Applied Data Mining for Forecasting Using SAS®

Tim Rey
Arthur Kordon
Chip Wells

§.sas.

The correct bibliographic citation for this manual is as follows: Rey, Tim, Arthur Kordon, and Chip Wells. 2012. *Applied Data Mining for Forecasting Using SAS®*. Cary, NC: SAS Institute Inc.

Applied Data Mining for Forecasting Using SAS®

Contents

Preface

It is utterly impossible that a mathematical formula should make the future known to us, and those who think it can would once have believed in witchcraft.

Jacob Bernoulli, in *Ars Conjectadi*, 1713

Curiosity about "what will happen next" is part of human nature, and thus the first attempts at forecasting are found rooted in history. In the ancient and medieval times, prophets like the Oracle of Delphi or Nostradamus had the status of demigods. The situation is significantly different in the 21st century, though, when predicting the future is not divine magic anymore but a necessity in contemporary business. Thousands of professionals are building forecasts in almost all areas of human activity. Since the global recession of 2008–2009, it has been much more widely understood that reliable forecasting is necessary.

The increased demand for forecasting triggered the development of new methods in addition to the "classical" time series statistical approaches, such as exponential smoothing and the Box-Jenkins AutoRegressive Integrated Moving-Average (ARIMA) models. One fruitful direction of development is that of nonlinear time series modeling, based on various computational intelligence methods, such as neural networks, support vector machines, and genetic programming. Other developments, of special importance to industrial applications, are the efforts for improving the time series forecasts by selecting the best potential drivers using various data mining methods. A short list of such methods includes but is not limited to the following: similarity analysis, sequential pattern matching, Principal Component Analysis (PCA), decision trees, co-integration analysis, variable cluster analysis, stepwise regression, and genetic programming.

Unfortunately, the available literature for integrating data mining methods in forecasting is very limited. The existing books on the market are either focused on forecasting methods or on data mining approaches. In addition, there are very few references that discuss the numerous practical issues of applying forecasting in a business setting. The practitioner needs a book that addresses the issues of applied industrial forecasting, gives a framework for integrating data mining and time series forecasting, and gives a methodology for large-scale multivariate industrial forecasting.

Applied Data Mining for Forecasting Using SAS is one of the first books on the market that fills this need.

Purpose of the Book

The purpose of the book is to give the reader an industrial perspective concerning applying data mining for forecasting different business activities using some of the most popular software—SAS Institute's range of SAS products including Base SAS, SAS Enterprise Guide, SAS Enterprise Miner, and SAS Forecast Server. The key topics of the book are as follows:

1. *What a practitioner needs to know to successfully apply data mining for forecasting* – The first main topic of the book focuses on the ambitious task of giving guidelines to practitioners about building the necessary framework for effective forecasting in a business setting. It covers the issues of justifying the need for industrial forecasting, offering a work process within the popular Six Sigma platform, and discussing the necessary infrastructure and application issues.

2. *How data mining improves forecasting* – The second key topic of the book clarifies the important question of using data mining for forecasting. Its main focus is on presenting the key data mining methods for variable reduction and selection and their implementation in SAS.

3. *How to apply data mining for forecasting in practice* – The third key topic of the book covers the central point of interest: the application strategy for business forecasting. It includes a short survey of

the key contemporary forecasting methods based on time series and illustrates them with appropriate examples from business practices.

Who This Book Is For

The targeted audience is much broader than the existing scientific communities in forecasting and data mining. The readers who can benefit from this book are described below:

- *Industrial practitioners* – This group includes forecasters in a number of different traditional company departments, such as strategy, sales, marketing, finance, supply-chain, purchasing, and so on. They will benefit from the book by understanding the impact of data mining on forecasting and using the discussed forecasting methods and application methodology to broaden and improve their forecast's performance.

- *Data miners and modelers* – This group consists of the large professional community of users of data mining technologies in different industries. This book will introduce them to contemporary forecasting methods and will demonstrate how they can leverage their data mining skills in the area of industrial forecasting.

- *Econometricians* – This group includes the key community driving the demand for development and application of time series statistical methods, which is at the basis of industrial forecasting. The book will give them substantial information about data mining methods related to time series forecasting, as well as important feedback from industry about the demand for corresponding methods for effective forecasting.

- *Six Sigma users* – Six Sigma is a work process for developing high-quality processes and solutions in industry. It has been accepted as a standard by the majority of global corporations. The estimated users of Six Sigma are tens of thousands of project leaders, called black belts, and hundreds of thousands of technical experts, called green belts. Usually, they use classical statistics in their projects. Data mining for forecasting is a natural extension to Six Sigma for solving complex problems, which both the black and green belts can take advantage of.

- *Academics* – This group includes a large class of academics in both fields (data mining and forecasting) who are not familiar with the research and technical details of the other. They will benefit from the book by using it to broaden their area of expertise and understanding specific requirements for successful practical applications as defined by industrial experts.

- *Students* – Undergraduate and graduate students in technical, economical, and even social disciplines can benefit from the book by understanding the advantages of using data mining in forecasting and its potential for implementation in their specific field. In addition, the book will help students gain knowledge about the practical aspects of forecasting and data mining and the issues faced in real-world applications.

How This Book Is Structured

The first four chapters of the book focus on the main topic of applying data mining for industrial forecasting. Chapter 1 clarifies the business forces that drive the use of data mining for forecasting while Chapter 2 presents a work process, akin to Six Sigma methodologies, that helps to integrate the proposed approach into corporate culture. Chapter 3 describes the critical efforts of building hardware, software, and organizational infrastructures that are needed for the successful application of business forecasting. Chapter 4 gives a systematic view of the key technical and nontechnical application issues as well as a complete checklist for applying data mining for forecasting. The next three chapters focus on presenting the necessary process and methods of data mining as it relates to forecasting. The focus of Chapter 5 is on data collection while Chapter 6 identifies the main data preprocessing steps and emphasizes their critical role for high-quality forecasting. Chapter 7 defines, from a practical perspective, the key data mining methods of forecasting, such as similarity analysis, varcluster analysis, principal component analysis, stepwise regression, decision trees, co-integration analysis, and genetic programming.

Chapters 8 through 11 cover the most important topic of the book—how to define an implementation strategy for successful real-world applications of data mining for forecasting. These chapters present a practitioner's guide of time series forecasting methods that details univariate, multivariate, hierarchical, and nonlinear models. Finally, Chapter 12 illustrates the key topics in applying data mining for forecasting on a real business example.

What This Book Is NOT About

- *Detailed theoretical description of data mining and forecasting approaches* – This book does not include a deep academic presentation of the various data mining and forecasting methods. The reader who is interested in more detailed knowledge on any individual approach is referred to the appropriate resources, such as books, critical papers, and Websites. The focus of the book is on the application of related data mining and forecasting methods. All methods are described and analyzed at the level of detail that will help their broad practical implementation.

- *Introduction of new data mining and forecasting methods* – The book does not propose new data mining and forecasting methods. The novelty of the book is on integrating both methodologies and on the application of data mining for forecasting.

- *Software manual of SAS products* – This is not an introductory manual of the SAS software products used in the application of data mining for forecasting. It is assumed that the interested reader has some basic knowledge on the specific SAS software used herein: Base SAS, SAS Enterprise Guide, SAS Enterprise Miner, and SAS Forecast Server.

Features of the Book

The key features that differentiate this book from other titles on data mining and forecasting are:

1. Integrating data mining and forecasting – One of the main messages in the book is that a critical factor for improving forecasting is using data mining methods. The synergetic benefits of both approaches are mostly in the area of variable reduction and variable selection for building multivariate forecasting models.

2. *A broader view of industrial forecasting* – Another important topic of the book is the proposed broadening of the forecasting approaches by using nonlinear predictions in addition to the existing time series methods. This allows handling cases with short time series and extraordinary business or process conditions.

3. *Emphasis on practical applications* – The third key feature of the book is the predominant practical view of all discussed topics. The examples given are from real industrial applications and the reader has the opportunity to "learn from the kitchen" regarding how data mining for forecasting works in an industrial setting.

Acknowledgments

The authors would like to thank Jan Baumgras and Terry Woodfield whose constructive comments substantially improved the final manuscript. The authors also highly appreciate the comments and clarifications of our technical reviewers Lorne Rothman, Abhijit Kulkarni, Sean Cai, Sara Vidal, and Udo Sglavo.

The staff of SAS Press has been most helpful, especially George McDaniel who successfully managed the project and responded to our requests. We gratefully acknowledge the contributions of our copyeditor Brad Kellam, production specialist Candy Farrell, designer Jennifer Dilley, and marketing specialists Aimee Rodriguez and Shelly Goodin.

x

Chapter 1: Why Industry Needs Data Mining For Forecasting

1.1 Overview

In today's economic environment there is ample opportunity to leverage the numerous sources of time series data that are readily available to the savvy decision maker. This time series data can be used for business gain if the data is converted first to information and then to knowledge—knowing what to make when for whom, knowing when resource costs (raw material, logistics, labor, and so on) are changing or what the drivers of demand are and when they will be changing. All this knowledge leads to advantages to the bottom line for the decision maker when times series trends are captured in an appropriate mathematical form. The question becomes how and when to do so. Data mining processes, methods and technology oriented to transactional type data (data that does not have a time series framework) have grown immensely in the last quarter century. Many of the references listed in the bibliography (Fayyad et al. 1996, Cabena et al. 1998, Berry 2000, Pyle 2003, Duling and Thompson 2005, Rey and Kalos 2005, Kurgan and Musilek 2006, Han et al. 2012) speak to the many methods and processes aimed at building prediction models on data that does not have a time series framework. There is significant value in the interdisciplinary notion of data mining for forecasting when used to solve time series problems. The intention of this book is to describe how to get the most value out of the host of available time series data by using data mining techniques specifically oriented to data collected over time. Previous authors have written about various aspects of data mining for time series, but not in a holistic framework: Antunes, Oliveira (2006), Laxman, Sastry (2006), Mitsa (2010), Duling, Lee (2008), and Lee, Schubert (2011).

In this introductory chapter, we help build the case for using data mining for forecasting and using forecasting as a competitive advantage. We cover the explosion of available economic time series data, the basic background on forecasting, and the limitations of classical univariate forecasting (from a business perspective). We also define what a time series database is and what data mining for forecasting is all about, and lastly describe what the advantages of integrating data mining and forecasting actually are.

1.2 Forecasting Capabilities as a Competitive Advantage

Information Technology (IT) Systems for collecting and managing transactional data, such as SAP and others, have opened the door for businesses to understand their detailed historical transaction data for revenue, volume, price, costs and often times even the whole product income statement. Twenty-five years ago IT managers worried about storage limitations and thus would design "out of the system" any useful historical detail for forecasting purposes. With the decline of the cost of storage in recent years, architectural designs have in fact included saving various prorated levels of detail over time so that companies can fully take advantage of this wealth of information. IT infrastructures were initially put in place simply to manage the transactions. Today, these architectures should also accommodate leveraging this history for business gain by looking at it from an advanced analytics view point. Various authors have discussed this framework in detail (Chattratichat et al. 1999, Mundy et al. 2008, Pletcher et al. 2005, Duling et al. 2008).

Large corporations generally have many internal processes and functions that support businesses—all of which can leverage quality forecasts for business gain. This is beyond the typical supply chain need for having the right product at the right time for the right customer in the right amount. Some companies have moved to a lean pull replenishment framework in their supply chains. This lean approach does not preclude the use of high-quality forecasting processes, methods, and technology.

In addition to those who analyze the supply chain, many other organizations in a corporation can use high-quality forecasts. Finance groups generally control the planning process for corporations and deliver the numbers that the company plans against and reports to Wall Street. Strategy groups are always in need for medium- to long-range forecasts for strategic planning. Executive sales and operations planning (ESOP) demand medium-range forecasts for resource and asset planning. Marketing and sales organizations always need short- to medium-range forecasts for planning purposes. New business development (NBD) incorporates medium- to long-range forecasts in the NPV (net present value) process for evaluating new business opportunities. Business managers themselves rely heavily on short- and medium-term forecasts for their own businesses data but also need to know about the market. Since every penny saved goes straight to a company's bottom line, it behooves a company's purchasing organization to develop and support high-quality forecasts for raw material, logistics, materials and supplies, and service costs.

Differentiating a planning process from a forecasting process is important. Companies do in fact need to have a plan to follow. Business leaders do in fact have to be responsible for the plan. But claiming that this plan is in fact a forecast can be disastrous. Plans are what we "feel we can do" while forecasts are mathematical estimates of what is most likely. These are *not* the same; but both should be maintained. In fact, the accuracy of both should be maintained over a long period of time. When reported to Wall Street, accuracy in the actual forecast is more important than precision. Being closer to the wrong number does not help.

Given that so many groups within an organization have similar forecasting needs, why not move towards a "one number" framework for the whole company? If finance, strategy, marketing and sales, business ESOP, NBD, supply chain and purchasing are not using the same numbers, tremendous waste can result. This waste can take the form of rework or mismanagement if an organization is not totally aligned with the same numbers. Such cross-organizational alignment requires a more centralized approach that can deliver forecasts that are balanced with input from the business and financial planning parts of the corporation. Chase (2009) presents this corporate framework for centralized forecasting in his book called *Demand Driven Forecasting*.

1.3 The Explosion of Available Time Series Data

Over the last 15 years, there has been an explosion in the amount of time series-based data available to businesses. To name a few, Global Insights, Euromonitor, CMAI, Bloomberg, Nielsen, Moody's Economy.com, Economagic—not to mention government sources such as www.census.gov, www.statistics.gov.uk/statbase, www.statistics.gov.uk/hub/regional-statistics, IQSS database, research.stlouisfed.org, imf.org, stat.wto.org, www2.lib.udel.edu, and sunsite.berkeley.edu. All provide some sort of time series data—that is, data collected over time inclusive of a time stamp. Many of these services are available for a fee, but some are free. Global Insights (www.ihs.com) contains over 30,000,000 time series. It has been the authors' collective experience that this richness of available time series data is not the same worldwide.

This wealth of additional time series information actually changes how a company should approach the time series forecasting problem in that new processes, methods, and technology are necessary to determine which of the potentially thousands of useful time series variables should be considered in the exogenous or multivariate in an X forecasting problem (Rey 2009). Business managers do not have the time to scan and plot all of these series for use in decision making. Statistical inference is a reduction process and data mining techniques used for forecasting can aid in the reduction process.

In order to provide some structure to data concerning various product lines consumed in an economy, there has long been a code structure used to represent an economies market. Various government and private sources provide this data in a time series format. This code structure is called NAICS (*North American Industry Classification System*) in North America (www.census.gov/naics). Various sources provide historical data in this classification system, but some also produce forecasts (Global Insights). For global product histories, an international system was recently deployed (ICIS—International Code Industry System). This system is at a higher level than the NAICS codes. For reference, there are cross-walk tables between the two (www.naics.com/). Both of these systems, among others, provide potential Y variables for a corporation's market forecasting endeavors. In some cases, depending on the level of detail being considered, these same sources may even be considered Xs.

Many of these sources offer databases for historical time series data but do not offer forecasts themselves. Other services, such as Global Insights and CMAI, do in fact offer forecasts. In both of these cases though, the forecasts are developed based on an econometric engine versus simply supplying individual forecasts. There are many advantages to having these forecasts and leveraging them for business gain. How to do so by leveraging both data mining and forecasting techniques will be discussed in the remainder of this book.

1.4 Some Background on Forecasting

A couple of important distinctions about time series models are important at this point. First, the one thing that differentiates time series data from transaction data is that the time series data contains a time stamp (day, month, year.) Second, time series data is actually related to "itself" over time. This is called serial correlation. If simple regression or correlation techniques are used to try and relate one time series variable to another, without regard to serial correlation, the business person can be misled. Therefore, rigorous statistical handling of this serial correlation is important. Third, there are two main classes of statistical forecasting approaches detailed in this book. First there are univariate forecasting approaches. In this case, only the variable to be forecast (the Y or dependent variable) is considered in the modeling exercise. Historical trends, cycles, and the seasonality of the Y itself are the only structures considered when building the univariate forecast model. In the second approach, where a multitude of time series data sources as well as the use of data mining techniques come in, various Xs or independent (exogenous) variables are used to help forecast the Y or dependent variable of interest. This approach is considered multivariate in the X or exogenous variable forecast model building. Building models for forecasting is all about finding mathematical relationships between Ys and Xs. Data mining techniques for forecasting become all but mandatory when 100s or even 1000s of Xs are considered in a particular forecasting problem.

For reference purposes, short-range forecasts are defined as one to three years, medium-range forecasts are defined as three to five years, and long-term forecasts are defined as greater than five years. Generally, the authors agree that anything greater than 10 years should be considered a scenario rather than a forecast. More often than not, in business modeling, quarterly forecasts are being developed. Quarterly data is the frequency that the historical data is stored and forecast by the vast majority of external data service providers. High-frequency forecasting might also be of interest even in finance where data can be collected by the hour or minute.

1.5 The Limitations of Classical Univariate Forecasting

Thanks to new transaction system software, businesses are experiencing a new richness of internal data, but, as detailed above, they can also purchase services to gain access to other databases that reside outside the company. As mentioned earlier, when building forecasts using internal transaction Y data only, the forecasting problem is generally called a univariate forecasting model. Essentially, the transaction data history is used to define what was experienced in the past in the form of trends, cycles, and seasonality to then forecast the future. Though these forecasts are often very useful and can be quite accurate in the short run, there are two things that they cannot do as well as the multivariate in X forecasts: They cannot provide any information about the "drivers" of the forecasts. Business managers always want to know what variables drive the series they are trying to forecast. Univariate forecasts do not even consider these drivers. Secondly, when using these drivers, the multivariate in X or exogenous models can often forecast further in time, with accuracy, then the univariate forecasting models.

The 2008–09 economic recession was evidence of a situation where the use of proper Xs in a multivariate in X "leading indicator" framework would have given some companies more warning of the dilemma ahead. Services like ECRI (Economic Cycle Research Institute) provided reasonable warning of the downturn some three to nine months ahead of time. Univariate forecasts were not able to capture these phenomena as well as multivariate in X forecasts.

The external databases introduced above not only offer the Ys that businesses are trying to model (like that in NAICS or ISIC databases), but also provide potential Xs (hypothesized drivers) for the multivariate in X forecasting problem. Ellis (2005) in "Ahead of the Curve" does a nice job of laying out the structure to use for determining what X variables to consider in a multivariate in X forecasting problem. Ellis provides a thought process that, when complemented with the data mining for forecasting process proposed herein, will help the business forecaster do a better job of both identifying key drivers and building useful forecasting models.

Forecasting is needed not only to predict accurate values for price, demand, costs, and so on, but it is also needed to predict when changes in economic activity will occur. Achuthan and Banerji—in their *Beating the Business Cycle* (2004) and Banerji in his complementary paper in 1999—present a compelling approach for determining which potential Xs to consider as leading indicators in forecasting models. Evans et al. (2002), as well as www.nber.org and www.conference-board.org, have developed frameworks for indicating large turns in economic activity for large regional economies as well as for specific industries. In doing so, they have identified key drivers as well. In the end, much of this work shows that, if we study them over a long enough time frame, we can see that many of the structural relations between Ys and Xs do not actually change. This fact offers solace to the business decision maker and forecaster willing to learn how to use data mining techniques for forecasting in order to mine the time series relationships in the data.

1.6 What is a Time Series Database?

Many large companies have decided to include external data, such as that found in Global Insights, as part of their overall data architecture. Small internal computer systems are built to automatically move data from the external source to an internal database. This practice, accompanied with tools like the SAS® Data Surveyor for SAP (which is used to extract internal transaction data from SAP), enables both the external Y and X data to be brought alongside the internal Y and X data. Often the internal Y data is still in transactional form that, once properly processed, can be converted to time series type data. With the proper time stamps in the data sets, technology such as Oracle, Sequel, Microsoft Access or SAS itself can be used to build a time series database from this internal transactional data and the external time series data. This database would now have the proper time stamp and Y and X data all in one place. This time series database is now the starting point for the data mining for forecasting multivariate in X modeling process.

1.7 What is Data Mining for Forecasting?

Various authors have defined the difference between "data mining" and classical statistical inference (Hand 1998, Glymour et al. 1997, and Kantardzic 2011, among others). In a classical statistical framework, the scientific method (Cohen 1934) drives the approach. First, there is a particular research objective sought after. These objectives are often driven by first principles or the physics of the problem. This objective is then specified in the form of a hypothesis; from there a particular statistical "model" is proposed, which then is reflected in a particular experimental design. These experimental designs make the ensuing analysis much easier in that the Xs are orthogonal to one another, which leads to a perfect separation of the effects therein. So the data is then collected, the model is fit and all previously specified hypotheses are tested using specific statistical approaches. In this way, very clean and specific cause-and-effect models can be built.

In contrast, in many business settings a set of "data" often contains many Ys and Xs, but there was no particular modeling objective or hypothesis in mind when the data was being collected in the first place. This lack of an original objective often leads to the data having multi-collinearity—that is, the Xs are actually related to one another. This makes building cause-and-effect models much more difficult. Data mining practitioners will mine this type of data in the sense that various statistical and machine learning methods are applied to the data looking for specific Xs that might predict the Y with a certain level of accuracy. Data mining on transactional data is then the process of determining what set of Xs best predicts the Ys. This is quite different than classical statistical inference using the scientific method. Building adequate prediction models does not necessarily mean that an adequate cause-and-effect model was built, again, due to the multi-collinearity problem.

When considering time series data, a similar framework can be understood. The scientific method in time series problems is driven by the economics or physics of the problem. Various structural forms can be hypothesized. Often there is a small and limited set of Xs that are then used to build multivariate in X times series forecasting models or small sets of linear models that are solved as a set of simultaneous equations. Data mining for forecasting is a similar process to the transaction data mining process. That is, given a set of Ys and Xs in a time series database, the goal is to find out what Xs do the best job of forecasting the Ys. In an industrial setting, unlike traditional data mining, a data set is not normally available for doing this data mining for forecasting exercise. There are particular approaches that in some sense follow the scientific method discussed earlier. The main difference here will be that time series data cannot be laid out in a "designed experiment" fashion. This book goes into much detail about the process, methods, and technology for building these multivariate in X time series models while taking care to find the drivers of the problem at hand.

With regard to process (previously discussed), various authors have reported on the process for data mining transactional data. A paper by Azevedo and Santos (2008) compared the KDD process, SAS Institute's SEMMA (Sample, Explore, Modify, Model, Assess) process and the CRISP data mining process. Rey and Kalos (2005) review the Data Mining and Modeling process used at The Dow Chemical Company. A common theme in all of these processes is that there are many Xs, and therefore some methodology is necessary to reduce the number of Xs provided as input to the particular modeling method of choice. This reduction is often referred to as variable or feature selection. Many researchers have studied and proposed numerous approaches for variable selection on transaction data (Koller 1996, Guyon 2003). One of the main concentrations of this book will be on an evolving area of research in variable selection for time series type data.

At a high level, the data mining process for forecasting starts with understanding the strategic objectives of the business leadership sponsoring the project. This is often secured via a written charter that documents key objectives, scope, ownership, decisions, value, deliverables, timing and costs. Understanding the system under study with the aid of the business subject matter experts provides the proper environment for focusing on and solving the right problem. Determining from here what data helps describe the system previously defined can take some time. In the end, it has been shown that the most time-consuming step in any data mining prediction or forecasting problem is the data processing step where data is defined, extracted, cleaned, harmonized and prepared for modeling. In the case of time series data, there is often a need to harmonize the data to the same time frequency as the forecasting problem at hand. Then there is often a need to treat missing data properly. This may be in the form of forecasting forward, backcasting or simply filling in missing data points with various algorithms. Often the time series database has hundreds if not thousands of hypothesized Xs in it. So, just as in data mining for transactional data, a specific feature or variable selection step is needed. This book will cover the traditional transactional feature selection approaches, adapted to time series data, as well as introduce various new time series specific variable reduction and variable selection approaches. Next, various forms of time series models are developed; but, just as in the data mining case for transaction data, there are some specific methods used to guard against overfitting, which helps provide a robust final model. One such method is dividing the data into three parts: model, hold out, and out of sample. This is analogous to training, validating, and testing data sets in the transaction data mining space. Various statistical measures are then used to choose the final model. Once the model is chosen, it is deployed using various technologies.

This discussion shows how and why it is important that the subject matter experts' knowledge of a company's market dynamics is captured in a form that institutionalizes this knowledge. This institutionalization actually surfaces through the use of mathematics, specifically statistics, machine learning and econometrics. When done, the ensuing equations become intellectual property (IP) that can be leveraged across the company. This is true even if the data sources are in fact public, since how the data is used to capture the IP in the form of mathematical models is in fact proprietary.

The core content of the book is designed to help the reader understand in detail the process described in the previous paragraphs. This will be done in the context of various SAS technologies, including SAS® Enterprise Guide®, SAS Forecast Server and various SAS/ETS® time series procedures like PROC EXPAND, PROC TIMESERIES, PROC ARIMA, PROC SIMILARITY, PROC Xll/12, as well as the SAS® Enterprise Miner™ time series data mining nodes, and others.

1.8 Advantages of Integrating Data Mining and Forecasting

The reason for integrating data mining and forecasting is simply to provide the highest-quality forecasts possible. Business leaders now have a unique advantage in that they have easy access to thousands of Xs, and the knowledge about a process and technology that enables data mining on time series data. With the tools now available through various SAS technologies, the business leader can create the best explanatory (cause and effect) forecasting model possible, and this can be accomplished in an expedient and cost efficient manner.

Now that models of this type are easier to build, they then can be used in other applications, including scenario analysis, optimization problems, and simulation problems (linear systems of equations as well as non-linear system dynamics). All in all, the business decision maker is now prepared to make better decisions with these advanced analytics forecasting processes, methods and technologies.

1.9 Remaining Chapters

The next chapter defines and discusses in detail the process of data mining for forecasting. In Chapter 3, details are given about how to set up an infrastructure for data mining for forecasting. Chapter 4 covers issues with data dining for forecasting applications. This then leads to data collection in Chapter 5 and data preparation in Chapter 6, which has an entire chapter dedicated to the topic since 60–80% of the work lies in this step. Chapter 7 discusses the foundation for the actually doing data mining by providing a practitioner's guide to data mining methods for forecasting. Chapters 8 through 11 present a practitioner's guide to time series forecasting methods. Chapter 12 finishes the book by walking through an example of data mining for forecasting from start to finish.

Chapter 2: Data Mining for Forecasting Work Process

2.1 Introduction

This chapter describes a generic work process for implementing data mining in forecasting real-world applications. By work process the authors mean a sequence of steps that lead to effective project management. Defining and optimizing work processes is a must in industrial applications. Adopting such a systematic approach is critical in order to solve complex problems and introduce new methods. The result of using work processes is that productivity is increased and experience is leveraged in a consistent and effective way. One common mistake some practitioners make is jumping to real-world forecasting applications while focusing only on technical knowledge and ignoring the organizational and people- related issues. It is the authors' opinion that applying forecasting in a business setting without a properly defined work process is a clear recipe for failure.

The work process presented here includes a broader set of steps than the specific steps related to data mining and forecasting. It includes all necessary action items to define, develop, deploy, and support forecasting models. First, a generic flowchart and description of the key steps is given in the next section, followed by a specific illustration of the work process sequence when using different SAS tools. The last section is devoted to the integration of the proposed work process in one of the most popular business processes widely accepted in industry–Six Sigma.

2.2 Work Process Description

The objective of this section is to give the reader a condensed description of the necessary steps to run forecasting projects in the real world. We begin with a high-level overview of the whole sequence as a generic flowchart. Each key step in the work process is described briefly with its corresponding substeps and specific deliverables.

2.2.1 Generic Flowchart

The generic flowchart of the work process for developing, deploying, and maintenance of a forecasting project based on data mining is shown in Figure 2.1. The proposed sequence of action items includes all of the steps necessary for successful real-world applications–from defining the business objectives to organizing a reliable maintenance program to performance tracking of the applied forecasting models.

Figure 2.1: A Generic flowchart of the proposed work process

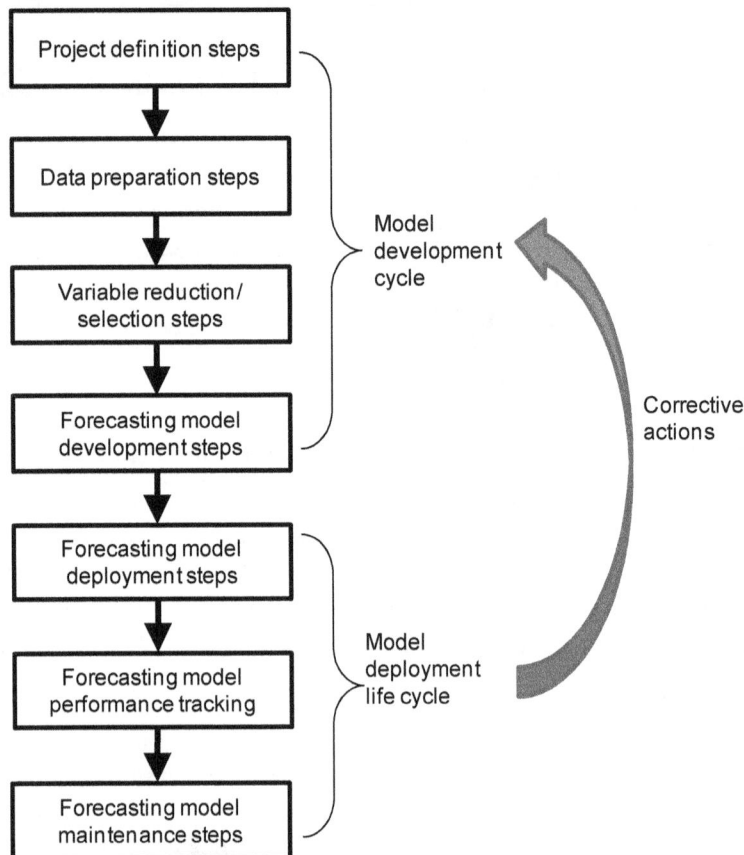

The forecasting project begins with a project definition phase. It gives a well-defined framework for approving the forecasting effort based on well-described business needs, allocated resources, and approved funding. As most practitioners already know, the next block—data preparation—often takes most of the time and the lion's share of the cost. It usually requires data extraction from internal and external sources and a lot of tricks to transfer the initial disarray in the data into a time series database acceptable for modeling and forecasting. The appropriate tricks are discussed in detail in Chapters 5 and 6.

The block for variable reduction and selection captures the corresponding activities, such as various data mining and modeling methods, that are used to take the initial broad range of potential inputs (Xs) that drive the targeted forecasting variables (outputs, Ys) to a short list of the most statistically significant factors. The next block includes the various forecasting techniques that generate the models for use. Usually, it takes several iterations along these blocks until the appropriate forecasting models are selected, reliably validated, and presented to the final user. The last step requires an effective consensus building process with all stakeholders. This loop is called the model development cycle.

The last three blocks in the generic flowchart in Figure 2.1 represent the key activities when the selected forecasting models are transferred from a development environment to a production mode. This requires automating some steps in the model development sequence, including the monitoring of data quality and forecasting performance. Of critical importance is tracking the business performance metric as defined by its key performance indicators (KPIs), and tracking the model performance metric as defined by forecasting accuracy criteria. This loop is called the model deployment cycle in which the fate of the model depends on the rate of model performance degradation. In the worst-case scenario of consistent performance degradation, the whole model development sequence, including project definition, might be revised and executed again.

2.2.2 Key Steps

Each block of the work process is described by defining the related activities and detailed substeps. In addition, the expected deliverables are discussed and illustrated with examples when appropriate.

Project definition steps

The first key step in the work process—project definition—builds the basis for forecasting applications. It is the least formalized step in the sequence and requires proactive communication skills, effective teamwork, and accurate documentation. The key objectives are to define the business motivation for starting the forecasting project and to set up as much structure as possible in the problem by effective knowledge acquisition. This is to be done well before beginning the technical work. The corresponding substeps to accomplish this goal as well as the expected deliverables from project the definition phase are described below.

Project objectives definition

This is one of the most important and most often mishandled substeps in the work process. A key challenge is defining the economic impact from the improved forecasts through KPIs such as reduced cost, increased productivity, increased market share, and so on. In the case of demand-driven forecasting, it is all about getting the right product to the right customer at the right time for the right price. Thus, the value benefits can be defined as any of the following (Chase 2009):

- a reduction in the instances when retailers run out of stock
- a significant reduction in customer back orders
- a reduction in the finished goods inventory carrying costs
- consistently high levels of customer service across all products and services

It is strongly recommended to quantify each of these benefits (for example, "a 15% reduction in customer back orders on an annual basis relative to the accepted benchmark").

An example of an appropriate business objective for a forecasting project follows:

> More accurate forecasts will lead to proactive business decisions that will consistently increase annual profit by at least 10% for the next three years.

Another challenge is finding a forecasting performance metric that is measurable, can be tracked, and is appropriate for defining success. An example of an appropriate quantitative objective that satisfies these conditions is the following definition:

> The technical objective of the project is to develop, deploy, and support, for at least three years, a quarterly forecasting model that projects the price of Product A for a two-year time horizon and that out-performs the accepted statistical benchmark (naïve forecasting in this case) by 20% based on the average of the last four consecutive quarterly forecasts.

The key challenge, however, is ensuring that the defined technical objective (improved forecasting) will lead to accomplishing the business goal (increased profitability).

Project scope definition

Defining the forecasting project scope also needs to be as specific as possible. It usually includes the business geography boundaries, business envelope, market segments covered, data history limits, forecasting frequency, and work process requirements. For example, the project scope might include boundaries such as the following: the developed forecasting model will predict the prices of Product A in Germany based on internal record of sales. The internal historical data to be used starts in January of 2001, uses quarterly data, and the project implementation has to be done in Six Sigma according to the standard requirements for model deployment and with the support of the Information Technologies department.

Project roles definition

Identifying appropriate stakeholders is another very important substep to take to ensure the success of forecasting projects. In the case of a typical large-scale business forecasting project, the following stakeholders are recommended as members of the project team:

- the management sponsor who provides the project funding
- the project owner who has the authority to allow changes in the existing business process
- the project leader who coordinates all project activities
- the model developers who develop, deploy, and maintain the models
- the subject matter experts—SMEs—who know the business process and the data
- the users (use the forecasting models on a regular basis)

System structure and data identification

The purpose of this substep is to capture and document the available knowledge about the system under consideration. This step provides a meaningful context for the necessary data and the data mining and forecasting steps. Knowledge acquisition usually takes several brainstorming sessions facilitated by model developers and attended by selected subject matter experts. The documentation may include process descriptions, market structure studies, system diagrams and process maps, relationship maps, etc. The authors' favorite technique for system structure and data identification is mind-mapping, which is a very convenient way of capturing knowledge and representing the system structure during the brainstorming sessions.

Mind-mapping (or concept mapping) involves writing down a central idea and thinking up new and related ideas that radiate out from the center.[1] By focusing on key topics written down in SME's words, and then defining branches and connections between the topics, the knowledge of the SMEs can be mapped in a manner that will help understanding and document the details of knowledge necessary for future data and modeling activities. An example of a mind-map[2] for system structure and data identification in the case of a forecasting project for Product A is shown in Figure 2.2.

The system structure, shown in the mind-map in Figure 2.2, includes three levels. The first level represents the key topics related to the project by radial branches from the central block named "Product A Price Forecasting." In this case, according to the subject matter experts, the central topics are: Data, Competitors, Potential drivers, Business structure, Current price decision-making process, and Potential users. Each key topic can be structured in as many levels of detail as necessary. However, beyond three levels down, the overall system structure visualization becomes cumbersome and difficult to understand. An example of an expanded structure of the key topic Data down to the third level of detail is shown in Figure 2.2. The second level includes the two key types of data – internal and external. The third level of detail in the mind-map captures the necessary topics related to the internal and external data. All other key topics are represented in a similar way (not shown in Figure 2.2). The different levels of detail are selected by collapsing or expanding the corresponding blocks or the whole mind-map.

Figure 2.2: An example of a mind-map for Product A price forecasting

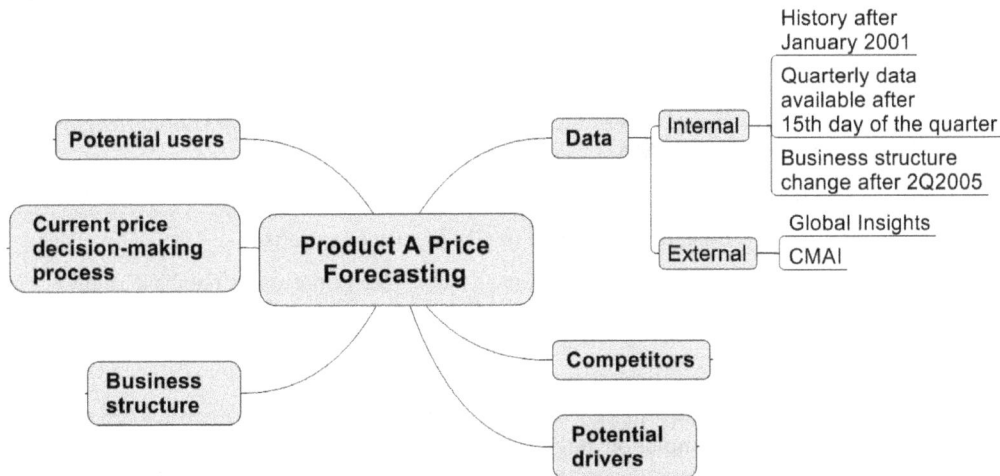

Project definition deliverables

The deliverables in this step are: (1) project charter, (2) team composition, and (3) approved funding. The most important deliverable in project definition is the charter. It is a critical document which in many cases defines the fate of the project. Writing a good charter is an iterative process which includes gradually reducing uncertainty related to objectives, deliverables, and available data. The common rule of thumb is this: the less fuzzy the objectives and the more specific the language, the higher the probability for success. An example of the structure of this document in the case of the Product A forecasting project is given in the Appendix at the end of this chapter.

The ideal team composition is shown in the corresponding charter section in the Appendix. In the case of some specific work processes, such as Six Sigma, the roles and responsibilities are well defined in generic categories like green belts, black belts, master black belts, and so on.

The most important practical deliverable in the project definition step is a committed financial support for the project since this is when the real project work begins. No funding—no forecasting. It is as simple as that.

Data preparation steps

Data preparation includes all necessary procedures to explore, clean, and preprocess the previously extracted data in order to begin model development with maximal possible information content in the data.[3] In reality, data preparation is time consuming, nontrivial, and difficult to automate. Very often it is also the most expensive phase of applied forecasting in terms of time, effort, and cost. External data might need to be purchased, which can be a significant part of the project cost. The key data preparation substeps and deliverables are discussed briefly below. The detailed description of this step is given in Chapters 5 and 6.

Data collection

The initial data collection is commonly driven by the data structure recommended by the subject matter experts in the system structure and data identification step. Data collection includes identifying the internal and external data sources, downloading the data, and then harmonizing the data in a consistent time series database format.

In the case of the example for Product A price forecasting, data collection includes the following specific actions:

- identifying the data mart that stores the internal data
- identifying the specific services and tags of the external time series available in Global Insights (GI), Chemical Market Associates, Inc. (CMAI), Bloomberg, and so on.
- collecting the internal data is generally conducted by the business data SMEs
- collecting the external data is done using local GI or CMAI service experts
- harmonizing the collected internal and external data as a consistent time series database of the prescribed time interval

Data preprocessing

The common methods for improving the information content of the raw data (which very often are messy) include: imputation of missing data, accumulation, aggregation, outlier detection, transformations, expanding or contracting, and so on. All of these techniques are discussed in separate sections in Chapter 6.

Data preparation deliverables

The key deliverable in this step is a clean data set with combined and aligned targeted variables (Ys) and potential drivers (Xs) based on preprocessed internal and external data.

Of equal importance to the preprocessed data set is a document that describes the details of the data preparation along with the scripts to collect, clean and harmonize the data.

Variable reduction /selection steps

The objective of this block of the work process is to reduce the number of potential economic drivers for the dependent variable by various data mining methods. The data reduction process is done in two key substeps: (1) variable reduction and (2) variable selection in static transactional data. The main difference between the two substeps is the relation of the potential drivers or independent variables (Xs) to the targeted or dependent variables (Ys). In the case of variable reduction, the focus is on the similarity between the independent variables, not on their association with the dependent variable. The idea is that some of the Xs are highly related to one another thus removing redundant variables reduces data dimensionality. In the case of variable selection, the independent variables are chosen based on their statistical significance or similarity with the dependent variables. The details of the methods for variable reduction and selection are presented in Chapter 7 and a short description of the corresponding substeps and deliverables is given below.

Variable reduction via data mining methods

Since there is already a rich literature for the statistical and machine learning disciplines concerning approaches for variable reduction or selection, this book often refers to and contrasts methods used in "non-time series" or transactional data. New methods specifically for time series data are also discussed in more detail in Chapter 7. In the transaction data approach, the association among the independent variables is explored directly. Typical techniques, used in this case, are variable cluster analysis and principal component analysis (PCA). In both methods, the analysis can either be based on correlation or covariance matrices. Once the clusters are found, the variable with the highest correlation to the cluster centroid in each cluster is chosen as a representative of the whole cluster. Another approach, used frequently, is variable reduction via PCA where a transformed set of new variables (based on the correlation structure of the original variables) is used that describes some minimum amount of variation in the data. This reduces the dimensionality of the problem in the independent variables.

In the time series-based variable reduction, the time factor is taken into account. One of the most used methods is similarity analysis where the data is first phase shifted and time warped. Then a distance metric is calculated to obtain the similarity measures between each two time series x_i and x_j. The variables below some critical distance are assumed as similar and one of them can be selected as representative. In the case of correlated inputs the dimensionality of the original data set could be significantly reduced after removing the similar variables. PCA can also be used in time series data, an example of such is the work done by the Chicago Fed wherein a National Activity Index (CFNAI), based on 85 variables representing different sectors of the US economy, was developed.[4]

Variable selection via data mining methods

Again, there is quite a rich literature in variable or feature selection for transactional data mining problems. In variable selection the significant inputs are chosen based on their association with the dependent variable. As in the case with variable reduction, there are different methods applied to data with a time series nature as compared to that of transactional data. The first approach uses traditional transactional data mining variable selection methods. Some of the known methods, discussed in Chapter 7, are correlation analysis, stepwise regression, decision trees, partial least squares (PLS), and genetic programming (GP). In order to use these same approaches on time series data, the time series data has to be preprocessed properly. First, both the Ys and Xs are made stationary by taking the first difference. Second, some dynamic in the system is added by introducing lags for each X. As a result, the number of extended X variables to consider as inputs is increased significantly. However, this enables you to capture dynamic dependences between the independent and the dependent variables. This approach is often referred to as the poor man's approach to time series variable selection since much of the extra work is being done to prepare the data and then non-time series approaches are being applied.

The second approach is more specifically geared toward time series. There are four methods in this category. The first one is the correlation coefficient method. The second one is a special version of stepwise regression for time series models. The third method is similarity as discussed earlier in the variable reduction substep but in this case the distance metric is between the Y and the Xs. Thus, the smaller the similarity metric the better the relationship of the corresponding input to the output variable. The fourth approach is called co-integration, which is a specialized test that two time series variables move together in the long run. Much more detail is presented in Chapter 7 concerning these analyses.

One important addition to the variable selection is to be sure to include the SME's favorite drivers, or those discussed as such in market studies (such as CMAI in the chemical industry) or by the market analysts.

Event selection

Specific class variables in forecasting are events. These class variables help describing big discrete shifts and deviations in the time series. Examples of such variables are advertising campaigns before Christmas and Mother's Day, mergers and acquisitions, natural disasters, and so on. It is very important to clarify and define the events and their type in this phase of project development.

Variable reduction and selection deliverables

The key deliverable from the variable reduction and selection step is a reduced set of Xs that are less correlated to one another. It is assumed that it includes only the most relevant drivers or independent variables, selected by consensus based on their statistical significance and expert judgment. However, additional variable reduction is possible during the forecasting phase. Selected events are another important deliverable before beginning the forecasting activities.

As always document the variable reduction/selection actions. The document includes a detailed description of all steps for variable reduction and selection as well as the arguments for the final selection based on statistical significance and subject matter experts approval.

Forecasting model development steps

This block of the work process includes all necessary activities for delivering forecasting models with the best performance based on the available preprocessed data given the reduced number of potential independent variables. Among the numerous options to design forecasting models, the focus in this book is on the most used practical approaches for univariate and multivariate models. The related techniques and development methodologies are described in Chapters 8−11 with minimal theory and sufficient details for practitioners. The basic substeps and deliverables are described below.

Basic forecasting steps: identification, estimation, forecasting

Even the most complex forecasting models are based on three fundamental steps: (1) identification, (2) estimation, and (3) forecasting. The first step is identifying a specific model structure based on the nature of the time series and modeler's hypothesis. Examples of the most used forecasting model structures are exponential smoothing, autoregressive models, moving average models, their combination–autoregressive moving average (ARMA) models, and unobserved component models (UCM). The second step is estimating the parameters of the selected model structure. The third step is applying the developed model with estimated parameters for forecasting.

Univariate forecasting model development

This substep represents the classical forecasting modeling process of a single variable. The future forecast is based on discovering trend, cyclicality, or seasonality in the past data. The developed composite forecasting model includes individual components for each of these identified patterns. The key hypothesis is that the discovered patterns in the past will influence the future. In addition to the basic forecasting steps, univariate forecasting model development includes the following sequence:

- Dividing the data into in-sample set (for model development) and out-of-sample set (for model validation)
- Applying the basic forecasting steps for the selected method on an in-sample set
- Validating the model through appropriate residuals tests
- Comparing the performance by applying the model to an out-of-sample set where possible
- Selecting the best model

Multivariate (in Xs) forecasting model development

This substep captures all the necessary activities to design forecasting models based on causal variables (economic drivers, input variables, exogenous variables, independent variables, Xs). One possible option is to develop the multivariate models as a time series model by using multiple regression. A limitation of this approach, however, is that the regression coefficients of the forecasting model are based on static relationships between the independent variables (Xs) and the dependent variable (Y). Another option is to use dynamic multiple regression that represents the dynamic dependencies between the independent variables (Xs) and the dependent variable (Y) with transfer functions. In both cases, the same modeling sequence, described in the previous section, is followed. However, different model structures, such as autoregressive integrated moving average with exogenous input model (ARIMAX) or unobserved components model (UCM), are selected. Note that the forecasted values for each independent variable, selected in the multivariate model, are required for calculating the dependent variable forecast. In most cases the forecasted values are delivered via univariate models for the corresponding input variables, that is, developing univariate models is a part of the multivariate forecasting model development substep.

Consensus planning

In one specific area of forecasting—demand-driven forecasting—it is of critical importance that each functional department (sales, planning, and marketing) reach consensus on the final demand forecast. In this case, consensus planning is a good practice. It takes into account the future trends, overrides, knowledge of future events, and so on that are not contained in the history.

Forecasting model development deliverables

The selected forecasting models with the best performance are the key deliverable not only of this step but of the whole project. In order to increase the performance, the final deliverable is often a combined forecast from several models, derived from different methods. In many applications the forecasting models are linked in a hierarchy, reflecting the business structure. In this case, reconciliation of the forecasts in the different hierarchical levels is recommended.

Another deliverable is the selected models performance. The document summarizing the model performance of the final models must include key statistics as well as a detailed description of the model validation and selection process. If sufficient data are available, it is recommended to test the performance robustness while changing key model process parameters, that is, test the size of in-sample and out-of sample sets.

The most important deliverable, however, is to convince the user to apply the forecasting models on a regular basis and to accomplish the business objectives. One option is to compare the model-generated and judgmental forecasts. Another option is to give the user the chance to test the model with different "What-If" scenarios. For final acceptance, however, a consistent record of forecasts within the expected performance metric for some specified time period is needed. It is also critical to prove the pre-defined business impact, that is, to demonstrate the value created by the improved forecasting.

Forecasting model deployment steps

This block of the work process includes the procedures for transferring the forecasting solution from development to production environment. The assumption is that beyond this phase the models will be put into the hands of the final users. Some users actively apply the forecasting models to accomplish the defined business objectives either in an interactive mode, by playing "What-If" scenarios, or by exploring optimal solutions. Other users are interested only in the forecasting reports delivered periodically or on demand. In both cases, a special version of the solution in a system-like production environment has to be developed and tested. The important substeps and deliverables for this block of the work process are discussed briefly below.

Production mode model deployment

It is assumed in production mode the selected forecasting models can deliver automatic forecasts from updated data when invoked by the user or by another program. In order to accomplish this, the necessary data collection scripts, data preprocessing programs, and model codes are combined in one entity. (In the SAS environment the entity is called a stored process.) In addition to the software during the model development cycle, some code for testing the data consistency in the future data collections has to be designed and integrated in the entity. Usually, the test checks for large differences between the new data sample and the current historical values in the data. By default, the new forecast is based on applying the selected models with the existing model parameters over the updated data. In most cases the user interface in production mode is a user-friendly environment.

Forecasting decision-making process definition

In the end, the results from the forecasting models are used in business decisions, which create the real value. Unfortunately, with the exception of demand-driven forecasting (see examples in Chase, 2009), this substep is usually either ignored or implemented in an *ad hoc* manner. It is strongly recommended to specify the decision-making process as precisely as possible. Then the quality of the decisions should be tracked in the same way as the forecasting performance. Using the method of forecast value analysis (FVA) is strongly recommended.[5] Even the perfect forecast can be buried by a bad business decision.

Forecasting model deployment deliverables

The ideal deliverable from this block of the work process is a user interface designed for the final user in an environment that he likes. In most of the cases that environment is the ubiquitous Microsoft Excel. Fortunately, it is relatively easy to build such an interface with the SAS Microsoft Add-in, as shown in Section 2.3.4.

Documenting the forecasting decision-making process is a deliverable of equal importance. The purpose of such a document is to define specific business rules that determine how to use the forecasting results. Initially the rule base can be created via brainstorming sessions with the subject matter experts. Another source of business rules definition could be a well-planned set of "What-If" scenarios generated by the forecasting models and analyzed by the experts. The end result is a set of business rules that link the forecasting results with specific actions and a value metric.

Training the user is a deliverable, often forgotten by developers. The training includes demonstrating the production version of the software. It is also expected that a help menu is integrated into the software.

Forecasting model maintenance steps

The final block of the proposed work process includes the activities for model performance tracking and taking proper corrective actions if the performance deteriorates below some specified critical limit. This is one of the least developed areas in practical forecasting in terms of available tools and experience. It is strongly recommended to discuss the model support issue in advance during the model definition phase. In the best-case scenario the project sponsor signs a service contract for a specified period of time. The users must understand that due to continuous changes in the economic environment forecasting models deteriorate with time and professional service is needed to maintain high-quality forecasts. A short description of the corresponding substeps and deliverables is given below.

Statistical baseline definition

The necessary pre-condition for performance assessment is to define a statistical baseline. The accepted baseline is called the naïve forecast, which assumes that the current observation can be used as the future forecast. It is also very important to explain to the final user the meaning of a forecast since non-educated users are looking only at the predicted number at the end of the forecast horizon as the only performance metric. A forecast is defined as the combination of: (1) predictions, (2) prediction standard errors, and (3) confidence limits at each time sample in the forecast horizon (Makridakis et al. 1998). The performance metric can be based on the difference between the defined forecast of the selected model and the accepted benchmark (naïve forecast).

Performance tracking

Performance monitoring is usually scheduled on a regular basis after every new data update. The tracking process includes two key evaluation metrics: (1) data consistency checks and (2) forecast performance metric evaluation. The data consistency check validates if the new data sample is not different from the most current data beyond some defined threshold. The forecast performance check is based on a comparison of the difference between the forecast of the selected model and the naïve forecast. Based on these two metrics, a set of decision rules is defined for appropriate corrective actions. The potential changes include either re-estimation of the model parameters and keeping the existing structure or complete model re-design and identifying a new forecast model structure.

Of critical importance is also tracking the business impact on KPIs of the forecast decisions. One possible solution for doing so is using business intelligence portals and dashboards (Chase 2009).

Forecasting model maintenance deliverables

The key deliverable in this final block of the work process is a performance status report. It includes the corresponding tables and trend charts to track the discussed metrics as well as the action items if corrective actions are taken.

2.3 Work Process with SAS Tools

The objective of this section is to specify how the proposed generic work process can be implemented with the wide range of software tools developed by SAS. A generic overview of the key SAS software tools related to data mining and forecasting is shown in Figure 2.3.

The SAS tools are divided in two categories depending on the requirements for programming knowledge: (1) tools that require programming skills and (2) tools that are based on functional blocks schemes and do not require programming skills. The first category consists of the software kernel of all SAS products—Base SAS with its set of operators and functions as well as specific toolboxes of specialized functions in selected areas. Examples of such toolboxes, related to data mining and forecasting, are SAS/ETS (includes the key functions for time series analysis), SAS/STAT (includes procedures for a wide range of statistical methodologies), SAS/GRAPH (allows creating various high resolution color graphics plots and chart), SAS/IML (enables programming of new methods based on the powerful Interactive Matrix Language IML), and SAS High-Performance Forecasting (includes a set of procedures for High-Performance Forecasting).

The second category of SAS tools, based on functional block schemes, shown in Figure 2.3, includes three main products: SAS Enterprise Guide, SAS Enterprise Miner, and SAS Forecast Server. SAS Enterprise Guide allows high-efficiency data preprocessing and development, basic statistical analysis, and forecasting by linking functional blocks. SAS Enterprise Miner is the main tool for developing data mining models based on build-in functional blocks and SAS Forecast Server is a highly productive forecasting environment with a very high level of automation. The business clients can interact with all model development tools via SAS Microsoft Add-in.

Figure 2.3: SAS software tools related to data mining in forecasting

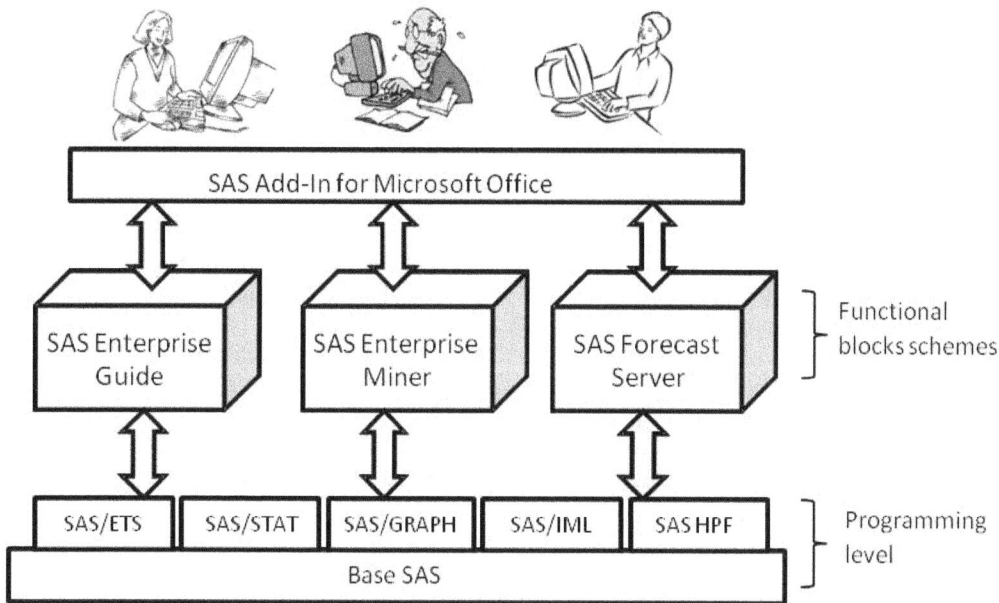

SAS also has another product with statistical, data mining, and forecasting capabilities. It is called JMP. However, because its functionality is similar to SAS Enterprise Guide and SAS Enterprise Miner, it is not discussed in this book. For those readers interested in the forecasting capabilities of JMP, a good starting point is *JMP Start Statistics: A Guide to Statistics and Data Analysis Using JMP* (Sall J., Creighton L., and Lehnan, A. 2009).

2.3.1 Data Preparation Steps with SAS Tools

The wide range of SAS tools gives the developer many options to effectively implement all of the data preparation steps. Good examples at the Base SAS level are procedures, such as DATA step for generic data collection or PROC SQL for writing specific data extracts.[6] The specific functions or built-in functional blocks for data preparation in the SAS tools that are related to data mining and forecasting are discussed briefly below.

Data preparation using SAS/ETS

The key SAS/ETS procedures for data preparation are as follows:

DATASOURCE provides seamless access to time series data from commercial and governmental data vendors, such as Haver Analytics, Standard & Poor's Compustat Service, the U.S. Bureau of Labor Statistics, and so on. It enables you to select the time series with specific frequency over a selected time range across sections of the data.

EXPAND provides different types of time interval conversions, such as converting irregular observations in periodic format or constructing quarterly estimates from annual data. Another important capability of this procedure is interpolating missing values for time series via the following methods: cubic splines, linear splines, step functions, and simple aggregation.

TIMESERIES has the ability to process large amounts of time-stamped data. It accumulates transactional data to time series and performs correlation, trend, and seasonal analysis on the accumulated time series. It also delivers descriptive statistics for the corresponding time series data.

X11 and X12 both provide seasonal adjustment of time series by decomposing monthly or quarterly data into trend, seasonal, and irregular components. The procedures are based on slightly different methods that were developed by the U.S. Census Bureau as the result of years of work by census researchers. X12 includes additional diagnostic tests to be run after the decomposition and the ability to remove the effect of input variables before the decomposition.[7]

Data preparation using SAS Enterprise Guide

SAS Enterprise Guide has built-in functional blocks that enable you to automate many data manipulation procedures (such as filtering, sorting, transposing, ranking, and comparing) without writing programming code. The two functional blocks for time series data preparation are Create Time Series Data and Prepare Time Series Data. Each block is a functional user interface to SAS/ETS procedures. Create Time Series Data is the user interface to TIMESERIES and Prepare Time Series Data is the corresponding user interface to EXPAND.

The advantage of using the functional block flows for implementing different steps of the proposed work process is clearly demonstrated with a simple example in Figure 2.4. The SAS Enterprise Guide flow shows the process of developing ARIMA forecasting models from the transactional data of 42 products. The original 42 transactional data are transformed as a time series of monthly data by the Create Time Series block, and the forecasting models are generated by the ARIMA Modeling functional block. The results with the corresponding graphical plots are summarized and output in a Word document.

Figure 2.4: An example of SAS Enterprise Guide flow for time series data preparation and |
modeling

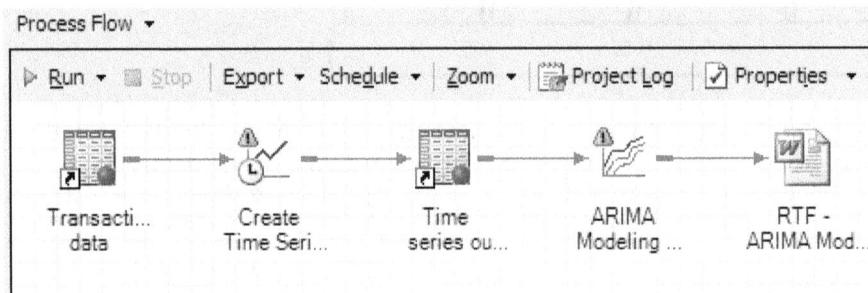

Another advantage of SAS Enterprise Guide is that it can access all SAS procedures either as separate blocks or as additional code within the existing blocks.

Data preparation using SAS Enterprise Miner

SAS Enterprise Miner is another SAS tool based on functional blocks, but its focus is on data mining. An additional advantage of this product is that it also imposes a work process. The work process abbreviation SEMMA (Sample-Explore-Modify-Model-Assess) includes the following key steps:

Sample
>the data by creating informational rich data sets. This step includes data preparation blocks for importing, merging, appending, partitioning, and filtering, as well as statistical sampling and converting transactional data to time series data.

Explore
>the data by searching for clusters, relationships, trends, and outliers. This step includes functional blocks for association discovery, cluster analysis, variable selection, statistical reporting and graphical exploration.

Modify
>the data by creating, imputing, selecting, and transforming the variables. This step includes functional blocks for removing variables, imputation, principal component analysis, and defining transformations.

Model

the data by using various statistical or machine learning techniques. This step includes the use of functional blocks for linear and logistic regression, decision trees, neural networks, partial least squares, among others, and importing models defined by other developers even outside SAS Enterprise Miner.

Assess

the generated solutions by evaluating their performance and reliability. This step includes functional blocks for comparing models, cutoff analysis, decision support, and score code management.

The data preparation functionality is implemented in the *Sample* and *Modify* sets of functional blocks.

Recently, a special set of SAS Enterprise Miner functional blocks related for Time Series Data Mining (TSDM) has been released by SAS. Its functionality covers most of the needed procedures for exploring forecasting data. The data preparation step is delivered by a Time Series Data Preparation node (TSDP), which provides data aggregation, summarization, differencing, merging, and the replacement of missing values.

2.3.2 Variable Reduction and Selection Steps with SAS Tools

Variable reduction and selection steps using specialized SAS subroutines

The key procedures for variable reduction and selection based on SAS/ETS and SAS/STAT are discussed briefly below.

AUTOREG (SAS/ETS) estimates and predicts linear regression models with autoregressive errors as well as stepwise regression. It also combines autoregressive models with autoregressive conditionally heteroscedastic (ARCH) and generalized autoregressive conditionally heteroscedastic (GARCH) models and generates a variety of model diagnostic tests, tables, and plots.

MODEL (SAS/ETS) analyzes and simulates nonlinear systems regression. It supports dynamic nonlinear models of multiple equations and includes a full range of nonlinear parameter estimation methods, such as nonlinear ordinary least squares, generalized method of moments, nonlinear full information maximum likelihood, and so on.

PLS (SAS/STAT) fits models by extracting successive linear combinations of the predictors, called factors (also called components or latent variables), which optimally address one or both of these two goals: explaining response or output variation and explaining predictor variation. In particular, the method of partial least squares balances the two objectives, seeking factors that explain both response and predictor variation. The contribution of the original variables to the factors is important to variable selection.

PRINCOMP (SAS/STAT) provides PCA on the input data. The results contain eigenvalues, eigenvectors, and standardized or unstandardized principal component scores.

REG (SAS/STAT) is used for linear regression with options for forward and backward stepwise regression. It provides all necessary diagnostic statistics.

SIMILARITY (SAS/ETS) computes similarity measures associated with time-stamped data, time series, and other sequentially ordered numeric data. A similarity measure is a metric that measures the distance between the input and target sequences while taking into account the ordering of the data.

VARCLUS (SAS/STAT) divides a set of variables into clusters. Associated with each cluster is a linear combination of the variables in the cluster. This linear combination can be generated by two options: as a first principal component or as a centroid component. The VARCLUS procedure creates an output data set with component scores for each cluster. A second output data set can be used to draw a decision tree diagram of hierarchical clusters. The VARCLUS procedure is very useful as a variable-reduction method since a large set of variables can be replaced by the set of cluster components with little loss of information.

Variable reduction and selection steps using SAS Enterprise Miner

The data mining capabilities in SAS Enterprise Miner for variable reduction and selection are spread in **Explore**, **Modify**, and **Model** tabs. It is not a surprise that the functional blocks are based on those SAS procedures, discussed in the previous section. The functional blocks or nodes of interest are the following:

In **Explore** tab:

Variable Clustering node implements the VARCLUS procedure in SAS Enterprise Miner—that is, it assigns input variables to clusters and allows variable reduction with a small set of cluster-representative variables.

Variable Selection node evaluates the importance of potential input variables in predicting the output variable based on R-squared and Chi-squared selection criterion. The variables that are not related to the output variable are assigned with rejected status and are not used in the model building.

In **Modify** tab:

Principal Components node implements the PRINCOMP procedure and in the case of linear relationship, reduces the dimensionality of the original input data to the most important principal components that capture a significant part of data variability.

In **Model** tab:

Decision Tree node splits the data in the form of a decision tree. Decision tree modeling is based on performing a series of if-then decision rules that sequentially divide the target variable into a small number of homogeneous groups that formulate a tree-like structure. One of the advantages of this block, in the case of variable selection, is that it automatically ranks the input variables, based on the strength of their relationship to the tree.

Partial Least Squares node implements the PLS procedure.

Gradient Boosting node uses a specific partitioning algorithm, developed by Jerome Friedman, called a gradient boosting machine.[8]

Regression node generates either linear regression models or logistic regression models. It supports stepwise, forward, and backward variable selection methods.

Two SAS Enterprise Miner nodes—TS Similarity (TSSIM) and TS Dimension Reduction (TSDR), which are part of the new **Time Series Data Mining** tab—can be used for variable reduction as well. The TS Similarity node implements the SIMILARITY based on four distance metrics: squared deviation, absolute deviation, mean square deviation, and mean absolute deviation and delivers a similarity map. The TS Dimension Reduction node applies four reduction techniques on the original data: singular value decomposition (SVD), discrete Fourier transformation (DFT), discrete wavelet transformation (DWT), and line segment approximations.

2.3.3 Forecasting Steps with SAS Tools

Forecasting using SAS/ETS

The key SAS/ETS forecasting procedures are described briefly below.

ARIMA generates ARIMA and ARIMAX models as well as seasonal models, transfer function models, and intervention models. The modeling process includes identification, parameter estimation, and forecast with generation of a variety of diagnostic statistics and model performance metrics, such as Akaike's information criterion (AIC) and Schwartz's Bayesian criterion (SBC or BIC).

ESM can generate forecasts for time series and transactional data based on exponential smoothing methods. It also includes several data transformation methods, such as log, square root, logistic, and Box-Cox.

FORECAST is the old version of ESM.

STATESPACE generates multivariate models based on different system representation by state space variables. It includes automatic model structure selection, parameter estimation, and forecasting.

UCM provides a development tool for unobserved component models. It generates the corresponding trend, seasonal, cyclical, and regression effects components, estimates the model parameters, performs model diagnostics, and calculates the forecasts and confidence limits of all the model components and the composite series.

VARMAX is very useful for forecasting multivariate time series, especially when the economic or financial variables are correlated to each other's past values. The VARMAX procedure enables modeling the dynamic relationship both between the dependent variables and between the dependent and independent variables. It uses a variety of modeling techniques, criteria for automatic determination of the autoregressive and moving average orders, model parameter estimation methods, and several diagnostic tests.

Forecasting using SAS Enterprise Guide

The forecasting capabilities of the SAS Enterprise Guide built-in blocks are very limited. However, all the SAS/ETS functionality can be used via SAS Enterprise Guide code nodes. The key built-in forecasting blocks in the Time Series Tasks are described briefly below.

Basic Forecasting
    generates forecasting models based on exponential smoothing and stepwise autoregressive fit of time trend.

ARIMA Modeling and Forecasting
    generates ARIMA models, but the identification and parameter estimation methods have to be selected by the modeler.

Regression Analysis with Autoregressive Errors
    provides linear regression models for time series data in the case of correlated errors and heteroscedasticity.

Forecasting using SAS Forecast Studio

SAS Forecast Studio is one of the most powerful engines for large-scale forecasting available in the market. It generates automatic forecasts in batch mode or executes custom-built models through an interactive graphical interface. SAS Forecast Studio enables the user to interactively set up the forecasting process, hierarchy, parameters, and business rules as well as to enter specific events. Another very useful feature is hierarchical reconciliation with the ability to reconcile the hierarchy bottom-up, middle-out, or top-down.

SAS Forecast Studio does not require programming skills, and the whole forecasting step can be done in an easy-to-use GUI interface where the model selection list includes exponential smoothing models with optimized parameters, ARIMA models, unobserved components models, dynamic regression, and intermittent demand models. It is possible also to define a model repository and events. The automatic model generation includes outlier detection, event identification, and automatic variable selection. The forecasting results are represented in numerous graphical reports. In the case of multivariate models, you can explore different "What if" scenarios to determine the influence of the key drivers in the dependent variable forecast. Forecast studio is a highly productive environment. It can generate thousands of time series forecasts in minutes.

2.3.4 Model Deployment Steps with SAS Tools

Model deployment on model development tools

Some forecasting applications use the development environment for model deployment. An obvious disadvantage of this option is that the user must be familiar with at least some of the capabilities of the development software. Using the development environment for model deployment is appropriate only in specialized cases with educated end users. In the case of large-scale industrial forecasting, however, this option is not recommended.

Model deployment via stored processes

One of the advantages of using various SAS tools is that you can communicate the results to the user via stored processes. A stored process is a SAS program that is stored centrally on a server. The final validated models are usually packaged as stored processes in the development environment (SAS Enterprise Guide, SAS Enterprise Miner, or SAS Forecast Server) and saved on a specified server. A client application can then execute the program, and then receive and process the results locally.

The most popular application for using the SAS stored processes on the client side is the SAS Add-In for Microsoft Office. After installing the Add-In, the user can select the corresponding stored process and invoke the forecasting application. In the case of Excel, the results are represented in spreadsheets with the rich graphic capabilities of this popular tool. Another option for model deployment is via SAS Web Report Studio.

2.3.5 Model Maintenance Steps with SAS Tools

One option for model maintenance is using SAS Model Manager, which manages and monitors analytical model performance in a central repository. It enables users to monitor the performance of a large number of models by defining different performance metrics and then generating model performance and comparison reports. Unfortunately, managing models generated by SAS Forecast Studio is currently not possible with SAS Model Manager.

Another option for forecasting model performance tracking is to develop corresponding stored processes that generate periodic reports. In the case of performance deterioration, the user can contact the developer to perform the proper corrective actions of parameter re-fitting or complete model re-development. These options are available in SAS Forecast Studio.

2.3.6 Guidance for SAS Tool Selection Related to Data Mining in Forecasting

At the end of this section are some generic guidelines on how to select the appropriate SAS tools so that you can apply the work process discussed.

SAS/ETS includes all generic functions for forecasting, such as: FORECAST, AUTOREG, ARIMA, VARMAX, X11, X12, SPECTRA, and so on.[9] It is a proper solution if the model developers have good programming skills in Base SAS and are knowledgeable in forecasting methods.

SAS/STAT provides the generic statistical functions such as REG, PRINCOMP, PLS, VARCLUS, and so on. From an implementation point of view, it has the same requirements as SAS/ETS. Both tools are appropriate for small-scale applications and prototype development and require skilled SAS programmers.

SAS Enterprise Guide enables fast system development based on a combination of built-in functional blocks using Base SAS procedures. It is a very good environment to integrate data preprocessing and some data mining and forecasting functions. SAS Enterprise Guide requires minimal Base SAS programming experience. Another advantage of SAS Enterprise Guide is its impressive reporting and graphical capabilities.

SAS Enterprise Miner is the ideal non-programming development tool for data preprocessing and data mining activities. An additional advantage in the case of data mining in forecasting is the currently developed node for Time Series Data Mining (TSDM), which enables fully functional time series preprocessing, variable reduction, and selection.

SAS Forecast Server provides automatic model and report generation for a wide range of forecasting algorithms. For large-scale industrial forecasting, this is the tool.

2.4 Work Process Integration in Six Sigma

The best-case scenario for implementing the proposed work process for data mining in forecasting is to integrate it into some existing work process. Doing so minimizes cultural change since the organization of interest has already introduced work processes according to its strategy. One popular work process in many organizations using demand-driven forecasting is Sales & Operation Planning (S&OP) in which the operations planning and financial departments match supply to a demand forecast and generate supply plans.[10]

In this book we discuss a more generic integration of the proposed work process into the most widespread work process in industry—Six Sigma. The advantages of this integration are as follows: fast acceptance in more than 50% of Fortune 500 companies (see below), reduced organizational efforts, established project management, well-defined stakeholders' roles, good opportunities for training and leveraging, and so on.

2.4.1 Six Sigma in Industry

What it is

A method or set of techniques, Six Sigma, has become a movement and a management religion for business process improvement.[11] It is a quality measurement and improvement program originally developed by Motorola in the 1980s that focuses on the control of a process to the point of ± Six Sigma (standard deviations) from a centerline. The Six Sigma systematic quality program provides businesses with the tools to improve the capability of their business processes. At the basis of Six Sigma methodology is the simple observation that customers feel the variance, not the mean. In other words, reducing the variance of product defects is the key to making customers happy. What is important for Six Sigma is that it provides not only technical solutions but a consistent work process for pursuing continuous improvement in profit and customer satisfaction. This is one of the reasons for the enormous popularity of this methodology in industry.

Industrial acceptance

According to the *iSixSigma Magazine*,[12] about 53% of Fortune 500 companies are currently using Six Sigma, and that figure rises to 82% for the Fortune 100. Over the past 20 years, use of Six Sigma has saved Fortune 500 companies an estimated $427 billion. Companies that properly implement Six Sigma have seen profit margins grow 20% year after year for each sigma shift (up to about 4.8 sigma to 5.0 sigma). Since most companies start at about 3 sigma, virtually each employee trained in Six Sigma will return on average $230,000 per project to the bottom line until the company reaches 4.7 sigma. After that, the cost savings are not as dramatic.

Key Roles

One of the key advantages of Six Sigma is its use of well-defined roles in project development. The typical roles are defined as: Champion, Black Belt, Green Belt, and Master Black Belt. The Champion is responsible for the success of the projects, provides the necessary resources and breaks down organizational barriers. The project leader is called a Black Belt. Project team members are called Green Belts and they do not spend all their time on projects. They receive training similar to that of Black Belts but for less time. There is also a Master Black Belt level. These are experienced Black Belts who have worked on many projects. They are the ones who typically know about more advanced tools, the business, and have had leadership training and often have teaching experience. A primary responsibility of Master Black Belts is mentoring new Black Belts.

2.4.2 The DMAIC Process

The classic Six Sigma methodology includes the following key phases known as Define-Measure-Analyze-Improve-Control (DMAIC) process:

Define: Understand the problem.

Measure: Collect data on the problem.

Analyze: Find root cause as to why the problem occurs.

Improve: Make changes to eliminate root causes.

Control: Ensure that the problem is solved.

A brief description of each phase is given below.

Define

The objective of the define phase is to identify clearly and communicate to all stakeholders the problem to be solved. The team members and the timelines are laid down. Project objectives are based on identifying the needs by collection of the voice of the customer. The opportunities are defined by understanding the flaws of the existing as-is process. The key document in this phase is the project charter, which includes the financial and technical objectives, assessment of the necessary resources, allocation of the stakeholders' roles, and a project plan.

Measure

The goal of the measure phase is to understand the problem in more detail by collecting all available data around it. The following questions need to be answered: what the problem really is, where it occurs, when it occurs, what causes it, and how does it occur. The key deliverables in this phase are identifying the necessary factors (inputs) that can influence the defect (output), and collecting and preparing the data for analysis. Another very important deliverable is the statistical measures from the data, such as the sigma level of the defect and process capability (the ability of a process to satisfy customer expectations), measured by the sigma range of a process's variation.

Analyze

The objective of the analyze phase is to analyze the collected data and to find out the root causes (critical Xs) of the problem. The analyses are mostly statistically based and are designed to identify the critical inputs affecting the defect or the output $Y = f(x)$. The potential cause-effect relationships or models are discussed and prioritized by the experts. As a result, several potential solutions of the problem are identified for deployment.

Improve

The goal of the improve phase is to prioritize the various solutions that were suggested during brainstorming and to explore the best solution with highest impact and the lowest cost effort. At the end of this phase, a pilot solution is implemented. The purpose of the improve phase is to remove the impact of the root causes by implementing changes in the process. Before beginning the implementation steps, however, the selected solution is tested with data to validate the predicted improvements. The initial results from pilot implementation are communicated to all stakeholders.

Control

The objective of this final phase is to complete the implementation of the selected solutions and to validate that the problem has gone away. A measurement system is usually set up to determine if the problem has been solved and the expected performance has been met. One of the key deliverables in the control phase is a process control and monitor plan. Part of the plan is the transition of ownership from the development team to the final user.

2.4.3 Integration with the DMAIC Process

Two options for integration with the DMAIC Six Sigma process are discussed briefly below: (1) the integration of a data mining work process and (2) the integration of the proposed data mining in a forecasting work process.

Data mining within DMAIC

The key blocks of a data mining process within the Six Sigma framework (defined in Kalos and Rey 2005) are shown in Figure 2.5.

The purpose of the first key block, Strategic Intent, is to ensure the relevance of the proposed data mining project with the strategic business goals, enterprise-wide initiatives, and management improvement plans. Another objective of this block is to identify the business success criteria, including various measurements (customer, process, and financial measurements).

The second key block, System & Data Identification, has objectives similar to those of the system structure and data identification substep in the Project Definition block. The third key block, Data Preprocessing, includes activities such as preliminary data analysis, variable selection, data transformation, and documenting the results. The fourth block, Opportunity Discovery, combines data analysis strategy development, exploratory data analysis, and model development and performance assessment. The last, fifth block of the data mining process within Six Sigma, Opportunity Deployment, is characterized by three main activities: (1) immediately using the developed models for business decisions, (2) integrating the developed models in other projects, and (3) triggering other projects based on the discovered opportunities and generating of preliminary Six Sigma project charters.

Figure 2.5: Key blocks of data mining in Six Sigma

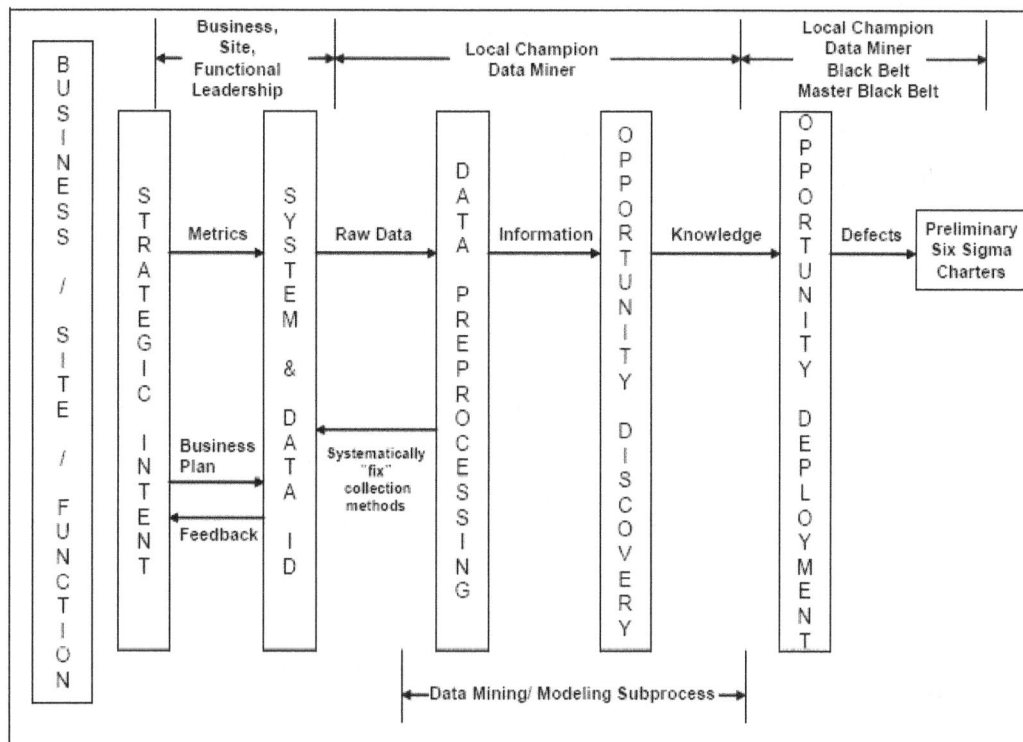

Figure 2.5 is from Alex Kalos and Tim Rey's paper, "Data mining in the chemical industry" (2005). The details of using this data mining process within the Six Sigma framework are also given in this paper.

Data mining in forecasting within DMAIC

The other option of integrating the proposed work process for data mining in forecasting within Six Sigma is illustrated in Figure 2.6 where we can see the corresponding links between the key blocks of both methodologies. The project definition steps, including system identification, are part of the define phase of DMAIC. The data preparation steps belong to the measure phase, and both variable selection and reduction and Forecasting model development steps are included in the analyze phase. The forecasting model deployment steps are part of the improve phase of DMAIC and the last part of the forecasting work process, the forecasting model maintenance steps, are linked to the control phase of DMAIC.

Figure 2.6: Correspondence of the Data Mining in Forecasting Work Process with DMAIC

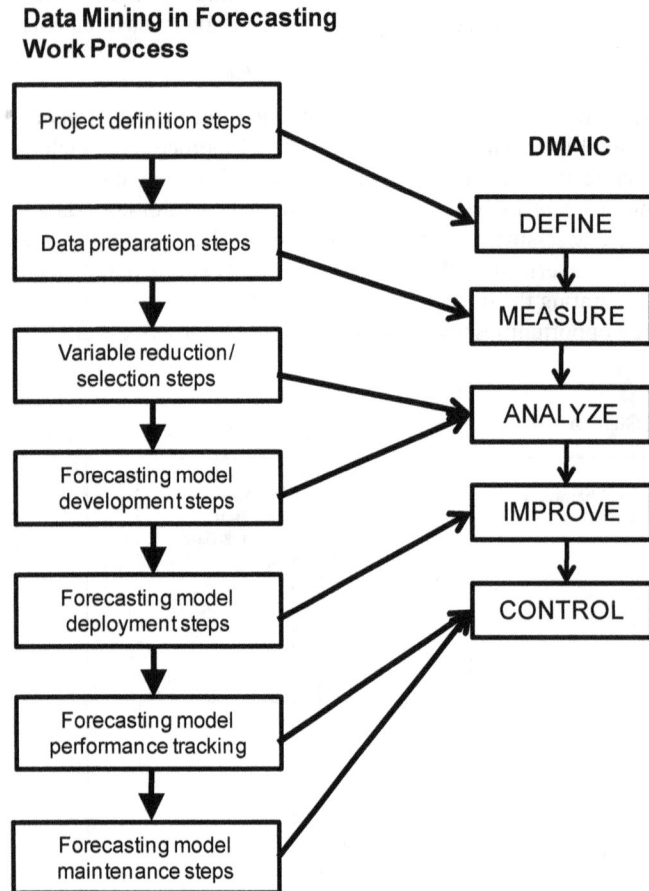

Because of the clear link between the proposed work process (based on the requirements for developing high-performance forecasting) and a work process such as Six Sigma (that is almost universally adopted in industry), you can integrate the two processes with minimal effort and cultural change. As a result, you have greater opportunities to introduce the proposed methodology and can more efficiently manage projects and develop forecast systems.

Appendix: Project Charter

Opportunity Statement:

- Current forecast is judgmental with an average Mean Average Percent Error (MAPE) of 16.5 for four quarterly forecasts.
- The opportunity is to improve the forecast by using statistical methods.
- The key hypothesis is that more accurate forecasts will lead to proactive business decisions that will increase consistently profit by at least 10%.

Project Goal and Objective:

The technical objective of the project is to develop, deploy, and support, for at least three years, a quarterly forecasting model that projects the price of Product A for a two-year time horizon and that outperforms the accepted statistical benchmark (naïve forecasting in this case) by 20% based on average of four consecutive quarterly forecasts.

Project Scope and Boundaries:

- The project will focus on Product A price in Germany.

Deliverables:

- a forecasting model with user interface in Excel
- a decision scheme with proactive action items to increase profits

Timeline:

Estimated duration of the key steps of the project:

Project definition:	40 hours
Data preparation:	80 hours
Model development:	60 hours
Model deployment:	20 hours
Model maintenance:	10 hours per year

Team Composition:

The ideal team includes:

Management sponsor

Project owner
Project leader
Technical subject matter experts
Model developers
End users

---

[1] A good starting point for developing mind-maps is Tony Buzan's *The Mind-map Book* (2003).

[2] The mind-maps in this book are based on the Mindjet product MindManager 8, available from http://www.mindjet.com/.

[3] A classic book about data preparation is Dorian Pyle's *Data Preparation for Data Mining* (1999).

[4] Evans, C., Liu, C. and Pham-Kanter, G. "The 2001 recession and the Chicago Fed National Activity Index: Identifying business cycle turning points," *Economic Perspectives* 26, no. 3 (2002): 26–43.

[5] The FVA method is described in Michael Gilliland's book, *The Business Forecasting Deal* (2010).

[6] A book with many examples of using different SAS solutions for data preparation is Gerhard Svolba's *Data Preparation for Analytics Using SAS* (2006).

[7] A good explanation of X11 and X12 is given by Spyros G. Makridacis et al. in *Forecasting: Methods and Applications* (1997).

[8] Friedman, J. H. "Greedy function approximation: A gradient boosting machine," *Annals of Statistics* 29 (2001): 1189–1232.

[9] A useful classification of the SAS/ETS functions is given in Table 1.1 in the book *SAS for Forecasting Time Series* (2003) by John Brocklebank and David Dickey.

[10] A detailed description of S&OP is given in Charles Chase's *Demand-Driven Forecasting: A Structured Approach to Forecasting* (2009).

[11] The reader can find more information about Six Sigma in *Implementing Six Sigma: Smarter Solutions Using Statistical Methods* (2003) by Forrest Breyfogle III.

[12] January/February 2007 Issue at http://www.isixsigma-magazine.com/

Chapter 3: Data Mining for Forecasting Infrastructure

3.1 Introduction

Applying data mining for forecasting in a business requires serious investments in hardware, software, and training, but a cultural change must also take place. It is very important to estimate the size of the investment based on technical requirements and the products that are available in the market. The four main components of any forecasting infrastructure are hardware, software, data, and organizational. The first three components build the technical basis to support applied data mining for forecasting, and the fourth component is critical to effectively change the culture of the organization. This chapter is focused on an enterprise-wide implementation strategy of data mining for forecasting. The importance of integrating the selected options into the existing corporate infrastructure is discussed at the end of the chapter.

3.2 Hardware Infrastructure

The objective of this section is to give the reader a condensed overview of the potential hardware architectures for implementing data mining for forecasting systems in an industrial setting. The following three options: (1) PC network, (2) client/server, and (3) cloud computing infrastructures are discussed briefly below. However, due to rapid technology changes today's recommendations can easily become obsolete tomorrow.

3.2.1 Personal Computers Network Infrastructure

The least expensive hardware solution for implementing data mining for forecasting systems in an industrial setting is to avoid any additional hardware expenses and use the existing information system infrastructure. Usually, this is based on a PC network. The key advantages of this option are as follows:

- low cost
- easy integration in the existing information system infrastructure
- minimal installation and maintenance efforts
- robust performance due to the decentralized architecture

The main limitations of the PC network infrastructure solution for implementing data mining for forecasting systems are as follows:

- limitations for large data set processing
- slower processing speed relative to servers
- limited operating systems options

3.2.2 Client/Server Infrastructure

The client/server model assumes a division of the computing resources between clients or workstations with local processing capabilities and servers with large memory and disk space and more powerful processors. The clients request services such as data, and the servers retrieve resources and deliver the requested information. The number of servers required depends on the number of clients, network speed and capacity, global and local operation, reliability, and so on.

An example of a minimal client/server infrastructure based on SAS is shown in Figure 3.1. The example includes four types of servers and two types of clients—modeler PC and final user PC. One server is allocated to handle metadata. A data mart server, based on Oracle, interacts with the large database cluster containing the corporate data. The third server includes the SAS server and is devoted to intensive computing tasks. Several clients can share the server resources either for developing new models or running developed models as stored processes.

The key advantages of the client/server infrastructure for implementing data mining for forecasting are given below:

- very powerful processing capabilities
- large memory and high-throughput disk
- the use of different operating systems
- capacity to process large data sets.

Figure 3.1: An example of client/server infrastructure based on SAS

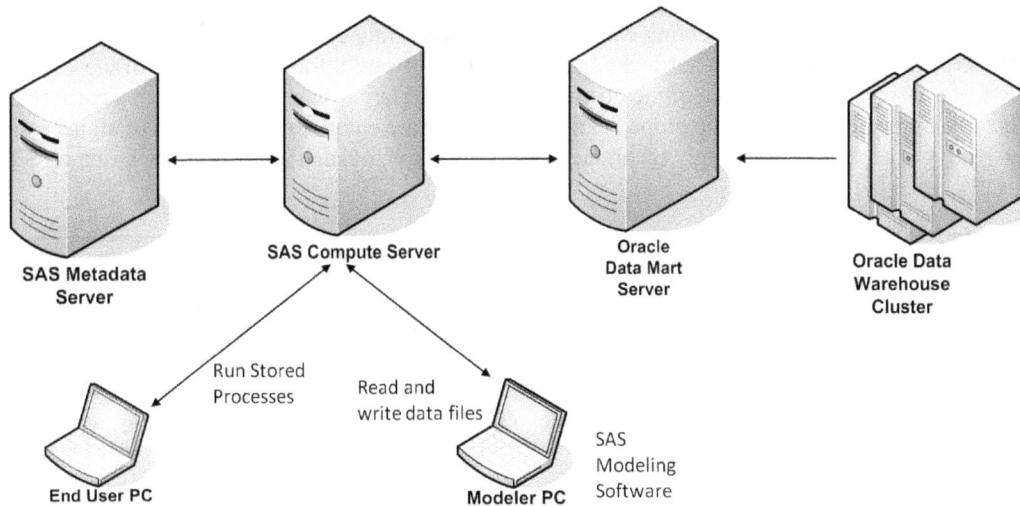

The disadvantages of this option are as follows:

- high cost
- more complex maintenance and support
- lower reliability if servers are down

The advantages, however, outweigh the disadvantages and the client/server infrastructure is the standard solution for large-scale industrial applications of data mining and forecasting.

3.2.3 Cloud Computing Infrastructure

Another potential solution, called cloud computing, uses powerful external and internal computing resources, and includes grid computing for parallel processing, multi-tiered computer architecture, and the capacity to handle super-large data sets. Such services are currently offered by a number of vendors including well-established industry leaders. Some of the advantages of using this option are as follows:

- low implementation and maintenance cost
- super-computer power, which is continuously upgraded by the cloud owner
- data consolidation in very large data sets
- increased reliability

The disadvantages of using a cloud computing infrastructure are summarized as follows:

- proprietary data security
- initial transfer of very large corporate data to the cloud
- limited software
- trust issues
- information technology (IT) management resistance

This option is still in an exploratory phase and has generated a lot of hype. However, if the technical and economic advantages are proved with more industrial applications, it could become a popular hardware infrastructure in the near future.

3.3 Software Infrastructure

The lion's share of the costs for implementing data mining for forecasting systems, especially for the PC network infrastructure, is not the cost of hardware but the cost of software infrastructure. One of the key decisions to make in advance is the scale of the efforts. In the case of large-scale forecasting on a corporate level that is to be implemented across the globe, an integrated software environment made up of all necessary components with global support is strongly recommended. An example of such infrastructure (based on SAS software) is discussed in this book.

3.3.1 Data Collection Software

This part of the infrastructure strongly depends on the existing corporate information system architecture. Unfortunately, it could be very diverse with different database platforms. In most cases, however, the data are organized in relational databases and stored in separate tables for each entity. The relationship between the tables is defined by two columns—primary key and foreign key columns (Svolba 2006). Data that are accessed from a relational database are usually extracted table by table and are merged according to the primary or foreign keys.

The software basis for handling data in relational database systems is the Structure Query Language (SQL). It includes the necessary operators for searching data pieces as well as different aggregations and joins of tables. The leading relational database systems include Oracle, SAP MaxDB and Sybase, Microsoft SQL Server, and IBM DB2. The good news is that the existing key software programs for data mining, such as SAS Enterprise Miner, IBM SPSS[1] and StatSoft STATISTICA Data Miner[2] include all necessary software interfaces to collect data from diverse sources.[3] For example, SAS offers a specialized tool, SAS/ACCESS, that has almost universal capabilities for access, retrieval, and integration with any available data source.[4]

3.3.2 Data Preparation Software

It is recommended that the selected software has the following functionality for data preparation:

- Data manipulation capabilities that include functions for summary tables generation, data split, concatenation, transposition, stacking, sorting, flexible filtering, joining tables, and so on.
- Missing data handling that includes different options to impute missing data.
- Data description capabilities that are usually based on basic descriptive statistics, frequency tables, histograms, and so on.
- Data visualization capabilities that include a broad spectrum of graphics, such as 3-D scatter plots, contour plots, parallel plots, and so on.
- Data pre-processing capabilities that include filtering, outlier detection and removal, data sampling, data partitioning, data transformation, and so on.

Examples of software tools with these capabilities are SAS Enterprise Guide, JMP, IBM SPSS, and StatSoft STATISTICA Data Miner.

3.3.3 Data Mining Software

From the broad range of available data mining methods and functions, the following capabilities for variable reduction and selection are needed for the forecasting applications:

- Basic statistical capabilities that include building and analyzing linear regression models with options for variable selection by forward and backward stepwise regression.
- Multivariate analysis capabilities that include cross-correlation analysis, PCA, and PLS.
- Clustering capabilities that include dividing variables in clusters by linear or nonlinear methods, similarity analysis, and building decision trees.
- Variable selection capabilities that include different algorithms for variable selection, such as stepwise regression, decision trees, gradient boosting, singular value decomposition (SVD), and so on.

The three most popular software options for industrial applications that offer most of these capabilities are SAS Enterprise Miner, IBM SPSS, and StatSoft STATISTICA Data Miner.

3.3.4 Forecasting Software

The recommended capabilities for effective development of forecasting models in industrial applications are as follows.

- Time series analysis capabilities that include generating time series, different time plots, correlations, seasonality adjustments, decompositions, and so on.

- Forecasting model generation capabilities that include the most popular methods, such as exponential smoothing, ARIMA, unobserved components, and so on with a variety of diagnostic statistics and model performance metrics.

- Forecasting modeling with events capabilities that enable the introduction of big discrete shifts in the model development.

- Hierarchical forecasting capabilities that include developing a model hierarchy at the desired level based on the existing business structure and reconciling this with the final forecast.

- Scenario generation capabilities for multivariate-based forecasting models—these different "what if" scenarios can show the impact of the key inputs on the final forecast.

The most powerful software tools that offer these capabilities are SAS Forecast Studio, Automatic Forecasting Systems Autobox and Business Forecast Systems Forecast Pro.

3.3.5 Software Selection Criteria

In addition to the specific technical capabilities of the key software components for a data mining for forecasting system, the following generic selection criteria are recommended:

- Cost depends mostly on the ease-of-use of the corresponding packages. Most of the time the tools based on building blocks (such as SAS Enterprise Miner) or the high-performance forecasting tools (such as SAS Forecast Server) cost more. However, the increased productivity they deliver is significant. An additional advantage is the shorter learning and product adaptation time, which lowers the total cost.

- Functionality—it is strongly recommended that you carefully check whether the necessary technical functionality is available, as described in the previous sections, and to avoid any compromises. The capability to add new methods is also recommended.

- Ease-of-use is enhanced by programming based on building blocks, a high level of automation for data pre-processing and model generation, an interactive graphic interface, and minimal programming necessary to deploy models (all features of SAS Enterprise Guide, for example).

- Report generation is a significant step during model development as well as during model deployment and when transferring ownership to clients. During the model building phase many detailed reports with time series analysis, model diagnostics, or variable selection results are needed for successful decision-making. For model deployment, good reporting capabilities for model performance and value tracking are critical in order to keep the client happy.

- The learning effort required depends on the software's ease-of-use, users' experience in statistics and forecasting within the organization, and the training courses and materials offered by the vendor. Products with a steep learning curve can significantly delay implementation efforts and reduce the impact of the technology for data mining in forecasting.

- Global support 24/7—a fast, professional response to model development and implementation issues that is available globally is critical for the success of data mining for forecasting in industry. This is one of the key factors to consider when selecting the proper software vendor. Very few have the capacity to provide this type of service.

3.4 Data Infrastructure

Developing and maintaining a data infrastructure that can reliably supply the data to the developed and apply forecasting models is a critical step for the final success. The data infrastructure for data mining in forecasting consists of two key parts: internal data from the business and external data from various sources, such as Global Insight, Bloomberg, CMAI, and so on. The essence of both cases is described briefly in the following sections.

3.4.1 Internal Data Infrastructure

Very often creating an internal data infrastructure for data mining in forecasting is the key bottleneck of the whole effort. There are several issues that contribute to this situation. The first issue is the diverse nature of data sources in different parts of the business. This issue is especially difficult to resolve during the transition period after mergers and acquisitions when various types of databases need to be integrated. The second issue is the different time interval and duration with which historical data are kept in the system. Very often the time interval (week, month, or quarter) is different and inconsistent for the historical periods of interest. A similar situation is observed with the duration of historical data. In many cases time history is too short to represent the patterns necessary to build and validate a good forecasting model. The third issue is the structural changes in the business since corresponding models need to be rebuilt with revised history after each significant change.

The internal data infrastructure depends on the corporate data infrastructure. One option to communicate and synchronize the extracts is by using a separate server. (See the example in Figure 3.1.) At the basis of data infrastructure design is the metadata (the data about the data) definition. The cost for maintenance and support of the internal data infrastructure depends on the internal cost structure derived by corporate IT.

3.4.2 External Data Infrastructure

Usually, the data about potential economic drivers are not available internally and need to be delivered by external sources. Examples of such sources are the Bloomberg services[5] with various types of financial data, such as equities, commodities, foreign exchange rates, and the Global Insight services[6] with more than 30 million time series of different nature across the globe, such as prices, economic indicators, and labor costs. The external data are generally consistent, collected in a timely manner, and some have forecast values for a given forecasting horizon. The last feature is very beneficial in the case of using these data as inputs in the multivariate in X forecasting models.

There are two options for delivering external data. The first one is based on accessing the necessary data by direct extracts from the key sources. The second option is based on building an internal database of the most frequently used external data. The advantage of the second approach is the synchronized update of all needed external data, fast search of the specific economic drivers, and more reliable maintenance of deployed models. However, this option requires allocating internal resources for the design and maintenance of the database and training of potential users.

An example of integrating different external and internal data sets in a data set that is appropriate for data mining in forecasting is shown in Figure 3.2. It includes three external data sets (Bloomberg, Global Insight, and CMAI) and two internal data sets. The different data are integrated in the forecasting data set based on a selected starting time and time interval (month, quarter, or year). Those time series with different time intervals are appropriately expanded or contracted in a previous step as described in Chapter 6.

Figure 3.2: Integrating external and internal data in a data set ready for data mining in forecasting analysis

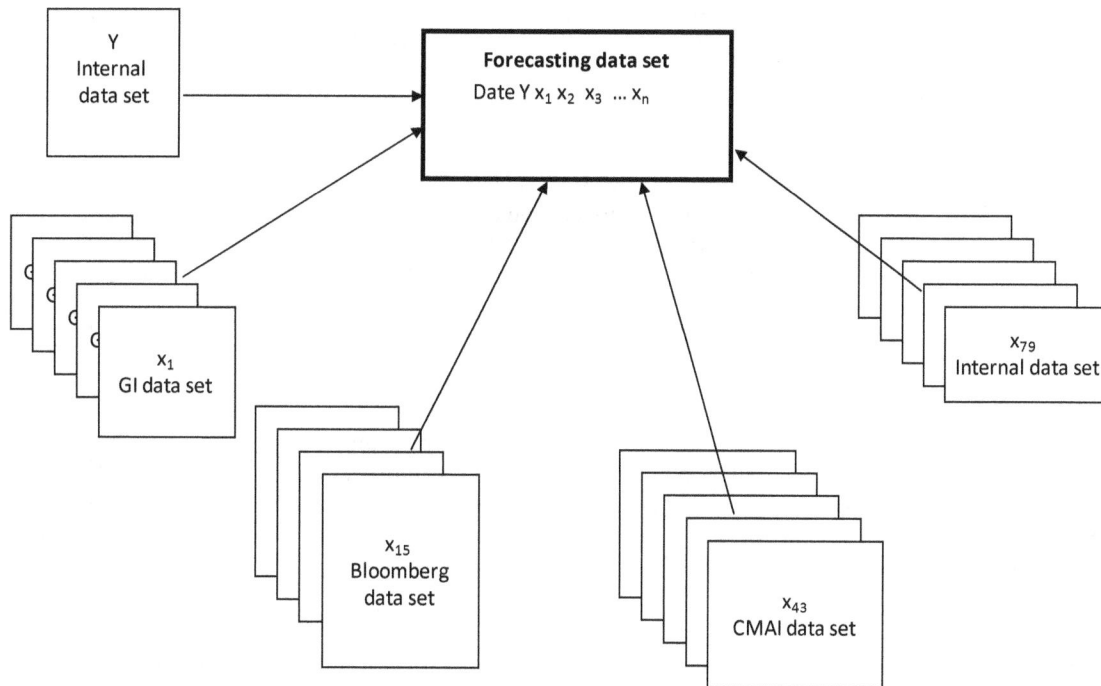

The cost of maintaining and supporting the external data infrastructure depends on the subscription services cost, the cost of developing and maintaining an internal database, and the internal cost of corporate IT.

3.5 Organizational Infrastructure

The objective of this section is to give the reader possible ways to build an organizational infrastructure for data mining in forecasting in a business. We briefly discuss organizing model developers and forecasting users, selecting a proper work process, and integrating everything into the corporate IT environment.

3.5.1 Developers Infrastructure

A key strategic business decision related to a forecasting organization is deciding how much to invest in people that can develop forecasting models. The type of the forecasting development effort and its size depend on the projected demand for forecasting projects in the organization. Other factors that have to be taken into account are as follows:

- the available internal personnel in corporate IT who can support forecasting models by managing the data, infrastructure, and operations
- the strategic commitment of key users for time and resources
- the available internal skills in the area of modeling, statistics, data mining, and forecasting
- the level of experience in applying forecasting projects

Below we briefly discuss three ways to organize developers: (1) external consultant services, (2) distributed developers in organizations (key users of forecasting services), and (3) a centralized group of developers.

External consultant services

This is the minimum-investment solution for when you have low expected demand, no strategic commitment, and a lack of internal resources. The only allocated internal resources are for project management and interaction with the external consultants. However, even in this case, some basic training for forecasting and statistics is recommended. It is preferable to have a well-prepared test case when you begin the working relationship with the external consultants. (Some suggestions on how to prepare an effective test case are given by Michael Gilliland in his book *The Business Forecasting Deal.*) The key advantage of this solution is the minimum cost. The key disadvantage is the total dependence on external resources.

Distributed developers

This organizational structure is appropriate in small or medium-size businesses when the demand for forecasting services is concentrated in several key users, such as marketing and sales, supply chain, and purchasing. Often they prefer to own the whole model development and deployment process and hire experts with forecasting knowledge. In many cases they do not invest in the high-end hardware and software infrastructure, such as SAS Forecast Studio. The key advantage of this solution is the availability to implement forecasting capabilities with internal resources in appropriate business functions at an affordable cost. The key disadvantage is the limited capacity for growth.

Centralized developers group

The best-case scenario for applying data mining in forecasting in larger organizations is by building a centralized group of developers. The group must have the capacity to respond fast to the growing demand of forecasting projects from various sections of a large corporation. The skill set of the developers' team must have a proper balance between system and data support expertise and modeling capabilities in the area of statistics, data mining, and forecasting. An example of key roles in a centralized group of data mining for forecasting is given below.

- The system administrator maintains servers, upgrades software, handles security issues, and interacts with IT.

- The data administrator maintains data integrity, identifies internal and external data sources, and collects and harmonizes data.

- The modeler interacts with clients, identifies system structure and data, pre-processes the data, performs variable reduction and selection, develops, validates, implements, and maintains forecasting models.

- The manager manages the group, delivers needed resources, and brings in projects.

The proper place of this group within a large organization is in the centralized corporate business services. This group serves all potential users so that the return of investment is maximized. The size of the group depends on projected demand. However, at least five to seven developers are needed to be efficient. It is assumed that a period of at least two to three years is needed for the group to establish itself by building infrastructure, hiring, learning, promoting to potential clients, and developing test projects. The funding during this period is centralized and gradually gives way to a self-support mode where projects are supported directly by their clients. The key issue that will determine the fate of this group is whether a sustainable project pipeline can be maintained.

3.5.2 Users Infrastructure

Forecasting users come from different parts of the organization. Typical clients for statistical forecasting services are the marketing, sales, financial, purchasing, and operations planning departments. Forecasting users can be classified in the following four categories, briefly discussed below: (1) forecasting reports users, (2) planners, (3) decision-makers, and (4) top level managers. (A similar user classification for demand-driven forecasting is described in detail in Charles Chase's book *Demand-Driven Forecasting: A Structured Approach to Forecasting.*)

Forecasting reports users

These are the users who passively use the delivered forecasts for information purposes only without making direct business decisions based on specific forecasting results or participating in judgmental forecasting or process planning. Most of the top managers are in this category. Recently many businesses have included forecasts in their regular performance tracking reports distributed to middle-to-top-level managers. The value of forecasting for this category of users is in giving them an awareness of the projected directions of the key performance indicators of interest.

Planners

In contrast to the previous category, planners actively use the delivered statistically based forecasting models in developing their sales, marketing, or operations plans. Very often they also have the right to override the statistical forecasts with their judgmental estimates. In the case of demand-driven forecasting, these are the users in marketing and sales who "shape the demand" based on analytics and domain knowledge. From all the categories of forecasting users, planners are the most educated and directly involved in the model development and deployment loops. They have the decisive role in introducing expert knowledge by defining events, evaluating model performance, and making the final forecasts adjustments. Planners also have the responsibility to recommend to the decision-makers which developed plans, based on the delivered and adjusted forecasts, get final approval.

Decision-makers

This category of forecasting users includes the middle-layer managers at the departmental level who are responsible for the results of the plans recommended by the planners. They also make the final decision for implementing the plans. Part of the decision-making process is balancing the recommended statistically-driven forecasts from the experts (planners and model developers) and the top management push. Often the decision-making process goes through several iterations until a consensus is reached. This category of users is critical for the success of specific forecasting projects and the overall forecasting activities in the business. Success for decision-makers is not based on model performance measured by forecasting accuracy but is based on the expected value measured by the key performance indicators (KPIs).

Top level managers

This category includes the top executives related to finances, IT, and operations. As users, they might have different roles. One critical role is to establish and support financially, for some period of time, the forecasting capabilities in the organization. Executives might request forecasting projects for developing a business strategy as well. It is expected that at any moment the top executives can access the forecasting reports at any level of the organization and keep track of the KPIs. And finally they can actively influence what decisions are made regarding the implementation of the action plans based on forecasting models.

3.5.3 Work Process Implementation

A key component in developing the organizational infrastructure is selecting and implementing an appropriate work process for data mining in forecasting. An example of such a work process is given in Chapter 2. It is also very important to integrate the selected work process with the existing corporate culture. The best-case scenario is to consolidate the data mining for forecasting work process with the existing standard work processes in the organization. If you can do so, the implementation cost and the time for integration into the corporate culture will be significantly reduced. An example of integration with the most popular work process in industry, Six Sigma, is described in Chapter 2. Another example of a popular work process in the case of demand-driven forecasting – Sales & Operation Planning (S&OP) is given in Chase 2009.

3.5.4 Integration with IT

An organizational issue of critical importance for the final success of applying data mining for forecasting is the smooth integration with corporate IT services. Unfortunately the integration process can be bumpy largely due to the different mode of operation of IT. The IT department is often focused primarily on implementing standard solutions across the business. The focus of data mining for forecasting is on delivering custom and, therefore, nonstandard solutions using specialized software. It is a well-known fact that maintenance and support of data mining for forecasting systems requires specialized expertise rather than the typical skill sets in corporate IT. One potential solution to this problem is allocating the specialized system support within the developers group. Part of the responsibilities of the developers' group system administrator is to coordinate all activities with IT. While establishing the developers group, however, support from top IT management is needed to promote the necessary changes beyond the IT standards.

---

[1] http://www.spss.com

[2] http://www.statsoft.com

[3] A good comparison between SAS Enterprise Miner, IBM SPSS, and StatSoft STATISTICA Data Miner is given in the *Handbook of Statistical Analysis & Data Mining Applications* (Nisbet et al. 2009).

[4] http://www.sas.com/technologies/dw/etl/access

[5] http://www.bloomberg.com/professional

[6] http://www.globalinsight.com

Chapter 4: Issues with Data Mining for Forecasting Application

4.1 Introduction

One of the big differences between mechanistic or statistical models applied in manufacturing and forecasting models that are applied mostly in a business context is that the latter is a continuously changing environment and performance is judged according to users' sometimes unrealistic expectations. The motivation for using data mining capabilities in the forecasting process is to mitigate these issues by improving model quality and gaining the user's trust by including economic drivers chosen by the user as model inputs. User acceptance is not automatically guaranteed by giving users knowledge about data mining, however. The final success of the application project depends on a good understanding of the key technical and nontechnical issues related to data mining and forecasting. These issues are discussed in this chapter.

Technical issues include data quality and the limitations of data mining and forecasting methods. Nontechnical issues include managing forecasting expectations, handling the politics of forecasting, and avoiding bad practices. A generic checklist—"Are we ready?"— which summarizes the level of preparedness to begin a successful industrial forecasting project, is given at the end of this chapter.

4.2 Technical Issues

The objective of this section is to clarify the technical limits of the applied forecasting systems. The emphasis is on the issues related to low-quality data and the limitations of the corresponding data mining and forecasting methods.

4.2.1 Data Quality Issues

Data quality is the key factor for success of any data-driven modeling technique. The requirements for time series data quality are even higher due to the nature of building forecasting models, which is based on discovering consistent patterns in time. Improper time alignment, data glitches, or high levels of noise significantly decrease the chance for developing high-performance forecasting models independently of the applied data mining or forecasting techniques. Some of the key data quality issues are discussed briefly below.[1]

Consolidating data from different sources

According to Ronald Pearson, a major source of data quality errors is related to manual entry and manual intervention (2006). Data consolidation is one of the data collection activities most vulnerable to manual intervention and human error. In almost any data mining for forecasting application, the data are collected from many sources and stitched together. Unfortunately, gathering data from different departments, companies, and external sources creates opportunities for harmful discrepancies during data integration. Some sources of potential issues in data consolidation are the following:

- Inexact soft keys—an important condition to integrate data from multiple sources is the availability of a common join key. If there is no definite key, like a credit card number, inexact joins, such as names or addresses, are used as a proxy. As a result, joining the different sources could be messy. To increase the matching rate, additional heuristic rules are added. Unfortunately, the potential for erroneous matches is not completely eliminated.

- Automatically generated data sets—it has to be taken into account that processing errors, such as truncating data, reverting to defaults, syntax errors, and numerical overflow, are more a rule than an exception in automatically generated data. It is strongly recommended that data consistency be checked after each automatic generation.

- Different metadata definitions—different departments in the business can have different definitions of the same entity. For example, a customer could be a person, a company, a service order, or a bill. Other examples are the different representations of dates (month/day/year versus day/month/year) and the definitions of different economic indexes (they might begin from different start dates).

- Different scales—integrating data with different conversion factors could create a mess. Especially vulnerable are the data related to volume as data can be measured in various ways depending on the nature of the specific products (liquid, material, service, and so on).

Noisy data

All real data are not ideal, and data can include some level of random noise caused by uncontrolled measurement errors and human factors. If the noise is at a low level, the data mining and forecasting algorithms can still generate proper variable selection and forecasting models. At a certain noise level, however, it's a challenge for the forecasting algorithms to capture time patterns (such as trend, seasonality, and cyclicality) in the noisy time series. Of special importance in this category are the efficient market systems with random walk-type behavior and low predictability. Another well-known case is demand volatility (Gilliland 2010). Demand volatility is the erratic behavior of sales driven by random promotions. It is recommended that the volatility be quantified by the coefficient of variation (CV), which is calculated by the ratio of standard deviation and the mean. The CV indicates the closeness of the data to its mean and the smoothness of the pattern. An example of time series is in Figure 4.1, where the CV of the data is 52.1%. In similar cases, it is strongly recommended that development efforts be stopped and the algorithms not be misused. The belief that the garbage in the data can be compensated with more sophisticated modeling algorithms is wrong. In the case of demand data, it is recommended that the volatility be reduced with less frequent promotions (Gilliland 2010).

Figure 4.1: An example of noisy time series with inherent randomness

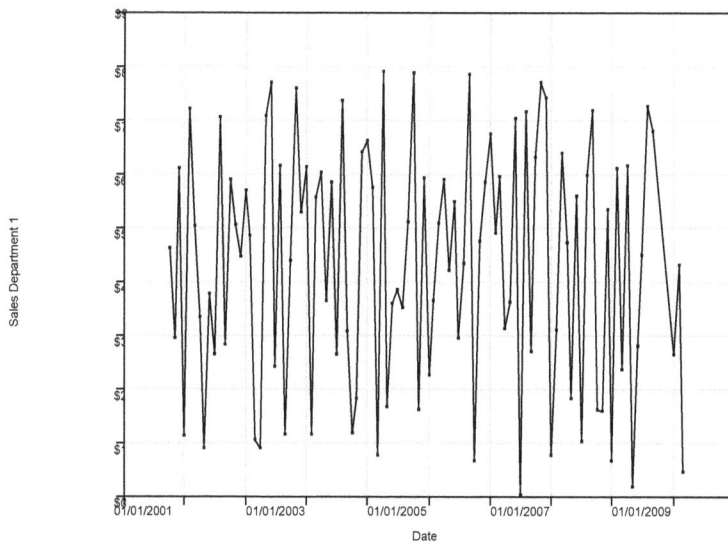

Data glitches

In contrast to data noise, a data glitch is any change introduced in the data by causes external to the process that generates the data (Dasu and Johnson 2003). Examples are unreported changes in layout, dropped data, switched data fields, duplicated records, and so on. The key issue with data glitches is that some of them are obvious and easy to detect, while others (for example, a swap of two numerical fields) are subtle and difficult to detect. Often, these glitches are observed only after they have been compounded several times, resulting in significant deviations from the true values.

Missing data

Missing data are defined as data values that should be present in a data set, but, for various reasons, are absent (Pearson 2005). Some sources of missing data are discussed briefly:

- Power failures in computerized data collection systems—these can result in the loss of data values that would normally be recorded. Fortunately, due to the low frequency of economic data collection and the increased hardware reliability, this potential source of missing data is a very rare event.

- Different metadata in data sources—this can result in the loss of data fields and values due to name inconsistencies. This scenario can happen during structural changes in the organization and after mergers and acquisitions.

- Different frequency of data collection—data collected with low frequency (for example, quarterly) can result in the loss of data in the higher frequencies (for example, monthly). However, the missing data can be imputed by time series expansion methods, which is discussed in Chapter 7.

- Different data collection history—some time series might have different histories, and this can result in the loss of data for the time period of interest. The frequency of internal historical data could be different as well. In some cases, the quality of historical data outside North America and Europe is low and critical indicators are missing.

Time series data with short history

In the case of a new product start-up or when establishing a new business, the available data history is very short, even for the highest possible data collection frequency. (See an example with monthly data in Figure 4.2.)

Figure 4.2: A short history data set example

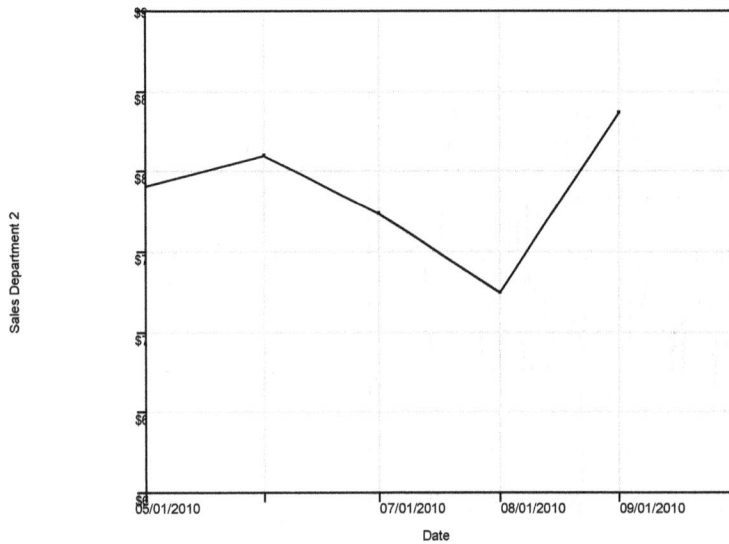

Answering the question, "What is the minimum number of time series observations to build a good forecasting model?" depends on the nature of the process, the selected forecasting method, and the targeted forecasting horizon. In the case of expected seasonality and cyclicality, at least a full cycle of collected data is expected, but multiple cycles are preferred.

Time series outliers

An outlier is defined as an entry in a data set that is anomalous with respect to the behavior seen in the majority of the other entries in the data set. A special category is time series outliers, which are of interest in forecasting. They can be defined as data points that violate the general pattern of smooth or otherwise regular (for example, discontinuous, but periodic) variation seen in the data sequence (Pearson 2005). An example of a time series outlier is shown in Figure 4.3.

Figure 4.3: A time series outlier example

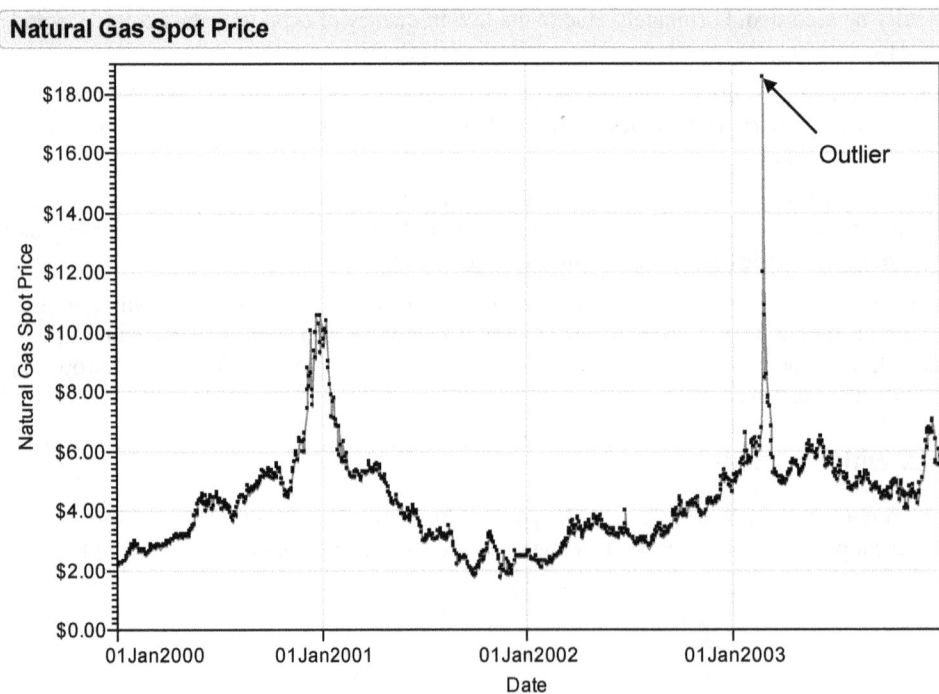

Figure 4.3 shows the daily natural gas spot prices in the US Gulf Coast from January 2000 to May 2004. The sudden jump in the price from $6.00 to $18.48 around February 25, 2003 is an obvious time series outlier. It is a short-term event not based on a real economic trend. An opposite example is the gradual peak of the price from between $5.00 and $6.00 to $10.00 around January 2001. It cannot be identified as a time series outlier because it fits consistently with the trends before and after the peak.

The proper identification and interpretation of time series outliers are of critical importance for model development because they can result in significant changes in dynamic data characterizations. On the one hand, time series outliers can be a detector of data glitches that can be removed in the analysis. On the other hand, time series outliers can be a detector of events, such as September 11, 2001 or a catastrophic failure in a key raw material production site. In this case, it is recommended that they be included as events in the model-building process by defining the proper shape of the event.

4.2.2 Data Mining Methods Limitations

The main hypothesis in this book is that business-related forecasting models can be improved by adding key economic drivers selected via data mining methods. This process, however, is not flawless, and it has its own limitations. The tasks of variable reduction and selection can become very challenging in the case of a large number of potential highly correlated inputs (hundreds to thousands). Often, the different data mining methods can give different results, thus creating confusion for the user or the less-experienced model developer. This is not a surprise because the methods are based on different scientific foundations. (See method descriptions in detail in Chapter 7.) One solution to the problem is not to rely on one method only, but to use several data mining approaches for variable reduction and selection and to compare and combine the results. It is a good practice for model developers to be aware of the limitations of the specific methods in advance. To fill this need, some of the limitations of the most-used approaches, such as stepwise regression, PCA, clustering algorithms, decision trees, and genetic programming, are briefly discussed below. The emphasis is mostly on the limitations related to data reduction and selection. In analyzing the limitations, one has to take into account that these methods were originally designed for transactional data, not time series data.

Stepwise regression methods limitations

The most widely used method for variable selection is stepwise regression, in which choosing predictive variables is carried out by an automatic procedure based on a sequence of various statistical tests, such as F-tests, t-tests, adjusted R-square, Akaike information criterion, and so on. The two key stepwise methods are forward selection and backward elimination. Forward selection starts with no variables in the model, and then tries out the variables one by one, including them if they are statistically significant. On the opposite end is backward elimination, which involves starting with all candidate variables and testing them one by one for statistical significance, deleting any that are not significant. The key issue of stepwise regression is that the different stepwise methods generate different variable selections from the same set of data. On top of that, none of the methods guarantee that the selected set of variables is optimal. Three problems with stepwise applications have been explored in some detail (Montgomery et al. 2006). First, computer packages use incorrect degrees of freedom in their stepwise computations, resulting in a greater likelihood of obtaining spurious statistical significance. Second, stepwise methods do not correctly identify the best variable set of a given size, especially in cases of multicollinearity (highly correlated inputs). Third, stepwise methods tend to capitalize on sampling error and, thus, tend to yield results that are not replicable. One of the most severe drawbacks of stepwise regression was discovered recently by Leo Breiman, who claimed that the stepwise regression algorithm lacks stability, in that a small perturbation on data might yield a very different model (1996).

As a result, stepwise regression should be applied with caution. It is recommended in statistical literature that the backward elimination algorithm be applied because it is less adversely affected by the correlative structure of the regressors than forward selection (Montgomery et al. 2006).

Principal Component Analysis (PCA) limitations

Another widely used approach for variable reduction and selection is PCA. It is based on the idea of constructing an independent linear combination of input variables in which the coefficients (or the eigenvectors) capture the maximum amount of variability in the Xs or input data itself (that is, PCA does not include the target variable). The linear combination with the largest variance in data is called the first principal component (PC). The second principal component is orthogonal to the first PC, and it captures the largest variation in the rest of the data. The next PCs are organized in a similar fashion. As a result, the system can be described by a low-dimensional set of PCs that capture the lion's share of variability in the data.

Some of the known PCA limitations are described briefly:

- Sensitivity to outliers—an outlier might result in a dramatic shift in the principal component line and bias the whole analysis.

- Limited to linear relationships—the PCA dimensionality reduction assumes linear relationships between the PCs and the original variables.

- Useful for correlated inputs only—if the original input variables are uncorrelated, then the PCA is useless.

- The assumption that observed data is independent and identically distributed (i.i.d.)—in general, it is invalid when modeling time series because some of the data are auto-correlated.

- The directions maximizing variance do not always maximize information—this drawback is closely connected to the fact that the PCA does not perform linear separation of classes, linear regression, or other similar operations, but it merely permits the input vector to be best restored on the basis of the partial information about it. All additional information pertaining to the vector (such as the identification of an image with one of the classes) is ignored.

Clustering algorithms limitations

Clustering algorithms are used mostly for variable reduction. Cluster analysis divides the data into groups (clusters) that are meaningful and useful. The groups of data objects are based only on information in the data. The idea is to define a cluster as a group of objects that are similar within the group and different from the objects in another group. (See more details in Chapter 7.) One of the key issues of this data mining approach is determining the correct number of clusters, which is still an ad hoc exercise. Usually, we are looking for the number of clusters at which there is a knee, peak, or dip in the plot of a quality metric when it is plotted against the number of clusters. Unfortunately, the clusters could be intertwined or overlapping and that could blur the selection.

Decision tree limitations

Decision trees consist of a set of rules for dividing a large population into smaller groups with respect to a particular target variable. In the data mining for forecasting process, decision trees are used for variable reduction. (See more details in Chapter 7.) The key technical issue with decision trees is that they can be unstable, and small variations in the data (such as those made by randomization) can cause the generation of very different trees. This is especially true in cases where the quality of splits for the different variables are close to each other in value. In this case, a small variation in the data is enough to influence the picking of one split over another.

Another issue is that decision tree models created from numeric data sets can be very large and complex. This can make the model difficult to interpret, understand, or justify.

Genetic programming limitations

Genetic programming (GP) is a novel data mining technique based on evolutionary computation. Of particular importance to data mining for forecasting is the application of GP, called symbolic regression, for automatically generating algebraic equations from the available data (Koza 1992). Symbolic regression has a built-in mechanism that selects the variables that are related to the target variable during the simulated evolution and gradually ignores variables that are not. The variable selection is based on the normalized frequency of appearance of the inputs in all equations and sub-equations at the end of the evolutionary process. One of the key advantages of this variable selection method is that it is, in principle, nonlinear because it is based on numerous nonlinear dependencies among input variables fighting for high fitness during simulated evolution. (For more details, see Chapter 7.)

Some limitations of GP-based variable selection are briefly described:

- Slow model generation—the longer time relative to the other modeling approaches is based on two factors:
    - Simulated evolution requires a prescribed number of generations to derive models with high quality.
    - Due to the stochastic nature of the algorithm, it is necessary to repeat it several times to achieve consistent results.
- Lack of statistical basis for variable selection—at this early phase of GP technology development, variable selection is based on ad hoc thresholds.
- Results are not repeatable—due to their stochastic nature, different GP model generation processes give slightly different variable rankings.

4.2.3 Forecasting Methods Limitations

In principle, no forecasting method is capable of accurately predicting the future when history does not repeat itself (Makridakis et al. 1998). However, it is well known that this is a very strong assumption. Due to social, technological, and structural economic changes, history never repeats itself exactly. Another limiting factor that has to be taken into account in forecasting is the level of randomness of complex social systems. For example, efficient markets have a random walk-type behavior, which reduces their predictability. The generic limitations of time series forecasting, the specific drawbacks of the key forecasting methods, and the constraints of the different forecasting horizons are discussed briefly.

Generic limitations of time series forecasting

- Requires a representative historical data set that captures the trend, seasonality, and cyclicality in the data—collecting these data sets is one of the biggest challenges in developing forecasting models. The most difficult requirement is representing the cyclicality, which is business specific and might require a long history containing at least several cycles. Very often, internal data sets are too short for representing cyclicality.

- Changes in fundamental conditions can cause the forecast to vary from actual results—the hypothesis that the identified patterns of the past will remain consistent into the future is tested with every new actual piece of data. When the subject being forecast undergoes fundamental changes, such as when revolutionary new technology is introduced or a significant business transformation occurs, time series forecasting cannot identify new patterns automatically.

- Assumes no change in the environment—in a similar fashion, changes in the economic environment, such as introducing new taxes or environmental regulations, cannot be captured automatically, and new model development is needed.

- Limited capability in anticipating rare events—adding events improves forecast accuracy after the event. Unfortunately, it is of limited value in forecasting a future event before it occurs.

Limitations of the key forecasting methods

A short list of the limitations of the key forecasting methods follows. (See Chapters 8–11 for more details.)

Moving average limitations

- does not respond to a rapid shift
- assumes no environmental changes
- equal weighting of all the data in the series

Exponential smoothing limitations

- still might not respond to a rapid shift
- assumes no environmental changes

ARIMA limitations

- requires statistical and domain knowledge
- more complicated to develop
- requires more data

Multivariate models limitations

- depends on the forecast of independent variables.
- depends on variable selection.
- depends on the strength of the statistical relationship between the economic drivers and the forecasted variable.
- model development is of an order of magnitude more complex.
- model maintenance is more costly.

Understanding the limitations of forecasting time horizon

Another important factor that has to be taken into account when defining realistic forecasting expectations is the forecasting time horizon length. Obviously, the longer the time horizon, the higher the uncertainty in the forecast and the more necessary it becomes to have a long historical data set. Forecasting time horizon can be defined in three categories—short-term (up to three time samples in the future), medium-term (up to 18 time samples in the future), and long-term (more than 18 time samples in the future). [2] The limitations for the short-term, medium-term, and long-term time horizons are discussed briefly.

- Limitations for short-term forecasting horizon—this is the best-case scenario for forecasting because the key economic phenomena that contribute to the forecasting models (trend and seasonality) do not change much or do not change frequently over a short time span (Makridakis et al. 1998). Most of the available forecasting methods can be used for short-term forecasting with varying success depending on their specific limitations. The main limitation in short-term forecasting is the appearance of an unexpected event, such as a weather disaster or an unplanned shutdown.

- Limitations for medium-term forecasting horizon—the main limitation of this forecasting category is the influence of business cycles and main economic moves, such as recessions and booms. Only limited forecasting methods, such as ARIMA, are usable in this case. Complementary business cycles estimation is strongly recommended. The recommended method is using multivariate ARIMAX-based forecasting methods.

- Limitations for long-term forecasting horizon—in most of the cases, the available time series forecasting methods are unreliable over a long time span. In addition to time series forecasting models, it is recommended that you identify and extrapolate megatrends, use analogies, and use scenario-based forecasting

4.3 Nontechnical Issues

The objective of this section is to provide guidelines on how to handle some of the important issues when dealing with the different users of forecasting systems. A careful analysis of potential nontechnical issues is as important for the success of the forecasting project as the knowledge of the methods' limitations. Some important issues, such as managing forecasting expectations, handling the politics of forecasting, and avoiding bad practices are discussed.

4.3.1 Managing Forecasting Expectations

One of the main nontechnical issues in forecasting is communicating clearly, to all stakeholders, what the realistic expectations are in terms of performance. We will briefly discuss the most important action items in this process, such as explaining the meaning of a forecast and the limitations of forecasting, avoiding model development with bad data, and defining realistic performance objectives.

Explain the meaning of a forecast

The first step in managing forecasting expectations is to educate the user about the key difference between model predictions that are based on first principles or statistics and a forecast. Although the accepted standard for the former is a number, a forecast includes three components: (1) a point forecast for each sample within the forecasting horizon, (2) low and high confidence limits around the point forecast, and (3) a forecasting horizon. Only when looking at these three components as a whole can you evaluate forecasting performance. Usually, non-educated users focus their attention on the point forecasts only, and ignore the meaning of the confidence limits and the forecasting horizon. Often, this ignorance creates confusion, and the modelers should use any opportunity to emphasize the importance of all three components in interpreting the forecasting results.

Explain forecasting limitations

The next step in managing forecasting expectations is explaining that forecasting is not a crystal ball and it has some limitations, as described in the previous section. Above all, users must understand that believing that statistical forecasts are infallible is wishful thinking. Because previous history never exactly repeats itself, they should realize that the economic environment is continuously changing and there is always the potential for unexpected events that can deteriorate model performance significantly.

Avoid the garbage in, gold out trap

The first necessary reality check when developing forecasting applications comes after the data collection and preprocessing phase. It is strongly recommended that model development not be started if the data has poor quality or a short history. The developers must resist the idea that some users have that poor quality data can be compensated for with more sophisticated algorithms (instead of garbage in, garbage out, idealistic users believe in the garbage-in-gold-out effect).

Define realistic performance objectives

According to Gilliland, the only reasonable expectations for forecasting performance is to beat a naïve forecasting model (or, at the least, not do worse), and to continuously improve the process (Gilliland 2010). The reasoning behind this idea is that by using a statistical metric alone, such as MAPE, you have no idea how efficient this metric is. By comparing it to the worst-case scenario forecast (based on a naïve model), at least you have a clear answer if the performance is improving and whether it makes sense to apply a forecasting model. Setting other arbitrary goals, based on absolute MAPE values or industry benchmarks, is not recommended (Gilliland 2010).

4.3.2 Handling Politics of Forecasting

Politics is an inevitable part of life in applying forecasting systems because the derived forecasts can influence the decision making of various stakeholders in an organization, including top management. It is practically impossible to offer complete guidelines for handling all political issues related to forecasting, but some important topics that might help the reader are discussed briefly below.

Define competitive advantage

The first step in opening the door to forecasting in an organization is convincing the key decision makers that forecasting capabilities gives them a competitive advantage in the contemporary business world. Convincing decision makers of this can create an environment in which there is strategic management support for applying forecasting systems and such activities are unaffected by future political swings. Some of the arguments that define the benefits of forecasting and data mining are given below.[3]

- Proactive management—forecasting enables decision making that anticipates future changes and reduces the risk of taking inadequate actions.

- Improved planning—a typical case is one of the most popular forecasting applications, demand-driven forecasting with significant economic impact. (See examples in Chase 2009.)

- Understanding business drivers—the unique combination of data mining and forecasting gives insight for the key drivers related to the specific business based on data and statistical relationships.

- Defining a winning strategy—long-term forecasting scenarios are critical for developing future business strategies.

Write a detailed project charter

A well-written project charter with clearly defined objectives, deliverables, responsibilities, and an action plan is the best way to avoid any future political games. It is strongly recommended that the model developers take the lead for this process and help the users in the project specification process. Usually, model developers have more experience in anticipating potential problems, and they try to avoid them through better planning. For example, it is especially important to include in the charter a detailed action plan for data collection with clear responsibilities defined for those who extract internal and external data. Another important topic that needs clarification in the charter is an agreement to define the performance objectives based on data quality after the preprocessing phase. Discussing the maintenance and support issues from the beginning and including them in the charter is strongly recommended.

Balance stakeholders' interests

Understanding the stakeholders' interests and concerns is a winning strategy for avoiding potential political issues. For example, typical business clients for forecasting projects expect trustworthy results that are similar to their own expert judgment and can help them make correct decisions and perform profitable planning. They prefer simple explanations and consistent performance. Their main concern is trying to avoid a forecasting fiasco and its negative impact on their careers. The business clients underestimate the issues of data collection and data quality. Most of them are strong believers in the garbage-in-gold-out effect. The interested reader can find a good analysis of the interests and concerns of the three most common stakeholders in advanced analytics—business clients, model developers, and IT and data experts in *Data Preparation for Analytics Using SAS* (Svolba 2006).

Handle biases

One of the realities in applying forecasting is that users have some preliminary opinions about how the future will look based on their knowledge. Handling this biased vision is one of the biggest challenges in managing forecasting projects. The most widespread bias is based on overconfidence in the power of modern forecasting algorithms and ignoring forecasting limitations. Most people are overly optimistic in their forecasts while they significantly underestimate future uncertainty. Often, people replace forecasting with their own wishful thinking.

Another bias is called the recent event bias (that is, the habit of humans to remember and to be greatly influenced by recent events and to forget the lessons from the past). For example, research findings have shown that the largest amount of flood insurance is taken out just after the occurrence of a flood and the smallest amount just before the next flood takes place (Makridakis et al. 1998).

Some subject matter experts are biased toward their hypothesis about potential economic drivers. If their list is not supported by the data and their favorite economic drivers have not been selected by the data mining algorithms, they want detailed explanations. It is strongly recommended that the statistical basis for the variable reduction be described carefully and data be shown. In some situations, it is possible to reach a compromise and include the expert-defined economic drivers in the inputs used for forecasting and let the forecasting algorithm do the final selection.

Avoid political overrides

In some cases, the statistical forecasts are corrected due to clear departmental political purposes. One extreme is sand-bagging when some sales departments lower the statistical forecast to reduce their sales quota in order to guarantee their bonuses. The other extreme is sales departments that have losses due to backorders. They prefer to raise the statistical forecast in hopes of managing inventory levels via the sales department forecast (Chase 2009).

Another typical case of political override is "evangelical" forecasting, defined by Gilliland. In this case, top-level management provides numbers, and the forecaster's task is to justify statistically the order given by upper management (Gilliland 2003).

4.3.3 Avoiding Bad Practices

Forecasting has been applied in different businesses for more than 30 years. A lot of lessons have been learned in this process. Of special importance for future applications is the negative experience due to implementation mistakes, biases, and organizational failures. Although success stories and best practices are well publicized by businesses and software vendors, there is very little in the literature related to analyzing the missteps and failures in applying forecasting projects.[4] Some of the key poor practices are discussed briefly below. It is strongly recommended that they be avoided in future forecasting applications.

Creating high expectations based on impressive historical data fit

A common mistake made mostly by inexperienced modelers is giving into the temptation of selling forecasting models based on their fit on historical data. It is especially dangerous if the fit does not include model validation performance measures on holdout data. This overly optimistic approach raises unrealistic expectations for users and can destroy a modeler's credibility when the forecasts significantly disagree with the actual data (as is the usual case with over-fitted models). To avoid this bad practice, it is recommended that, before you communicate the model performance based on available data history, you carefully validate the model on different holdout periods, if possible. An advantage of this approach is that comparing the model validation fit on different holdout periods demonstrates model performance robustness. It is strongly recommended that you officially communicate to users only the robust performance metric on the selected holdout periods with a warning that the expectations for performance are limited to the assumption that the system will not change in the future.

Obsession with high accuracy

Another observed bad practice in applied forecasting is the obsession with high accuracy of modelers and users. It demonstrates a one-dimensional view of the goals of forecasting efforts by ignoring all related factors. The following quote from one of the leading researchers in forecasting, Spyros Makridakis, explains the substance of this bad practice: "Concentrating on accuracy is like trying to melt an iceberg by heating the tip: when forecasting accuracy is slightly improved, other managerial problems of implementation rise to the surface to prevent the full realization of forecasting's promise" (Makridakis et al. 1998).

Forecasting has little or no impact on decision making

One of the worst-case scenarios in applying forecasting is when the delivered forecasts are not used for making business decisions. There are various reasons for this bad practice. It could be due to communication problems between model developers and forecasts users. Another situation is when the users are at a low level of the organization and have no impact on decision making. A possible explanation is users' distrust of the delivered forecasts. This could be based either on initial skepticism about the potential of statistical forecasting (typical for some "super" experts in the field) or consistently disappointing disagreements between the forecasts and the actual data. To avoid this situation, it is recommended that the actions to be taken in response to bad model performance be clarified and raised up front in the project charter.

Low priority on model maintenance and performance tracking

A common practice for model developers is to pay less attention to deployed models and to outsource maintenance to less-experienced personnel. Unfortunately, this practice can lead to bad relationships with the users due to growing disappointment about the results and the service. Of special importance is the case of consistent performance degradation with corresponding economic loss due to bad forecasting. Often, the solution is model re-development, which requires the involvement of highly skilled model developers. To avoid this bad practice, it is recommended that a service contract be negotiated with the user for a specific time period that includes the model developers in the performance tracking and analysis.

4.3.4 Forecasting Aphorisms

Often phenomena and issues are best captured by the short satirical style of aphorisms. In *The Business Forecasting Deal* (2010), Mike Gilliland defines several forecasting aphorisms that are easy to remember and represent the essence of the discussed forecasting issues. The forecasting aphorisms with their corresponding corollaries are given below (reprinted with permission). The interested reader can find more details in Chapter 9 of Gilliland's book.

Aphorism 1:

Forecasting is a huge waste of management time.

Aphorism 2:

Accuracy is determined more by the nature of the behavior being forecast than by the specific method being used to forecast it.

Corollary

Do not set arbitrary forecasting performance goals.

Aphorism 3:

Organizational policies and politics can have a significant impact on forecasting effectiveness.

Aphorism 4:

We cannot control the accuracy achieved, but we can control the process used and the resources we invest.

Corollary

Do not overspend in pursuit of unrealistic accuracy goals.

Aphorism 5:

The surest way to get a better forecast is to make the demand more forecastable.

Corollary

Any knucklehead can forecast a straight line.

Aphorism 6:

Minimize the organization's reliance on forecasting.

Aphorism 7:

Before investing in a new system or process, put it on the test.

4.4 Checklist "Are We Ready?"

A checklist that summarizes the steps necessary to begin a successful forecasting project, and avoid bad practices and potential technical and nontechnical issues, is proposed in this section. It is assumed that the organization has already invested in the necessary infrastructure for data mining and forecasting (as discussed in Chapter 3). It is assumed that there is a business request for developing a forecasting project that includes a commitment for allocating internal resources. Under these assumptions, the following checklist is recommended for successful forecasting project management.

- Clarify expectations regarding the value created by using forecasts in business decisions.
- Include key business stakeholders on the project team.
- Do not begin project work without an approved charter.
- Approve first version of the charter only if the following are true:
  o   A value statement is included.
  o   Responsibilities among stakeholders are clearly defined.
  o   A realistic estimate of resources needed is made.
  o   Deliverables are defined in realistic terms.
  o   A statement for maintenance support for a specific time period is included.

- Identify potential economic drivers and data sources.
- Collect and preprocess data.
- Revise first version of the charter with the results from the data reality check (if the available data are of low quality).
- Begin data mining and forecasting steps according to the work process defined in Chapter 2.

---

[1] Data quality issues are discussed in detail in *Exploratory Data Mining and Data Cleaning* (Dasu and Johnson 2003) and *Mining Imperfect Data: Dealing with Contamination and Incomplete Records* (Pearson 2006).

[2] The forecasting time horizon classification is according to Spyros Makridakis (Makridakis et al. 1998).

[3] Good examples of benefits of advanced analytics are given in *Analytics at Work* (Davenport et al. 2010).

[4] An exception is Gilliland's book (2010).

Chapter 5: Data Collection

5.1 Introduction

The importance of data collection is well understood by modelers, and even praised in a popular data mining joke that goes, "In God we trust. All others bring data." As many practitioners know, bringing high-quality data is not a trivial task, and it requires good knowledge of the target forecasting variables and the available internal and external data. The main issues related to extracting the necessary knowledge about the data from the experts and organizing an effective data collection effort for implementing data mining for forecasting are discussed in this chapter. The chapter begins with a brief description of the process for identifying the key economic drivers based on expert knowledge and through an analysis of the system structure and the available data. The topics of defining the internal and external data sources and the metadata follow. Special attention is given to the different methods for data extraction and alignment. The importance of automating data collection for model deployment is discussed at the end of the chapter.

5.2 System Structure and Data Identification

One of the key differences between experienced and inexperienced modelers is how they approach data collection. Inexperienced modelers ignore the existing knowledge about the system. They jump to start collecting the data based on their perceptions of the problem. In contrast, experienced modelers try to extract as much knowledge as possible from the experts. They organize data collection based on the recommended list of potential economic drivers. Unfortunately, the process for defining the system structure with its corresponding data is not well discussed in the data mining and forecasting literature.[1]

The objective of this section is to partially fill this gap and give the reader a condensed overview of this important topic. It includes a brief discussion of the issues of representing the captured knowledge with mind maps, system structure knowledge acquisition, and data structure identification. These issues are all focused on potential economic drivers.

5.2.1 Mind-Mapping

According to one of the gurus of mind-mapping, Tony Buzan, "A mind map harnesses the full range of cortical skills—word, image, number, logic, rhythm, color, and spatial awareness—in a single, uniquely powerful technique. In doing so, it gives you the freedom to roam the infinite expanse of your brain" (Buzan 2003). Developing a mind map is very intuitive. It requires minimal training using most available software tools.[2] An example of a mind map of the work process for data mining in forecasting (presented in Chapter 2) is shown in Figure 5.1.

Figure 5.1: An example of mind-mapping

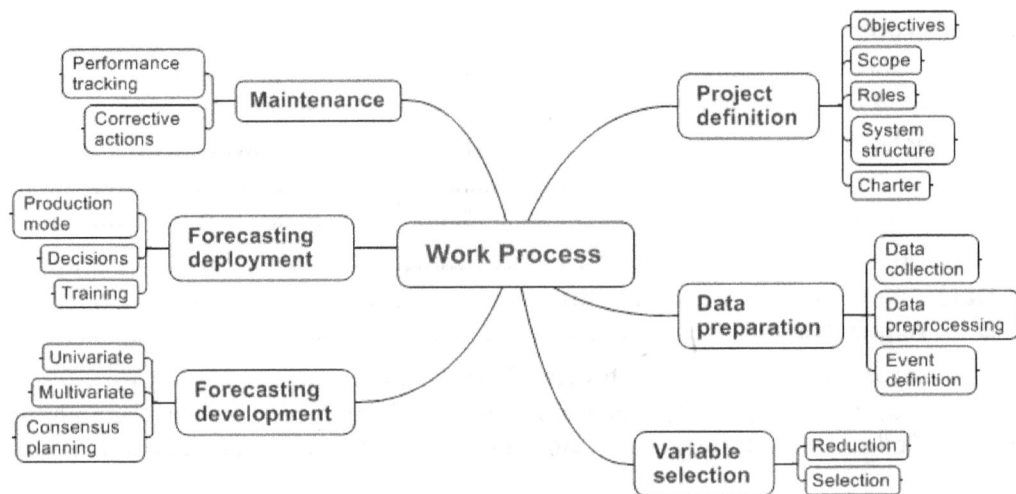

The creation of the map begins in the center with a block representing the key topic or theme of the map. In this specific case, this is the work process. The next elements in the mind map are the main branches that radiate from the central block. These branches signify the major subject areas. In this specific case, they are the key steps in the work process. Because our brains can typically hold about seven pieces of information in our short-term memory, the recommended number of main branches per mind map is between five and nine (Nast 2006). If there is an order of information in the mind map, it is read clockwise, beginning with the main branch at the one o'clock position (project definition branch in Figure 5.1). Each main branch can include a hierarchy of sub-branches that refine the corresponding main topic. In Figure 5.1, the sub-branches include only one level down, which are the substeps of the corresponding key steps of the work process for data mining in forecasting. To avoid creating complex mind maps and to improve their visual impact, it is recommended that short names and branches be used with no more than three hierarchical levels.

One of the advantages of mind-mapping is that it is possible to look at the system at different hierarchical levels by collapsing or expanding to a level. For example, by selecting the level one view of the mind map in Figure 5.1, you can see only the main branches, which represent the key work process steps. There are other options to capture knowledge in mind maps, such as adding notes, timing, prioritizing, relationships, resources, and so on. If you want to increase your proficiency in developing mind maps, you can find more details in *The Mind Map Book* (Buzan 2003) and *Idea Mapping* (Nast 2006).

5.2.2 System Structure Knowledge Acquisition

One of the key advantages of mind-mapping is being able to represent expert knowledge shared during brainstorming sessions when the substantial information about the system is captured and classified in real time. It requires good facilitating skills such as listening to the experts, making decisions about the number of details to be included in the map, reaching consensus on the proposed key words, and so on. To help the participants, it is strongly recommended that a mind map on a similar topic be shown at the beginning of the session. Ideally, the facilitator can prepare a prototype of the mind map with an idea of the initial structure illustrated with main branches.

An example of system structure knowledge acquired and captured in a mind map is shown in Figure 5.2.

Figure 5.2: An example of a system structure mind map

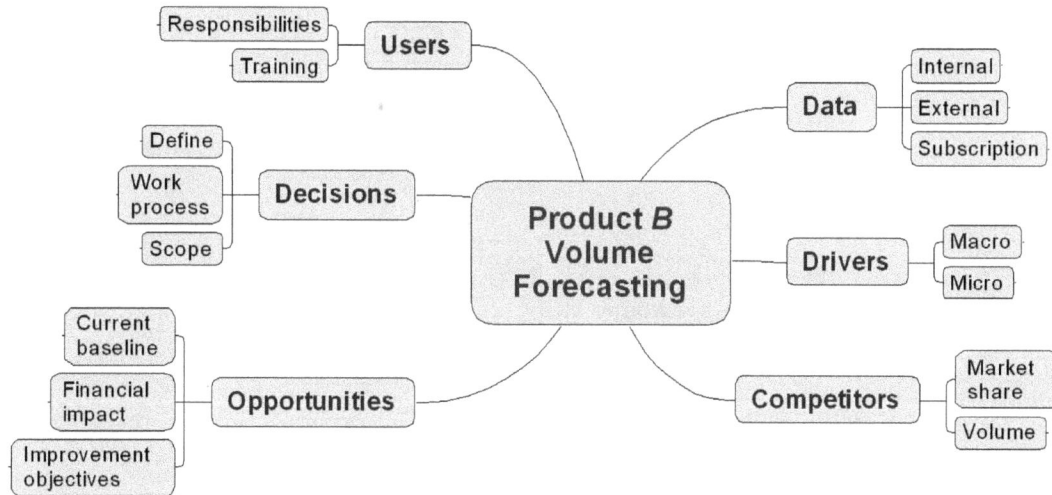

The mind map relates to a specific project for product *B* volume forecasting, represented in the central block. The key branches, defined by the experts, show the main issues of interest when developing this forecasting model, such as available data, potential economic drivers, competitors, opportunities from using the forecast, decisions driven by the forecast, and potential users. The next level of detail for each main branch is captured in the second level of sub-branches. For example, the opportunities main branch includes three sub-branches, related to creating a baseline for the performance of current volume planning using judgmental forecast, defining the potential financial impact due to more accurate statistical forecasts, and clarifying the technical objectives for improvement. The other sub-branches are organized in a similar mind map.

Representing the system structure in a mind map similar to the one shown in Figure 5.2 helps all project stakeholders focus on the key issues. It provides a condensed view of the system and reduces the risk for potential miscommunication. It pushes the experts to organize their knowledge in a structured way. In addition, typical template mind maps can be developed for more generic use and leveraged in future projects.

5.2.3 Data Structure Identification

One of the key objectives of the system structure and data identification step in data mining for forecasting is identifying potential economic drivers. Usually, this is one of the main branches of the system structure mind map and requires more detailed brainstorming and a separate mind map. An example is shown in Figure 5.3.

The mind map in Figure 5.3 is organized in three hierarchical levels. The first hierarchical level represents the experts' decision to divide the potential economic drivers for the targeted volume of the product *B* forecast into two key groups—macroeconomic and microeconomic. The second hierarchical level includes the main categories of economic drivers for each group. For example, the key macroeconomic groups of interest include GDP, exchange rate, population growth, consumer confidence, leading indexes, and governmental policies. The third hierarchical level includes the corresponding specific data for each main category. Because the scope of the project is limited to the United States, Europe, and Japan, most of the required data for the main macroeconomic categories are for these geographic regions. In Figure 5.3, the third-level branches for microeconomic potential drivers are based on more specific sectors within the second-level categories. For example, the personal care category includes data for soap, shampoo, and shaving cream volumes. It is assumed that for each of these third-level potential drivers, data for the three geographic regions of interest are needed.

The experts gave their rankings of the potential economic drivers in category level. Rankings are represented with numbers in the corresponding blocks of the mind map in Figure 5.3. According to the experts, the most influential categories for the volume of product *B* are personal care, oil and gas, construction, population growth, and cleaning.

Figure 5.3: An example of a key economic drivers mind map

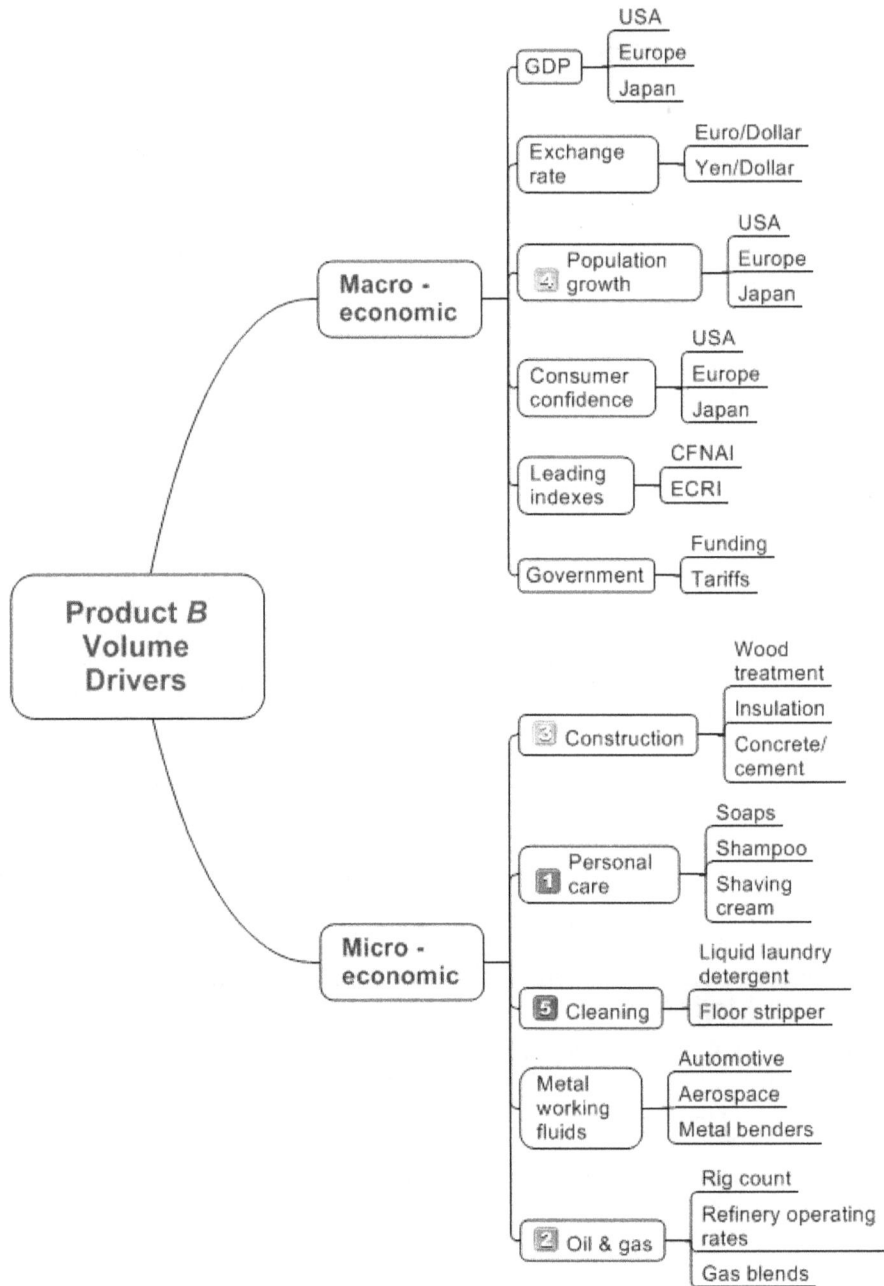

5.3 Data Definition

From the broad area of data definition, the focus in the book is on two key topics—identifying internal and external data sources and capturing the context of the data by defining the corresponding metadata.

5.3.1 Data Sources

The objective is to identify all sources of data recommended from the system structure and data identification phase, which was discussed in the previous section. The data sources are classified into two separate classes—internal and external data sources. The specific issues related to both classes are discussed briefly.

Internal data sources

In most of the cases, the origin of the targeted forecasting variables data is inside the business. It is a common practice that a person in the business who is familiar with the specific sources extracts the internal data during model development. Usually, the final tables with the time series of the targeted variables are delivered as Microsoft Excel spreadsheets. However, these are aggregated data collected by scripts from the transaction tables in the corresponding corporate relational databases.

External data sources

One of the advantages of the Internet is the availability of many services that offer almost any data of interest for building multivariate forecasting models. However, not all of these services are equal in quality or in the number of available economic indicators. Some of the best known sources are discussed below.

Global Insight

The most well-known external source with an almost universal collection of economic and financial data is IHS Global Insight (http://www.ihsglobalinsight.com/). The list of the key data in the main categories of global economic data, global financial data and financial indicators, and industry and sector data for more than 200 countries and more than 170 industries is given below.

Global Economic Data

- Balance of payments
- Cyclical indicators
- Finance and financial markets
- Government finance
- Housing and construction
- Output, capacity, and capacity utilization
- Merchandise trade
- National accounts
- Labor market
- Population
- Prices
- Wholesale and retail trade

Global Financial Data and Financial Indicators

- Commodities and futures
- Derivatives
- Equities
- Money markets
- Exchange rates

- Fixed income
- Indexes
- Stock markets

Industry and Sector Data

- Total sales and all of its components
- Sources and uses of production
- Production prices
- Producer price indices
- Consumer price indices
- Labor cost indices
- Agriculture, hunting, forestry, and fishing
- Automotive
- Mining and quarrying
- Manufacturing
- Fabricated metal, machinery, transport, and equipment
- Electricity, gas, and water
- Construction
- Wholesale and retail trade, restaurants, and hotels
- Transport, storage, and communications
- Finance, insurance, business services, and real estate
- Community, social, and personal services
- Public administration, personal, and household services

Chemical Market Associates, Inc. (CMAI)

Another service with reliable external data that is specialized for the chemical industry is CMAI (http://www.cmaiglobal.com/). It includes worldwide data for the chemical, plastics, fibers, and chlor-alkali industries. Most of the data have forecasts.

Bloomberg

The most famous service for financial data is Bloomberg (http://www.bloomberg.com/markets/). It offers time series of stocks, exchange rates and bonds, currencies, and mutual funds.

International Monetary Fund (IMF)

IMF, with its service International Financial Statistics, provides approximately 32,000 time series, covering more than 200 countries and areas (http://www.imf.org/external/data). The focus is mostly on current data that is needed for the analysis of problems with international payments and inflation and deflation. In other words, the data is useful for analyzing exchange rates, international liquidity, international banking, money and banking, interest rates, prices, production, international transactions, government accounts, and national accounts.

Indexes

Special types of external data are economic indexes. In the same way that the most famous stock market indexes (such as Dow Jones and S&P 500) represent the state of the stock market in the United States, the economic indexes capture the state of a national economy. Among the numerous available economic indexes, the focus in this section is on two—the Chicago Fed National Activity Index (CFNAI) and the economic indexes delivered by the Economic Cycle Research Institute (ECRI). CFNAI is the de facto standard economic index for the state of the American economy. One of the advantages of the indexes published by ECRI is that they include leading economic indicators.

The CFNAI

The CFNAI is a monthly index that is based on a weighted average of 85 existing monthly indicators of the US national economic activity (http://www.chicagofed.org/webpages/publications/cfnai/index.cfm). The 85 economic indicators that are included in the CFNAI are composed of four broad categories of economic sectors: production and income; employment, unemployment, and hours; personal consumption and housing; and sales, orders, and inventories. It is a normalized index, with an average value of 0 and a standard deviation of 1. Because economic activity tends toward trend growth rates over time, a positive index reading corresponds to growth above trend, and a negative index reading corresponds to growth below trend. A critical threshold on the negative side could be used as a warning that we are entering a recession when the index crosses –0.7.

ECRI indexes

ECRI indexes, such as the US Weekly Leading Index (WLI), Future Inflation Gauge (FIG), and Leading Home Price Index (LHPI), are based on leading indicators (http://www.businesscycle.com/). The leading index provides early signals about the direction in which the economy is going. One of the earlier signals that an ongoing expansion might start to decelerate is a sustained decline in the leading index growth rate. In contrast to CFNAI, however, the structure of ECRI indexes is proprietary.

5.3.2 Metadata

The key procedure in data definition is clarifying the structure and content of the data. This data about the data is called metadata. It provides generic information about key aspects of the data, such as the following:

- time and date of creation
- data source
- data description
- purpose of the data
- data type

Above all, metadata is data. As such, metadata can be stored and managed in a <u>database</u>. Some details about the main structural components of metadata are briefly discussed below.

For time series metadata, it is a must to have as much information as possible for each time sample, the frequency of data collection, start date, end date, and the time of last update. The data source has to include all necessary details to identify the specific time series, such as short name, specific bank of data, and its corresponding mnemonic. The data description section includes a long descriptive name, geographic area, units, series type, and more. An example of a metadata structure for external data that originated from Global Insight is shown in Table 5.1. This specific time series is an aggregated price of the US Producer Price Index (PPI) for iron and steel with a basic index year of 1982. The data are a blend of quarterly historical data from the first quarter of 1960 and a ten-year forecast until the fourth quarter of 2020, with the last update on June 3, 2010. The Global Insight bank name is CISSIM, and the corresponding DRI and WEFA mnemonic is WPIWP101.Q (the last letter means quarterly data).

Table 5.1: Metadata for Global Insight data

Source	Global Insight
Concept	Aggregate Prices
Short Label	PPI, Iron and Steel
Long Label	United States PPI, Iron and Steel Source: BLS/Global Insight Units: 1982=100 Historic
Geography	United States
Unit	(1982=100)
Frequency	QUARTERLY
SeriesType	Forecast - Cost Services Pricing and Purchasing 10 year
Start Date	1960 Q1
End Date	2020 Q4
Last Update	6.3.2010
Bank Name(s)	CISSIM
DRI Mnemonic	WPIWP101.Q
WEFA Mnemonic	WPIWP101.Q

The purpose of the data section of metadata is illustrated with the built-in options in SAS Enterprise Miner. The most frequently used options for different data roles in SAS Enterprise Miner are as follows:

- Target variables—dependent variables, predicted by the input or independent variables
- Input variables—independent variables that are used in estimating, classifying, or predicting the target variables
- Rejected variables—variables that are rejected or omitted from the model
- Time ID variables—variables that identify time
- Predict variables—variables that include predicted values required for a modeling node

An example of selected data roles in SAS Enterprise Miner is shown in Figure 5.4, where the target variable is Volume_B; the time ID variable is Data; the input variables are ICPCTOT, IP_Fabricated_Products, and Indust Production; and the rejected variables are Demand_Sheet_Metal_Work and ICPLODTOT.

Figure 5.4: An example of data roles in SAS Enterprise Miner

Name	Role	Level	
Date	Time ID	Interval	N
Demand_Sheet_Metal_Work	Rejected	Interval	N
ICPCTOT	Input	Interval	N
ICPLODTOT	Rejected	Interval	N
IP_Fabricated_Products	Input	Interval	N
Indust_Production	Input	Interval	N
Volume_B	Target	Interval	N

The type of data section of metadata is illustrated with the built-in options in SAS Enterprise Miner. The data types in SAS Enterprise Miner (called levels) are as follows:

- Binary type—categorical variables with only two distinct class levels
- Interval type—numeric variables that have more than 10 distinct levels
- Nominal type—character variables with more than two and less than 10 distinct nonmissing levels
- Ordinal type—numerical variables with more than two but no more than 10 distinct levels
- Unary type—variables with one nonmissing level or with assigned date formats

The Input Data Source node in SAS Enterprise Miner reads the data source and creates a data set called a metadata sample. This metadata sample automatically defines the variable attributes for processing within the process flow. By default, a metadata sample takes a random sample of 2,000 observations from the source data set. If the data set is smaller than 2,000 observations, then the entire data set is used to create the data mining data set. In the metadata sample, each variable is automatically assigned a data type (level of measurement) and a variable role.

5.4 Data Extraction

In the case of developing forecasting models based on data mining, data extraction includes two phases— collecting the data from the original sources and importing data to the data mining and forecasting software environments (SAS, in this case). Some details about how to collect data from internal and external sources are given below.

5.4.1 Internal Data Extraction

Internal data can originate from diverse sources such as simple text files, spreadsheets, relational databases, or Enterprise Resource Planning (ERP) systems. Bringing internal data is usually the responsibility of the team member in the corresponding business. Importing the data into SAS is the responsibility of the data miner or modeler. Fortunately, SAS offers various access modules for almost any data format and database. For example, SAS/ACCESS products can bring data from relational databases such as DB2, Informix, Microsoft SQL Server, MySQL, Oracle, Sybase, and Teradata. SAS/ACCESS products provide an interface to the most popular ERP systems, such as SAP, SAP BW, or PeopleSoft. Examples for implementing SAS/ACCESS products for various sources are given in *Data Preparation for Analytics Using SAS* (Svolba 2006).

The most common way to bring internal data into SAS is from Excel spreadsheets. This process is trivial in SAS Enterprise Guide. It is accomplished by applying an Import Data block, in which the developer has full control over the data attributes in a dialog mode. An example is shown in Figure 5.5.

Figure 5.5: An example of importing an Excel file in SAS Enterprise Guide

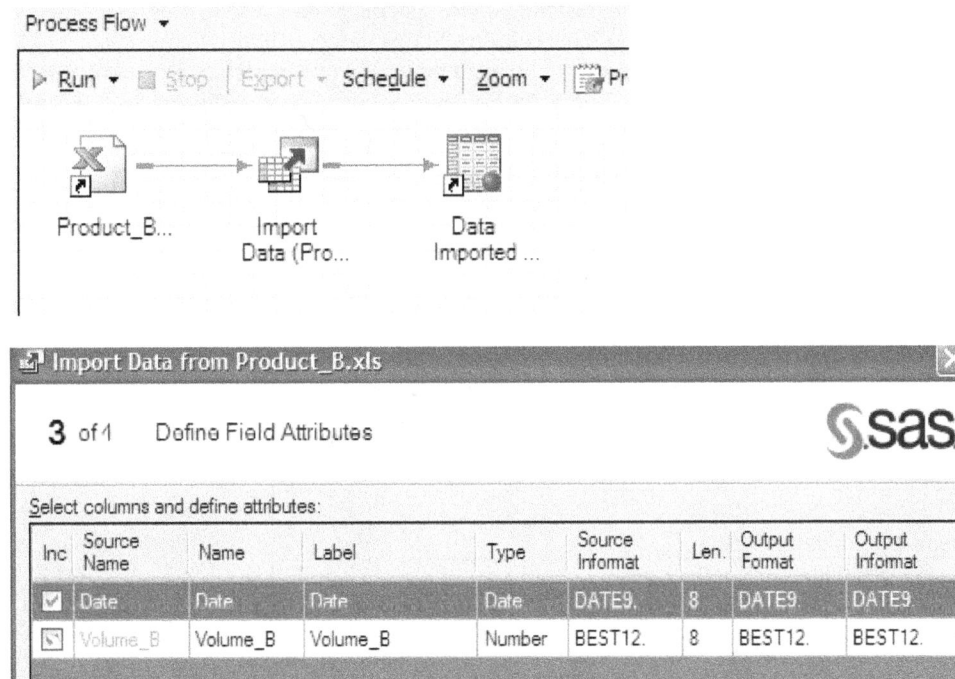

5.4.2 External Data Extraction

Most of the external data sources discussed in the previous section offer a variety of services to deliver data. For example, IHS Global Insight provides several options to navigate, retrieve, and automatically update the data that you need from its database. It is possible to set up fast data downloads or to construct complex data tables directly in Excel. One of the services, DataInsight, provides access to millions of time series in IHS Global Insight, with many time series dating back before the 1960s. There is fast and easy navigation via data tables, tree structures, and keyword searches. The data are delivered either directly to the client's PC or to a specific place in the corporate network. If using one of these services, it is strongly recommended that the metadata structure in Table 5.1 be implemented.

A potential solution for the problem of having many forecasting projects in a business is the development of an internal database that contains all necessary external data. This solution will lead to better coordination of all data collection activities and will reduce maintenance costs.

5.5 Data Alignment

An important issue for the quality of time series data that is collected from different internal and external sources is their proper synchronization. Two main types of data alignment—aligning to appropriate business structures and aligning to a selected time basis—are discussed briefly.

5.5.1 Data Alignment to a Business Structure

The first question after identifying the potential economic drivers is, "What is their alignment to a specific business structure or economic category?" In the case of the United States, Canada, and Mexico, you can use the industry classification system, the North American Industry Classification System (NAICS).[3] It is an economic classification system in which economic units that use similar processes to produce goods or services are grouped together. This production-oriented system permits statistical agencies in the United States, Canada, and Mexico to produce data that can be used for measuring productivity, unit labor costs, and the capital intensity of production. In addition, it can produce data to construct input-output relationships, estimate employment-output relationships, and to create other statistics that require that inputs and outputs be used together. The total number of economic units is 1843.

NAICS is a six-digit hierarchical numeric classification system, with the first digit or first two digits designating the broad industry sector, and subsequent digits representing more specific industry categories. For example, the first sector with the lowest first digit is agriculture. This is followed by mining, and then construction. Manufacturing begins with the third digit. Wholesale trade, retail trade, and transportation are industry sectors in the middle of the list. Codes beginning with 6, 7, or 8 are for services, and a code beginning with 9 is for public administration. An example of the NAICS code hierarchy in the construction industrial sector is shown in Table 5.2.

Table 5.2: An example of NAICS code hierarchy

NAICS	Description
23	Construction
236	Construction of Buildings
2361	Residential Building Construction
23611	Residential Building Construction
236115	New Single-Family Housing Construction (except Operative Builders)
236116	New Multifamily Housing Construction (except Operative Builders)
236117	New Housing Operative Builders
236118	Residential Remodelers
2362	Nonresidential Building Construction
23621	Industrial Building Construction
236210	Industrial Building Construction
23622	Commercial and Institutional Building Construction
236220	Commercial and Institutional Building Construction

An example of aligning some of the identified economic drivers shown in the mind map in Figure 5.3 is in Table 5.3.

Table 5.3: An example of aligning economic drivers

Economic driver	NAICS	Description
wood treatment	321911	Wood Window and Door Manufacturing
insulation	42333	Roofing, Siding, and Insulation Material Merchant Wholesalers
concrete/cement	3273	Cement and Concrete Product Manufacturing
soap	325611	Soap and other Detergent Manufacturing
shampoo	81211	Hair, Nail, and Skin Care Services
shaving cream	325620	Toilet Preparation Manufacturing
liquid laundry detergent	333312	Commercial Laundry, Drycleaning, and Pressing Machine Manufacturing
floor stripper	561740	Carpet and Upholstery Cleaning Services

5.5.2 Data Alignment to Time

Aligning internal and external data to a common timestamp is critical for analyzing time series. The first step in integrating the data from the different sources is to align data relative to the specific time window and frequency of the output (dependent) variable. This process requires expanding or contracting data. (See details in Chapter 6.)

Aligning indexes is of special interest. An unpleasant situation is when the external data provider changes the basis year for an index. To blend the historical data from the old index with the new index, the data of the old index must be aligned to the new basis. An example is shown in Figure 5.6, where a new basis has been established in May 2009. The old index is rescaled with the new basis.

Figure 5.6: An example of index change data alignment

Date	Old Index	New Index
1/9/2008	304.6	132.9
1/10/2008	278.3	121.4
1/11/2008	274.2	119.6
1/12/2008	266.7	116.4
1/1/2009	277.3	121.0
1/2/2009	264.4	115.4
1/3/2009	236.1	103.0
1/4/2009	239.2	104.4
1/5/2009	229.2	100.0
1/6/2009		98.6
1/7/2009		96.8
1/8/2009		96.2
1/9/2009		94.2
1/10/2009		93.8
1/11/2009		92.9
1/12/2009		94.0
1/1/2010		92.5

Old index aligned data

New index basis

New index aligned data

5.6 Data Collection Automation for Model Deployment

One of the biggest challenges in applying forecasting systems is the proper organization and implementation of deployed models. A critical process in this phase of the project is data collection automation. The focus here is on two main issues related to this topic—the differences between collecting data for model development and deployment, and describing the automation sequence.

5.6.1 Differences between Data Collection for Model Development and Deployment

The key difference between collecting data for model development and deployment is the participation of the model developer in every step of the modeling process. In the model deployment phase, the modeling process is executed by an automatic sequence of data collection, preparation, and model execution. To make this process more reliable, the model structure is fixed. A short list of the key differences between collecting data for model development and deployment (also called production mode) is given.

- Deployed models require only inputs that have been used in the final selected forecast models.
- Data collection of deployed models assumes automatic integration and alignment of internal and external data.
- Data collection of deployed models requires a data consistency check before applying the data to the forecasting model.

5.6.2 Data Collection Automation for Model Deployment

The ideal in automatic data collection in production mode is when the whole process is fully automated as a stored process and is run, on demand, by the user. A flowchart of this process is shown in Figure 5.7.

A requirement for the success of automatic data collection is the reliable and timely extraction of internal data. One potential issue that causes problems for the models themselves is when the business structure changes. In this case, the person responsible for internal data collection must communicate this change, in advance, to the project team and discuss the possible action to take in response. This might include rebuilding the forecasting models. Automation of external data collection, even if it requires several sources, can be done by the subscribed services. Integration of both data sets is usually done by using the proper SAS Enterprise Guide blocks or SAS functions.

The most important part of automatic data collection is checking data consistency. This task requires looking at the deviations from the previously collected periods, and comparing the deviations of the new data from the last data extraction. An example of a data consistency check is shown in Table 5.4. It includes a comparison of three GI quarterly time series from two consecutive extractions. For each corresponding input, a column with the percentage change for each quarter after 3Q2008 is given, as well as the average deviation for the last year. As shown in Table 5.4, the three time series demonstrate different patterns. The history of Input 1 is continuously changed, even back in 2008. Input 2 has minor changes for the last two data extractions. The history of Input 3 shows a pattern of increased deviation in the last three extractions that might raise a flag if this trend continues.

Figure 5.7: A flowchart for automatic data collection for model deployment

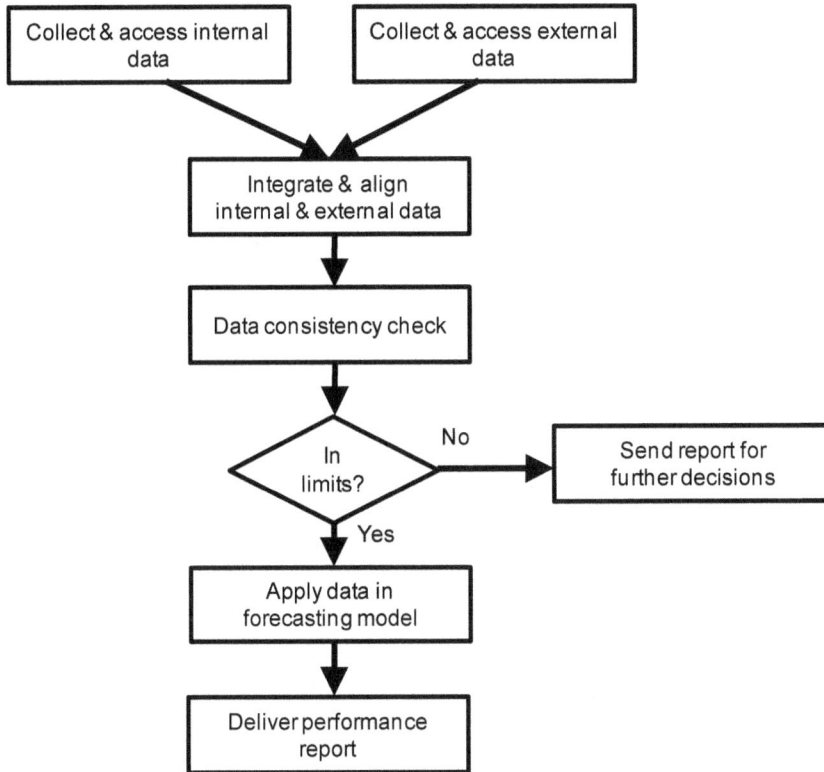

Table 5.4: An example of differences in two consecutive data collections

Date	input1_4Q	input1_1Q	%delta	input2_4Q	input2_1Q	%delta	input3_4Q	input3_1Q	%delta
7.1.2008	129,83	129,86	-0,02%	13223,5	13223,5	0,00%	287,17	287,17	0,00%
10.1.2008	111,13	111,16	-0,03%	12993,7	12993,7	0,00%	213,67	213,67	0,00%
1.1.2009	108,01	107,99	0,01%	12832,6	12832,6	0,00%	185,60	185,60	0,00%
4.1.2009	113,94	113,91	0,03%	12810	12810	0,00%	169,73	169,73	0,00%
7.1.2009	120,44	120,45	-0,01%	12860,8	12860,8	0,00%	186,30	186,30	0,00%
10.1.2009	124,29	124,31	-0,02%	13019	13019	0,00%	194,37	194,37	0,00%
1.1.2010	117,92	117,89	0,03%	13138,8	13138,8	0,00%	212,33	212,33	0,00%
4.1.2010	110,20	110,17	0,03%	13194,9	13194,9	0,00%	232,23	233,67	-0,62%
7.1.2010	112,93	112,94	-0,01%	13260,7	13278,5	-0,13%	221,20	223,73	-1,15%
10.1.2010	122,15	119,58	2,11%	13341,88	13382,6	-0,31%	204,61	223,49	-9,22%
Average %delta for last year		0,54%			-0,11%			-2,75%

---

[1] One of the few references is from Alex Kalos and Tim Rey (2005).

[2] The mind-maps in the book are based on the Mindjet product MindManager 8, with the website: http://www.mindjet.com/

[3] http://www.census.gov/eos/www/naics/

Chapter 6: Data Preparation

6.1 Overview

As might be expected, when data is received to be used in data mining for forecasting, it is not necessarily in the right format or clean enough for analysis. Almost always, the Y variable of interest comes from a different data source or system than the exogenous or X data. More often than not, the Y data might be in what is called a transactional format, which is typical of most ERP systems. This transaction data has to be converted to time series data. Generally, the exogenous data comes from a source that is in a format ready for time series data mining and forecasting. Often, not all of the data is in the same interval. Some might be annual, others quarterly, and yet others monthly. Thus, they all have to be brought into the same interval.[1] Once all of the data is put into a proper time series format and aligned to the same interval, then it has to be merged into one data set. Once it is in this format and is a single data set for analysis, the final steps in the data preparation process can be conducted. As is typical in any data analysis endeavor, both the Y and X data sources might have bad data. The data might need to be analyzed for outliers. There might also be missing data. Missing data can be from the beginning of the data set, throughout the data set, or at the end of the data set. Each missing data issue is handled differently. Data might have to be transformed mathematically before being used.

6.2 Transactional Data Versus Time Series Data

Data used for time series data mining and forecasting is not always in an appropriate time series format. First, all data needs a proper time stamp with each row. For use in SAS, this time stamp needs to be in an appropriate date and time format. Each column represents the dependent (Y) and exogenous (X) variables. Each row has to be "accumulated" to the same frequency (hour, day, week, month, quarter, year, and so on). Figure 6.1 (a SAS Enterprise Guide data set) represents a valid time series data mining and forecasting data set. Note that the data has an appropriate date and time stamp for each row—MMDDYY10. In this case, this is quarterly data designated by the months 1, 4, 7, and 10. The day is represented as the first day of the month, 1. There are always four quarters in each year for all data between the first and last year. The Y is designated, and the Xs have SAS names and are self-explanatory. Note that there is no missing data. This data is called a time series database.

Figure 6.1: Proper Time Series Data Mining and Forecasting Format

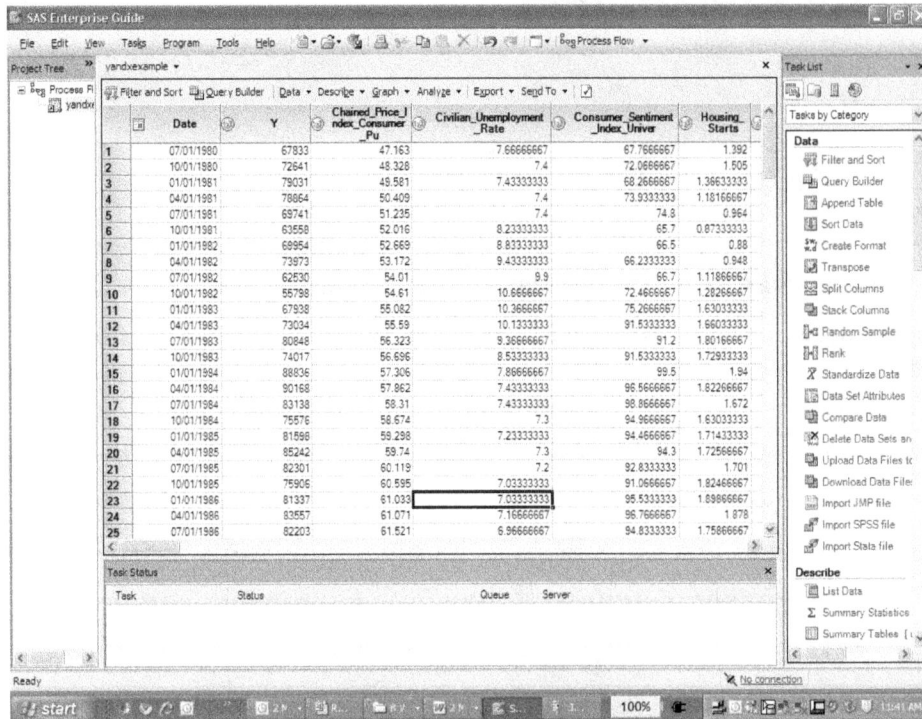

In any data mining endeavor, 60 to 80% of the time is getting the data ready for modeling. In business databases, the data might be in a transaction format and not accumulated to a time-stamped time series format. Figure 6.2 is an example. Transactional data has a time stamp on it, but each row is a "transaction." The variables shown in Figure 6.2 include the time stamp, order type, and so on. The variables in Figure 6.3 are in the same data set. They represent the Y that has to be aggregated or accumulated over the particular time frequency that the data mining and forecasting project is designed for.

Figure 6.2: Raw Transactional Data

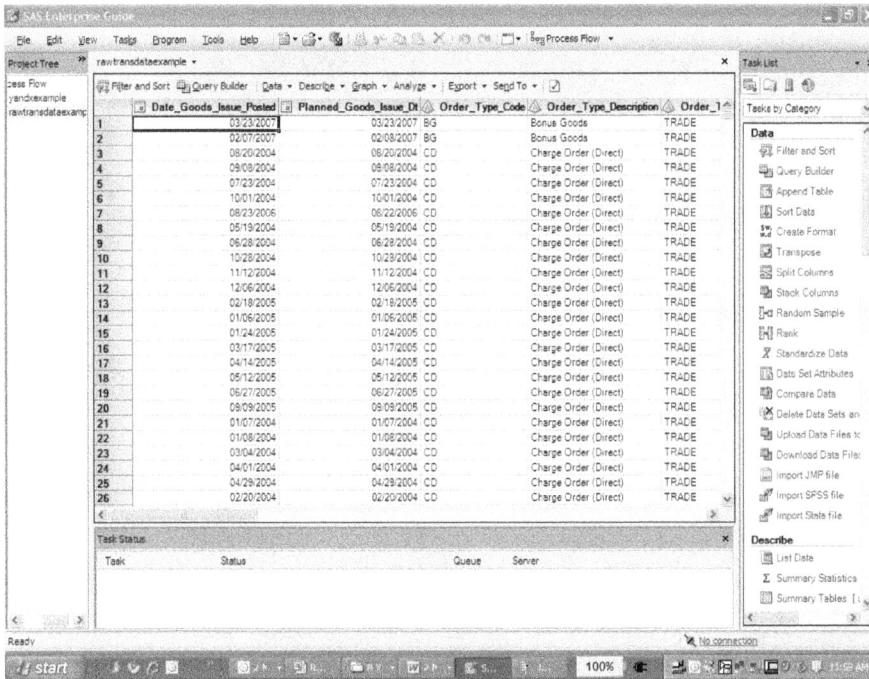

Figure 6.3: Transaction Variables, Inclusive of Y, Present in the Transactional Data Set

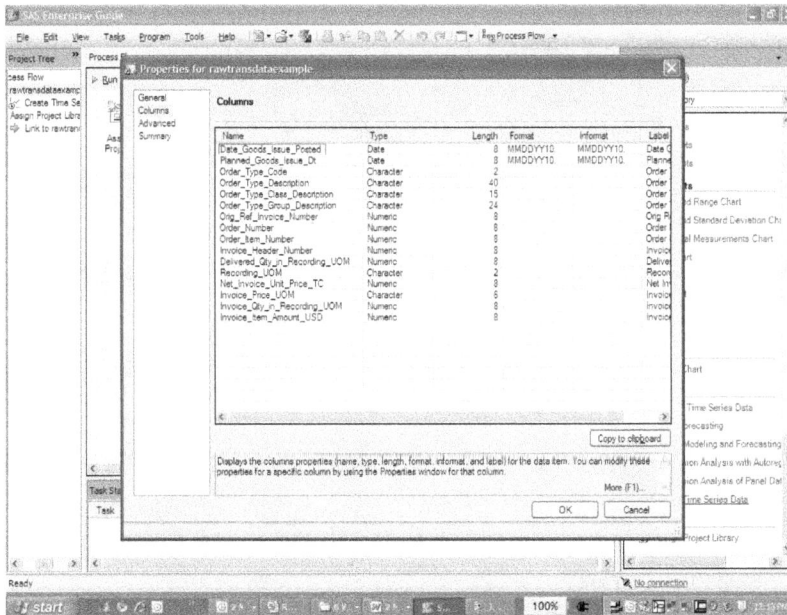

There are various approaches to getting the transaction data accumulated to the proper time series data format. A standard SAS DATA step code sequence can be used. But, both SAS Enterprise Guide (through PROC TIMESERIES) and SAS Forecast Studio do it automatically. In each case, it should be done properly. The following figure (Figure 6.4) shows an example of how to do this task in SAS Enterprise Guide. The key SAS code (generated by SAS Enterprise Guide) is shown. Note that the transaction data is the input source and that the Create Time Series Data task in SAS Enterprise Guide is used.

Figure 6.4: SAS Enterprise Guide Flow for Converting Transaction Data to Time Series Data

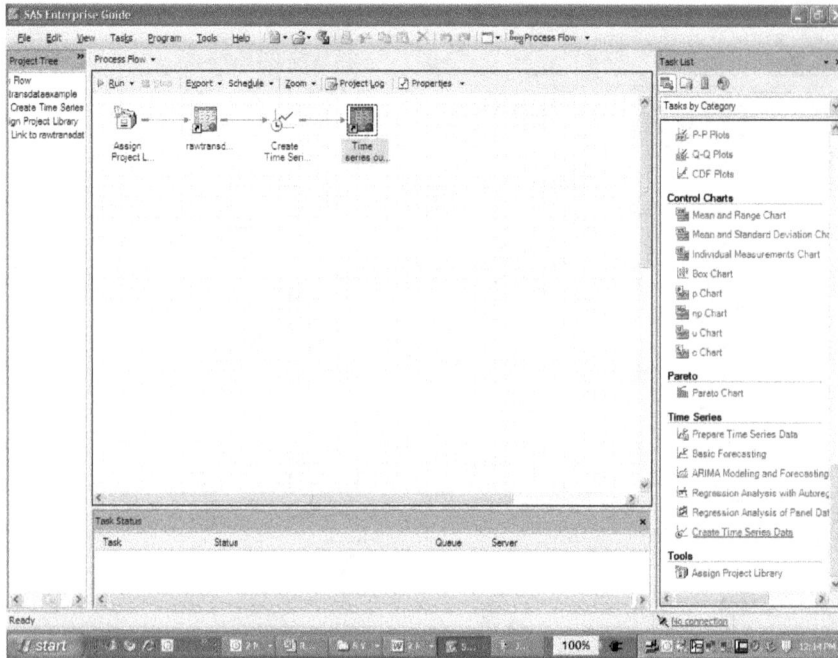

In step 2 of the Create Time Series Data task in SAS Enterprise Guide, Figure 6.5 shows where the Time ID has to be identified with the Y variable (time series).[2]

Figure 6.5: Step 2 of the Create Time Series Data Task

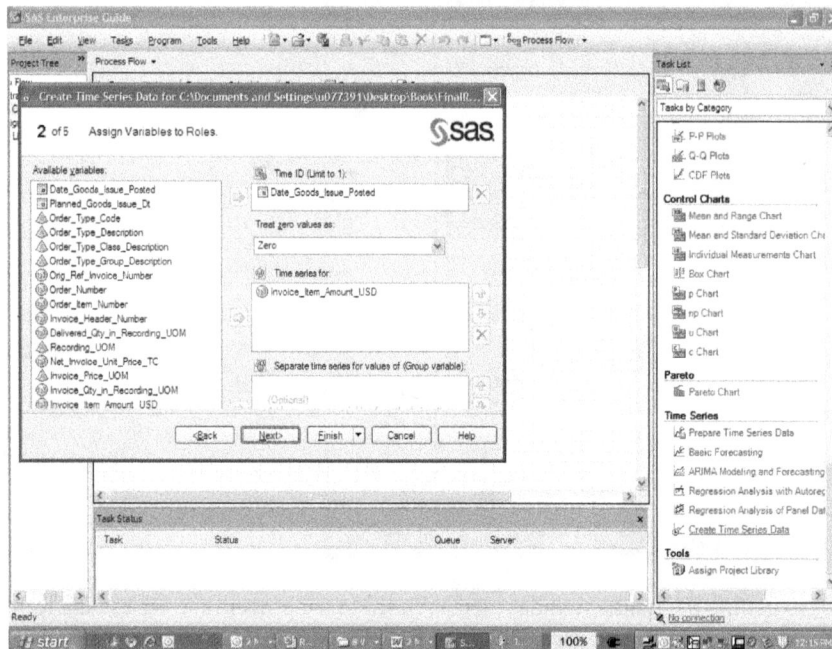

In step 3 of the Create Time Series Data task (Figure 6.6), three options are presented: what time interval (frequency) to accumulate up to, what statistic to use, and how to handle missing data in the transactional data set.

Figure 6.6: Step 3 of the Create Time Series Data Task

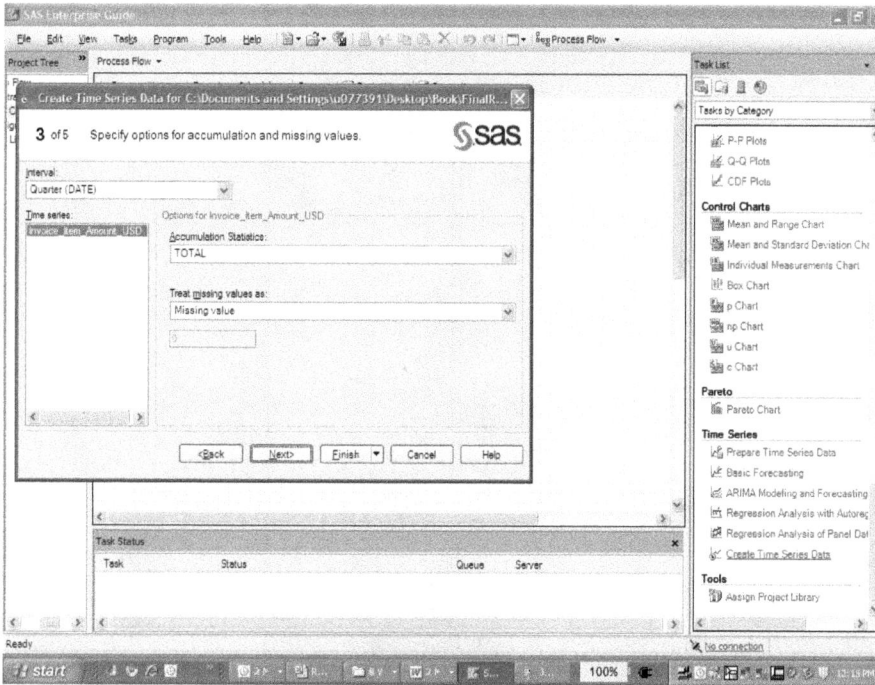

If you are interested in raw SAS coding, Figure 6.7 shows the PROC TIMESERIES code generated by SAS Enterprise Guide. PROC TIMESERIES is available in SAS/ETS. The same accumulation could be obtained with raw SAS code using the SUM option in PROC MEANS.

Figure 6.7: PROC TIMESERIES Code for Converting Transaction Data to Time Series Data

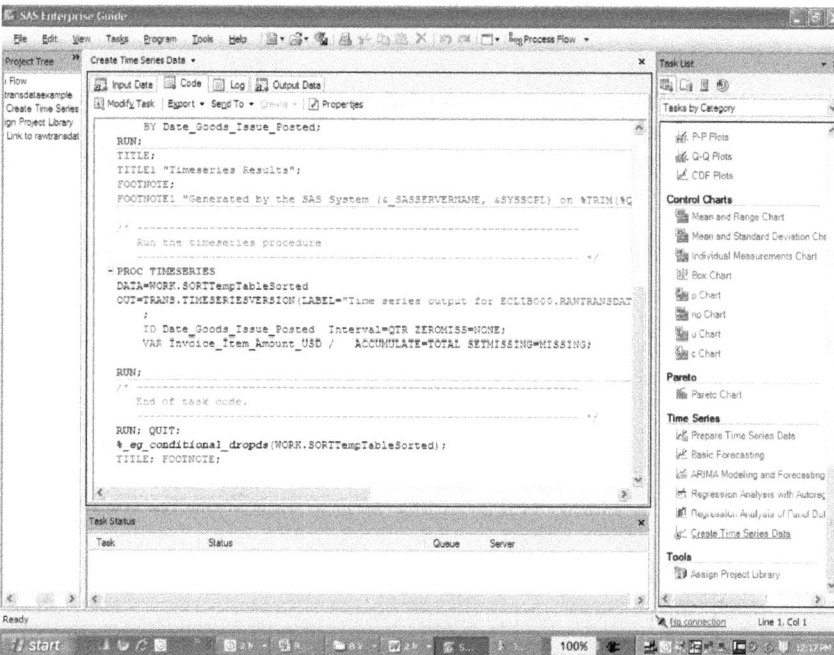

Figure 6.8: Final Time Series Data

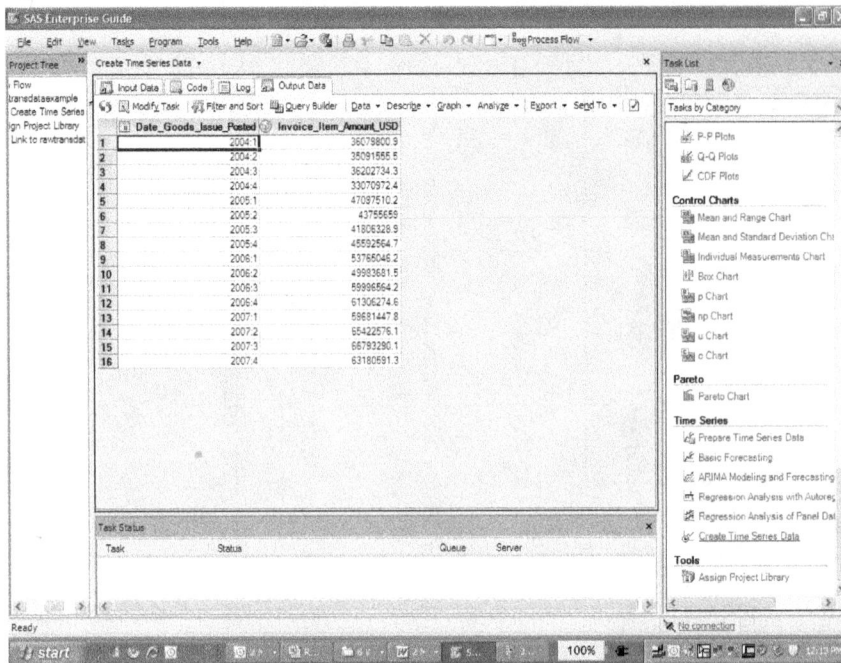

Processing can also be accomplished in SAS Forecast Studio. The complication is in mixing different transactional sources. Thus, only Y and X variables collected in the same transactional framework can be done simultaneously in SAS Forecast Studio. Though there are various approaches to converting transaction data to time series data, experience shows that using raw SAS code (such as PROC MEANS with the SUM option) offers the modeler the most freedom and ability to check that accumulation is done correctly. Most often, multiple sources of Y and X data are involved. The most effective process is to accumulate each of these data sets over the transaction details separately first, then get them in a common time interval and date format, and then merge them.

The following example shows how to import a raw transaction data set and set it up as a time series data set in SAS Forecast Studio. Figure 6.9 is the second step in the process, which is identifying the input data set for SAS Forecast Studio. The first step is naming the project. Step 3 is identifying whether there are classification variables for a hierarchical forecasting problem. In this case (illustrated in Figure 6.10), this is not a hierarchical problem.

Figure 6.9: Identifying the Input Data Set

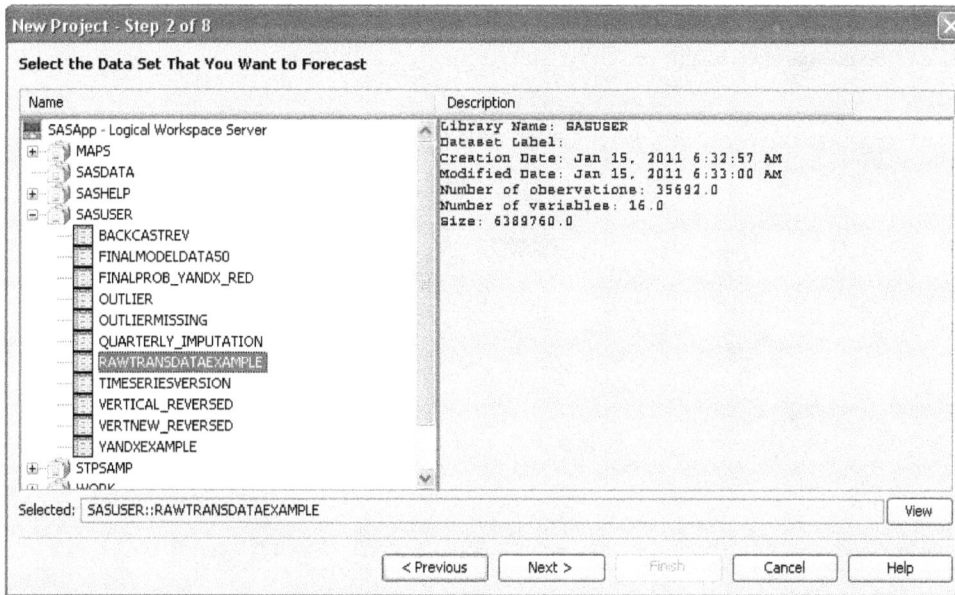

Figure 6.10: Identifying Hierarchical Variables

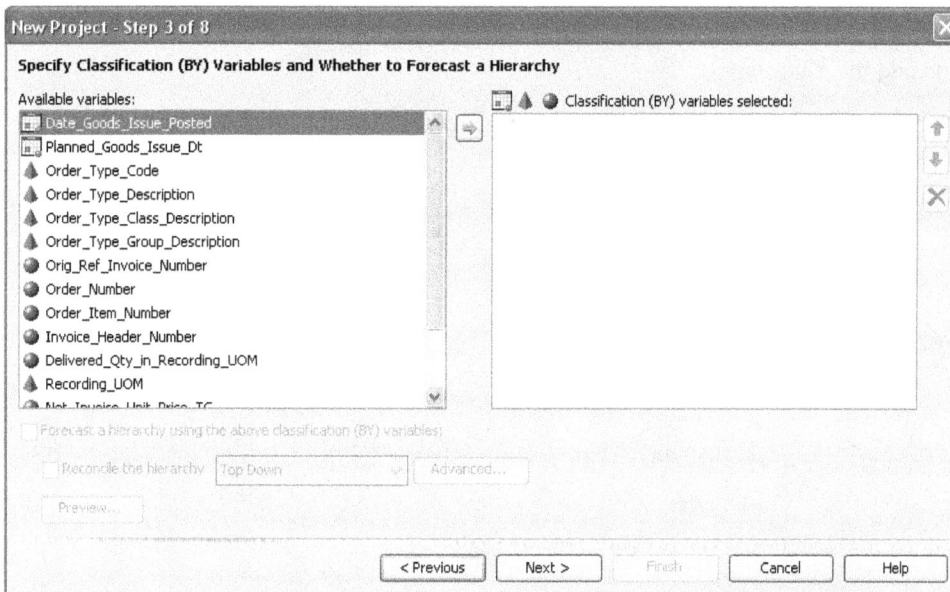

To properly set up the variables in SAS Forecast Studio, it is helpful to have a date variable that is already set up at the right frequency. In this case, the correct frequency is quarterly. Figure 6.11 shows that the time variable used is Date Goods Issue Posted. Once this variable is chosen as the date variable, and Quarter has been chosen as the interval, SAS Forecast Studio identifies the seasonal cycle length as 4.

Figure 6.11: Choosing the Date Variable

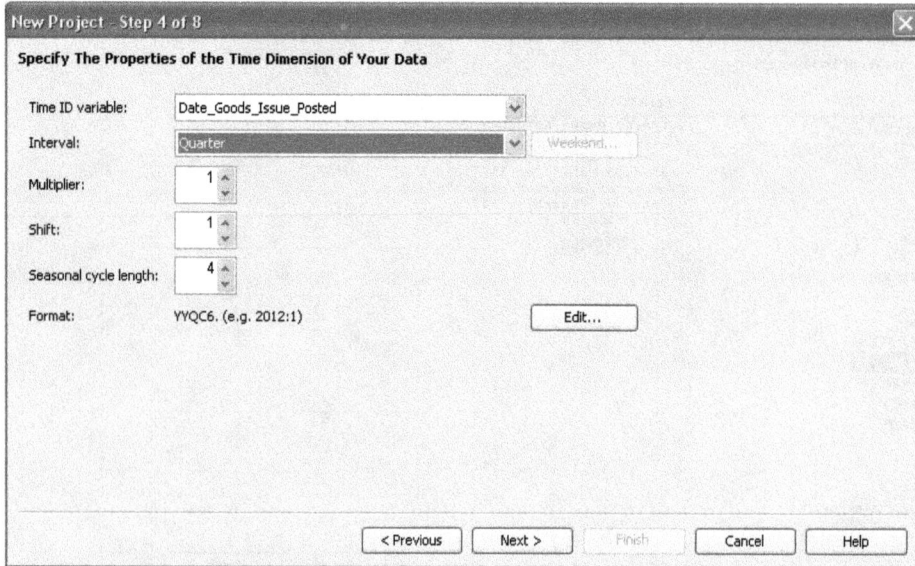

The last step in the process is to choose the proper Y variable. In this case, the Y variable is Invoice Item Amount USD. The accumulation option is Sum of values. Figure 6.12 shows this selection. The plot in Figure 6.13 shows that the data was accumulated correctly.

Figure 6.12: Choosing the Y Variable

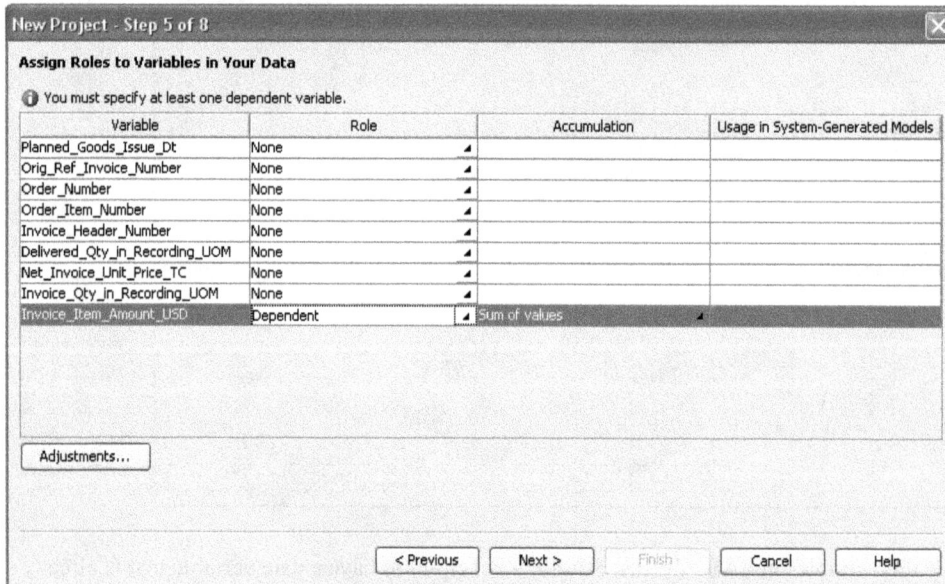

Figure 6.13: Final Accumulation

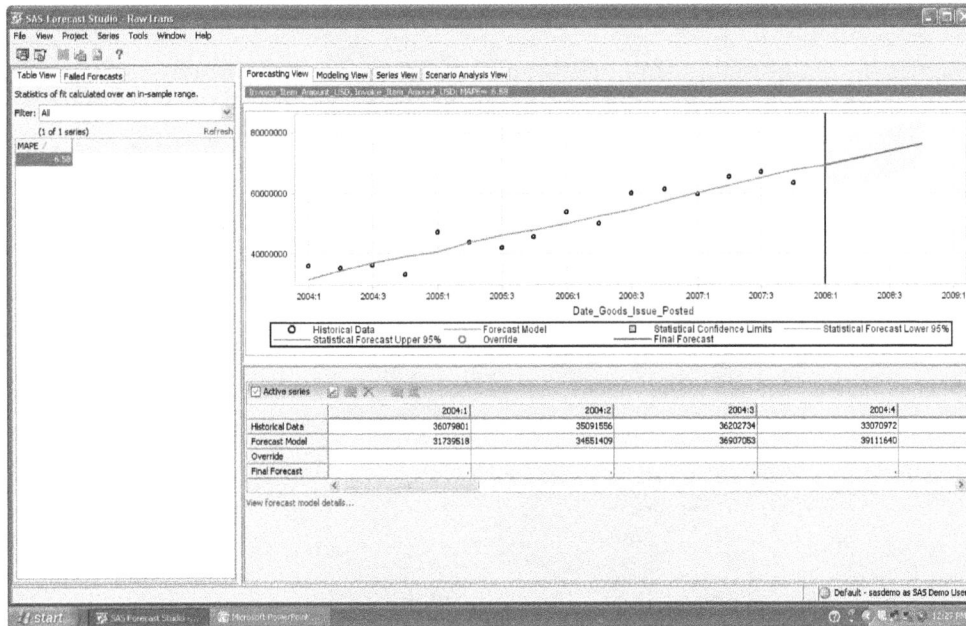

6.3 Matching Frequencies

Given the original objectives of the forecasting project, an analysis unit or interval was determined in an earlier phase of the project. For business processes such as Executive Sales, Operations, and Planning (ES&OP), the interval might be quarterly. All of the time series data has to be converted to this common interval for use. Sometimes, the time series data has to be converted from a finer interval to a coarser interval (monthly to quarterly). This is called contracting. Other times, an annual interval has to be converted to a quarterly interval. This is called expanding.

6.3.1 Contracting

Converting time series data that is at a finer interval, such as monthly or weekly, can be simply a matter of summing or averaging data to get a coarser interval. Time series data such as demand or sales should be summed. Data such as indices should be averaged. This processing step is basic and can be done with SAS Enterprise Guide using the Summary Statistics and Sum or Mean selection for each variable, or with Base SAS using PROC MEANS. Figure 6.14 shows a SAS Enterprise Guide example of converting a monthly sales figure to quarterly using the SAS Enterprise Guide framework.

Figure 6.14: SAS Enterprise Guide Flow for Converting Quarterly to Annual Sales

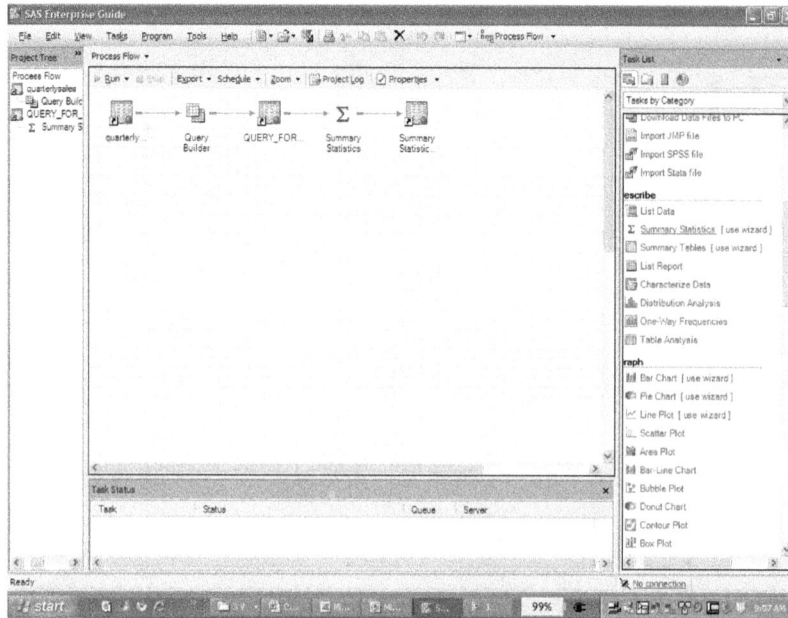

There are two processing steps in this flow. After bringing the data in, a second Date variable must be assigned (the Query Builder step). Then, the data has to be summed to annual (the Summary Statistics step).

Figure 6.15 shows the quarterly input data set.

Figure 6.15: Quarterly Input Data Set

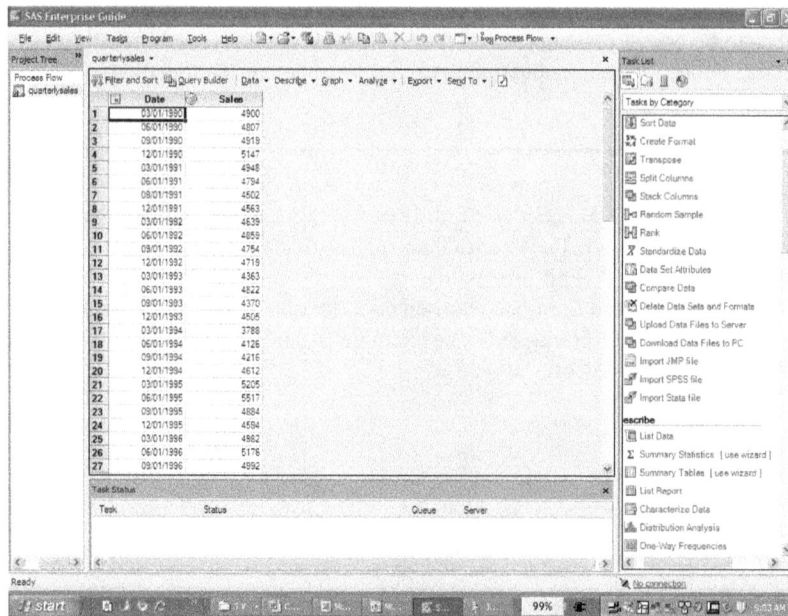

Figure 6.16 shows the new Date variable Year in the proper format. The Query/Filter step is not shown.

Figure 6.16: Date Variable Year

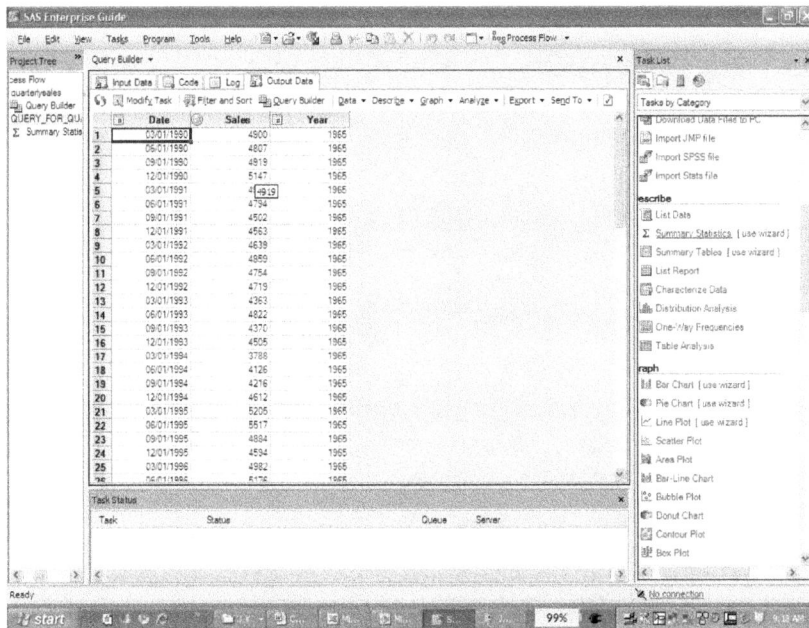

Figure 6.17 shows the SAS Enterprise Guide Summary Statistics options to obtain the sum.

Figure 6.17: Summary Statistics Options

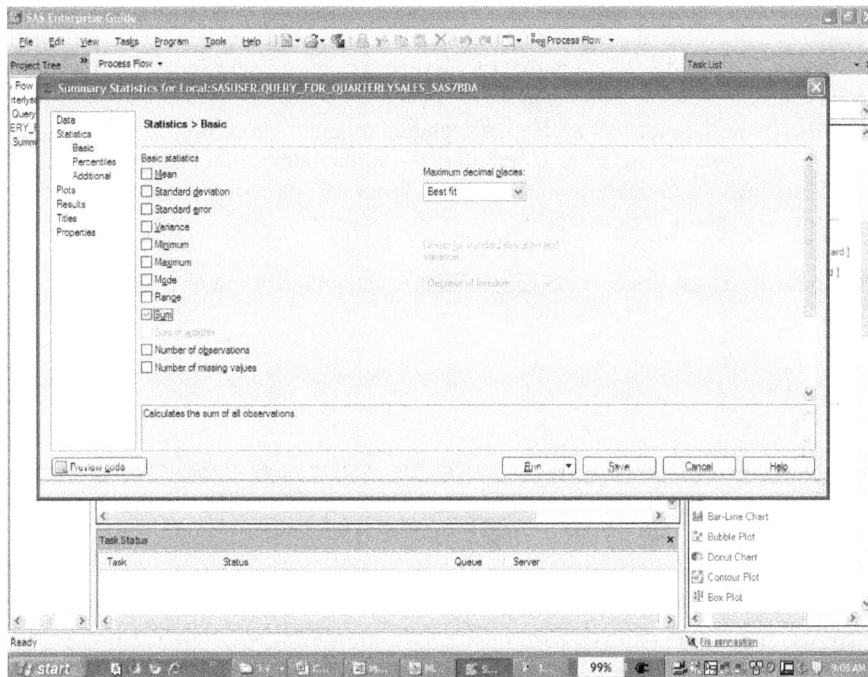

Figure 6.18 shows the new annual summed data set. Note that the new data variable, Year, is a proper date variable.

Figure 6.18: Annual Summarized Data

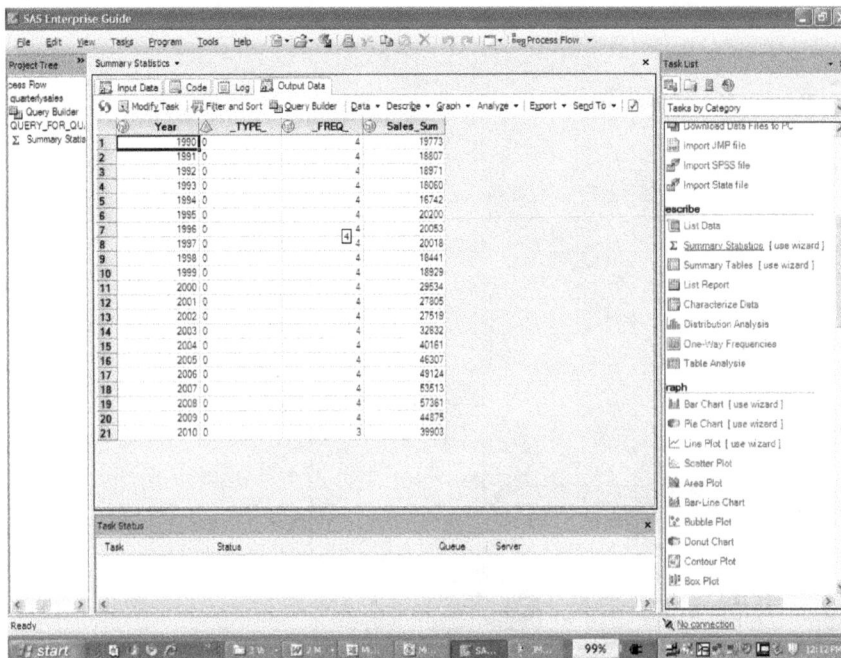

6.3.2 Expanding

Sometimes, data for the exogenous variables is available only at a more aggregated level than the Y variable. This means that the data has to be converted to a more frequent interval, such as going from annual to quarterly. SAS Enterprise Guide can do this. Knowledge of the nature of the time series being expanded from one interval to another is definitely useful. The simplest approach is to divide by 4 (4 quarters in a year), but this is generally not too informative. Expanding using some form of spline interpolation might be more useful. This approach would accommodate a trend from one year to the next. Combining the spline interpolation approach with seasonal adjustments when the exogenous variable is known to follow similar seasonal patterns of other exogenous variables might provide the best of both worlds.

Let's take the simpler case first—no seasonality and starting with the annual data in Figure 6.19.

Figure 6.19: Annual Data for Expanding

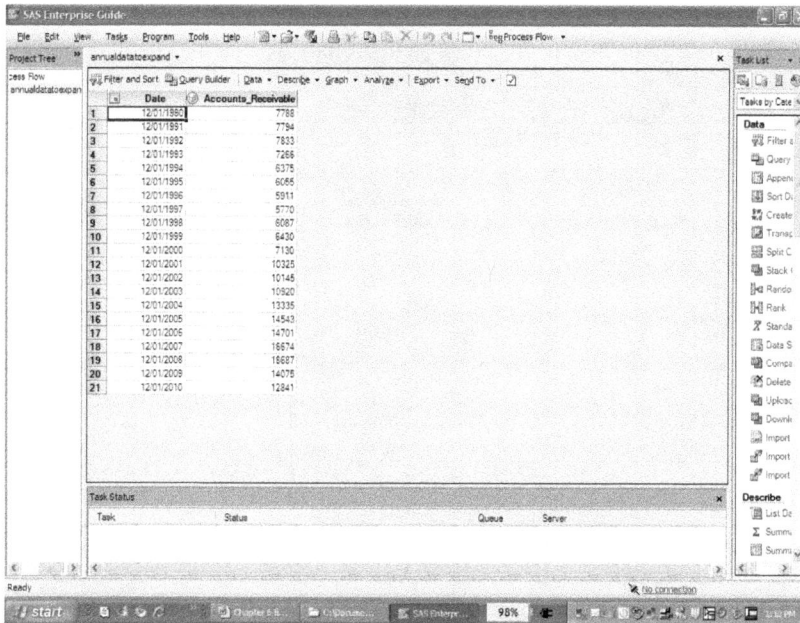

We apply the Prepare Time Series Data task in SAS Enterprise Guide shown in Figure 6.20.

Figure 6.20: SAS Enterprise Guide Flow for Expanding

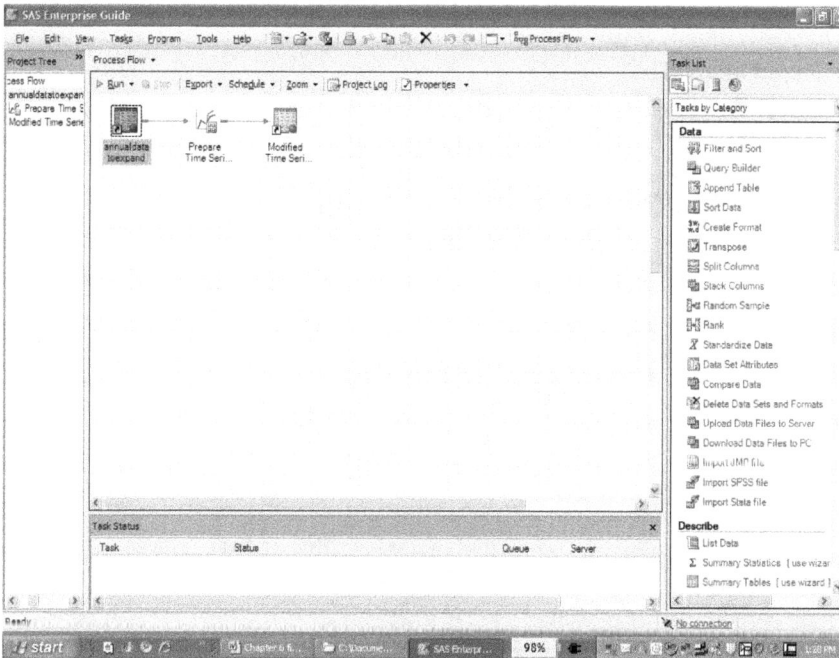

Figure 6.21 shows the first step in the process.

Figure 6.21: Choosing the Input Variables and Intervals

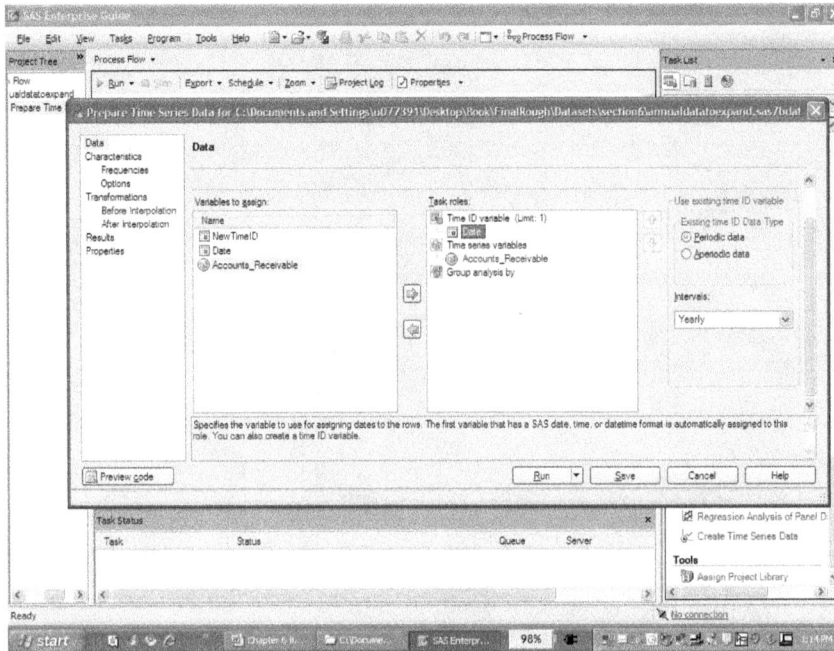

Here, we identify the date variable (Date). We tell the system what the interval the input data is in. In this case, the interval is yearly (annual). We identify the Time series variable that we are working on, Accounts Receivable. Next, we identify the output frequency that we want and have chosen Quarterly. (See Figure 6.22.)

Figure 6.22: Choosing the Output Interval

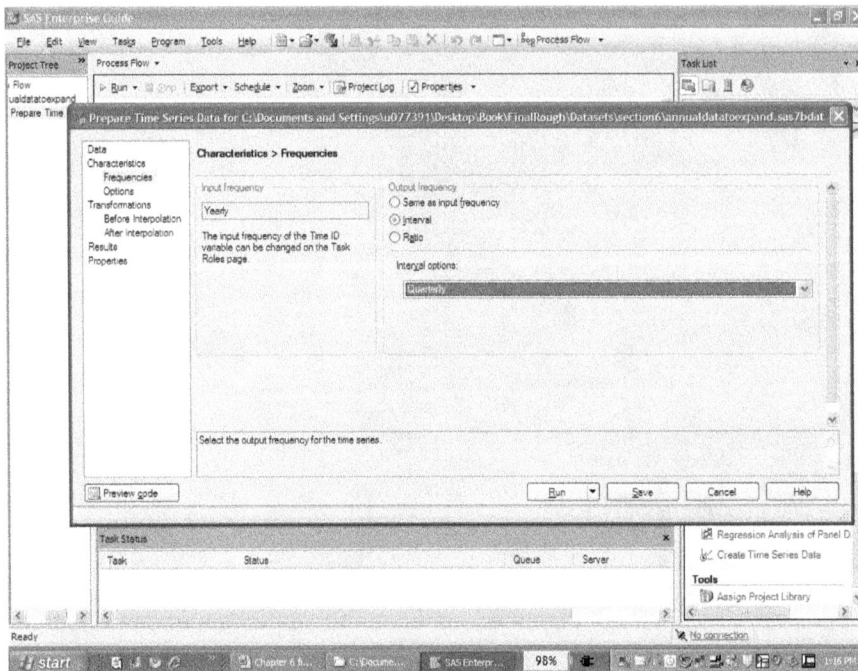

Next, we choose the interpolation method to use. In this case, we have chosen Cubic spline and have used the default end point anchoring methods (Not-a-knot). We have to choose both the Observation characteristics and the Date Alignment Interval. Observation characteristics indicate how the data was compiled (Total, Average, Beginning, Middle, or End). We have selected Beginning. The Date Alignment Interval is when you would like the dates to be set up and aligned (Beginning, Middle, or End). We have chosen Beginning. (See Figure 6.23.)

Figure 6.23: Choosing Output Details

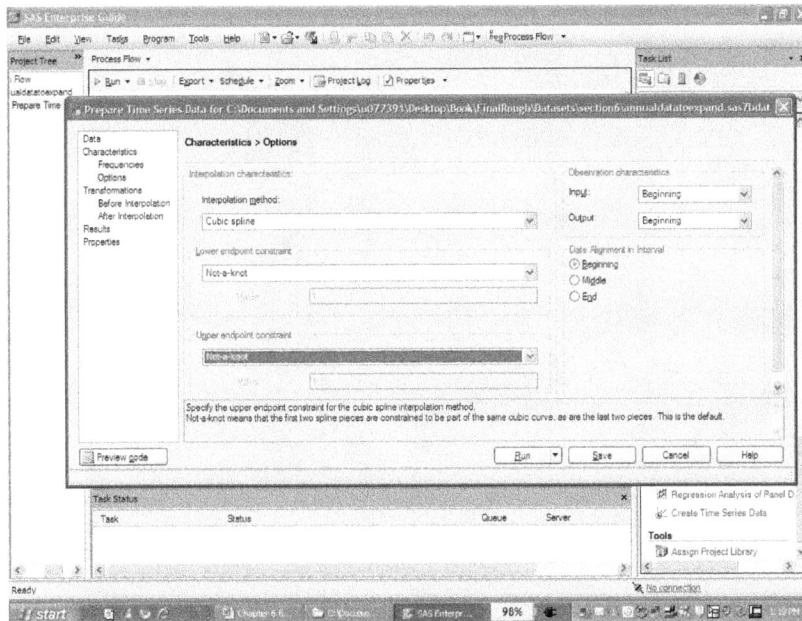

We have to choose a transformation in order to answer the question, "Is there a formula to correctly convert the annual data to quarterly?" (See Figures 6.24 and 6.25.)

Figure 6.24: First Step in Selecting Transformation

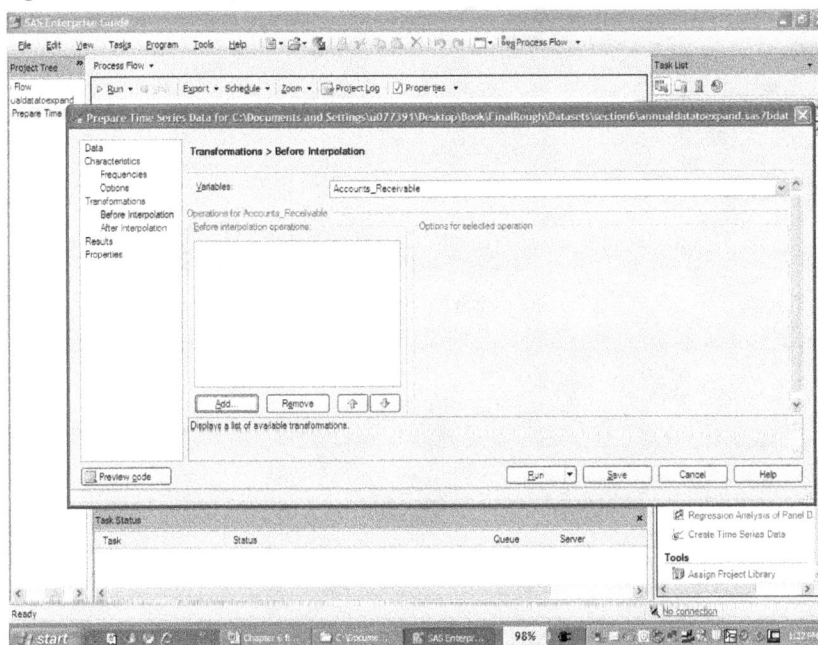

In this case, we choose to divide by 4. The combination of dividing by 4 and using the cubic spline, without any other knowledge of the time series or seasonality, is a reasonable way to capture the year-on-year trends properly.

Figure 6.25: Second Step in Choosing Transformation

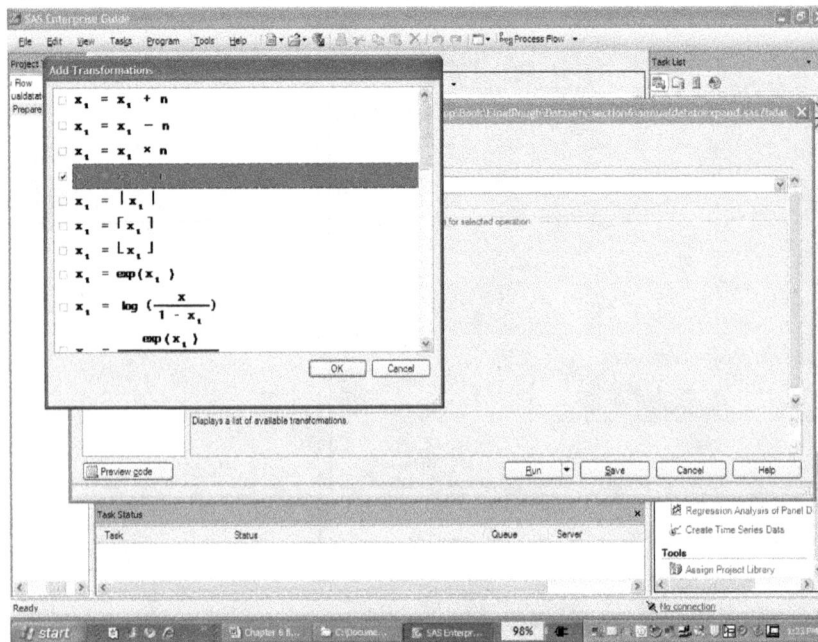

Figure 6.26: Choosing the Divisor for the Transformation

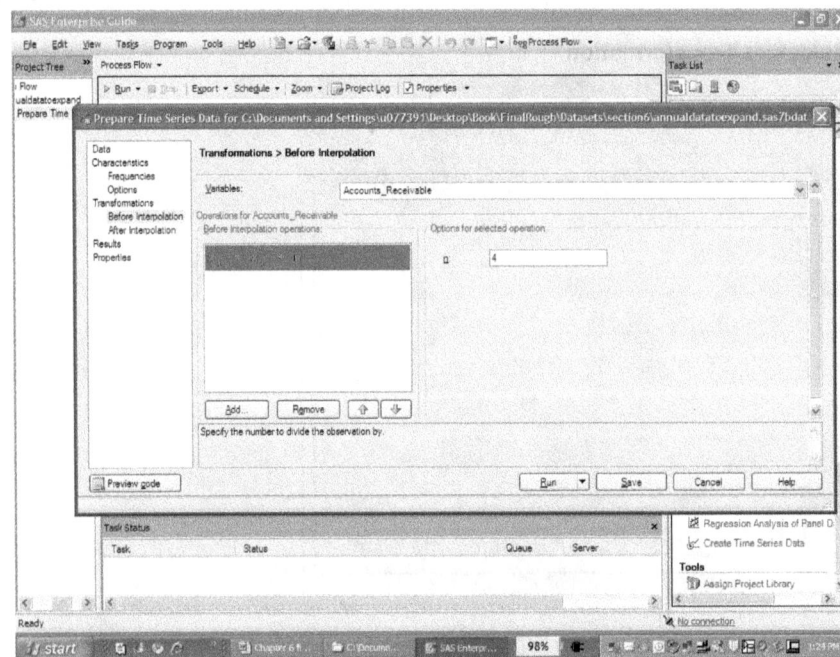

In Figure 6.27, we see the final expanded data. Note that the variable maintains its original name.

Figure 6.27: Final Expanded Data

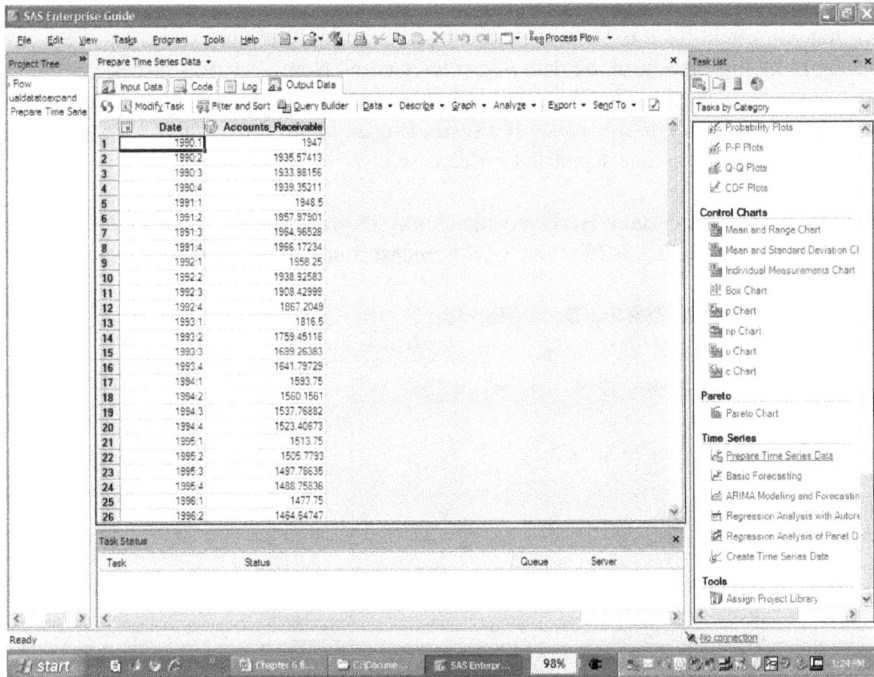

As a final check, the two series should be plotted together. (See Figure 6.28.)

Figure 6.28: Annual and Quarterly Expanded Accounts Receivables

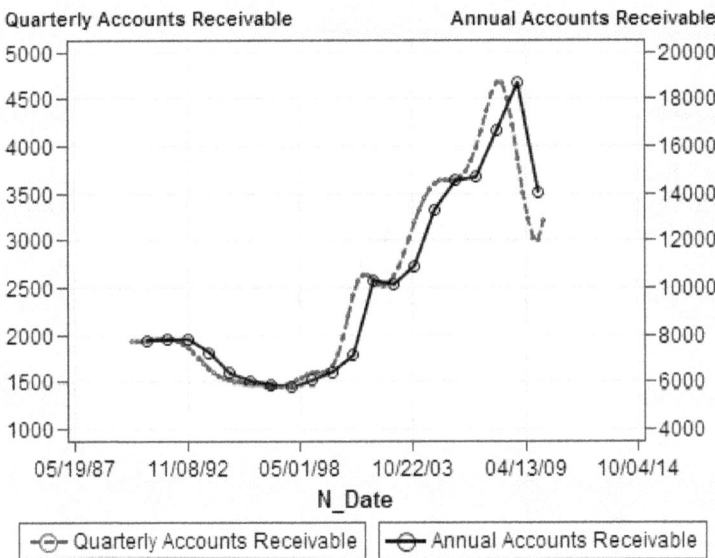

6.4 Merging

Once each of the data sources (Xs and Ys of varying intervals) is converted to a proper times series data set with matching intervals, they simply need to be merged. A standard Base SAS DATA step, using pre-sorting and a BY statement, is sufficient. If preferred, the Query Builder task in SAS Enterprise Guide can be used. In either case, this is the last step before treating the data set as the final time series data set. Because each is a basic SAS manipulation, it is up to the reader to decide how to accomplish this particular task.

6.5 Imputation

There are various reasons that you might have to "impute" missing data in the final time series data set. Some data simply might be missing randomly. Because the data might have come from varying sources, each time series might not start at the same time. Also, some series might have stopped and, as a result, is shorter than others. This last situation is the most detrimental because if a series is used in the final analysis, it needs to be as complete as possible. Each of these missing data scenarios is discussed.

The simplest scenario is the random missing data. Here, we can simply let SAS Forecast Studio do its job because it has missing data capabilities. Figure 6.29 is the SAS Forecast Studio view of the data in a project.

Figure 6.29: Time Series Data Input with Missing Data (Random)

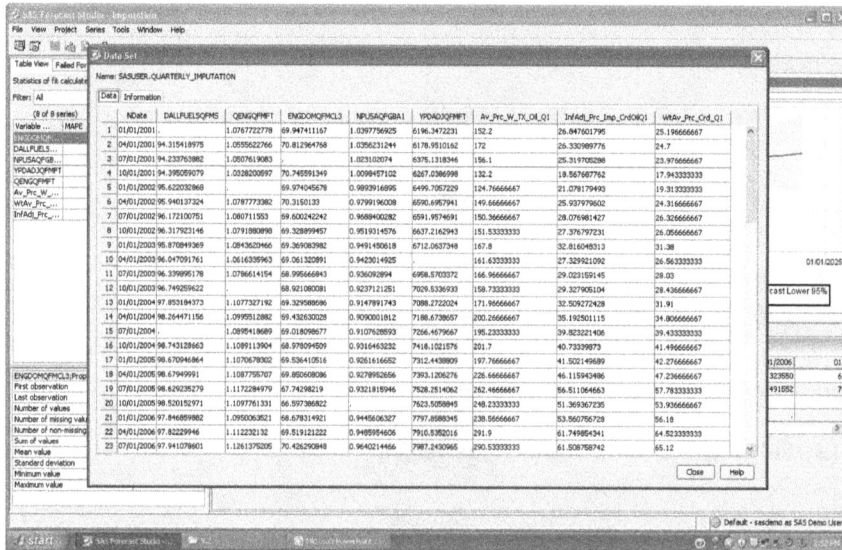

In the New Project wizard (specifically, step 6), we are asked how to handle missing data. The first decision is about how to handle embedded or random missing data. There are various choices for handling missing data, including simply leaving it as "missing." Each choice has ramifications that are unique to the forecasting project. The options are as follows: 0, Average, First, Last, Maximum, Median, Minimum, Missing, Next, and Previous. In situations where there is an evident trend or cycle, Previous, Next, or Missing seems to be the best choice. When no trend or cycle exists, Average or Median seems to be the best choice. When choosing 0, First, Last, Maximum, or Minimum, you need to make sure that the context is appropriate for the problem.

Figure 6.30: Selecting the Missing Data Option

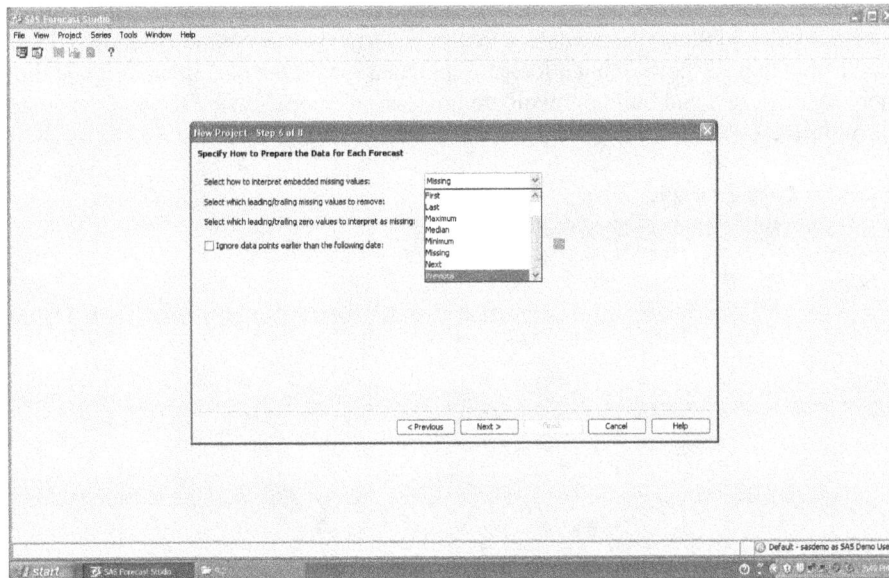

There are multiple choices for leading or trailing missing values. There are fewer choices for imputation. We can "remove" missing values by choosing Both Left and Right or Left or Right. We can also choose None (which leaves them in). These same choices apply to leading or trailing zero values.

You can actually truncate the series on the left side for data earlier than a given date.

Figure 6.31: Truncating Series

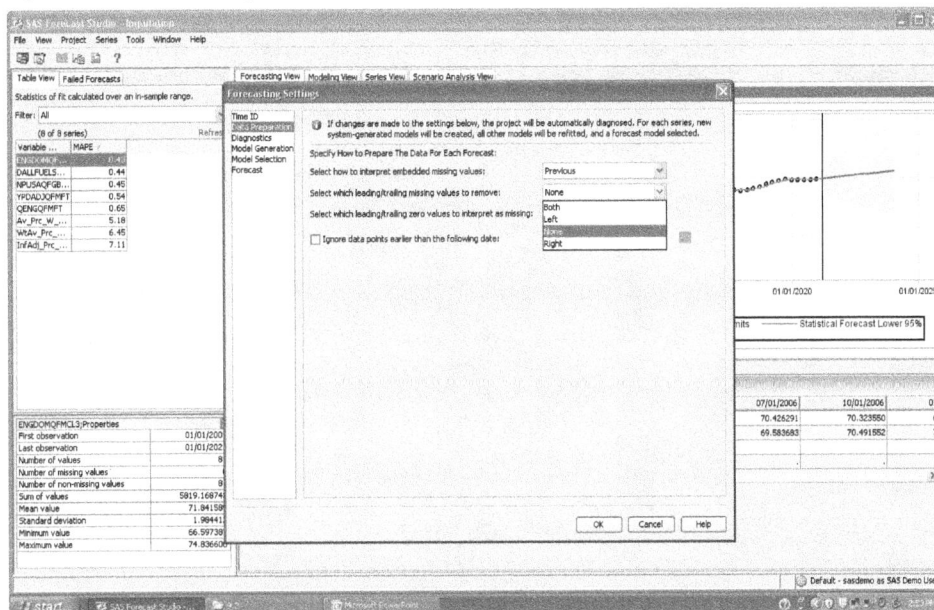

Each of these options might be useful for an extremely long time series or when doing early exploratory modeling. But in general time series data comes at a premium, and deleting trailing or leading missing data is not necessarily a good option. This can lead to *backcasting* for leading missing data and forecasting for trailing missing data. An example of leading missing data is shown in Figure 6.32.

Backcasting is the process of forecasting backward in time. Often, time series data from different sources do not start at the same time as each another. Backcasting to handle missing data is normally restricted to exogenous variables. The simplest method for handling missing data is to sort the data in reverse time order. (See Figure 6.33.) Use simple univariate methods for developing a forecast, and then re-sort the data in the original time order. You could get more sophisticated and build multivariate time serious models using the other exogenous variables as Xs for the X of interest. This would introduce specific collinearity between the Xs.

Figure 6.32: Leading Missing Data Example

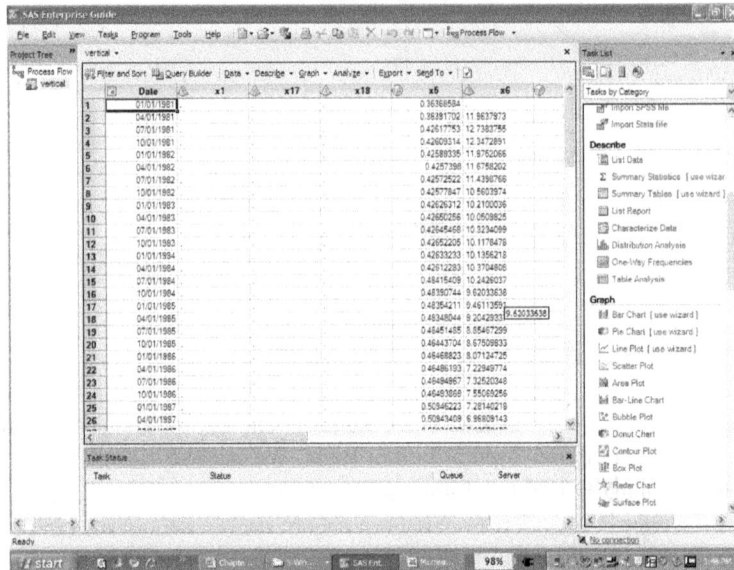

Figure 6.33: Reverse Date and Sorting for Backcasting

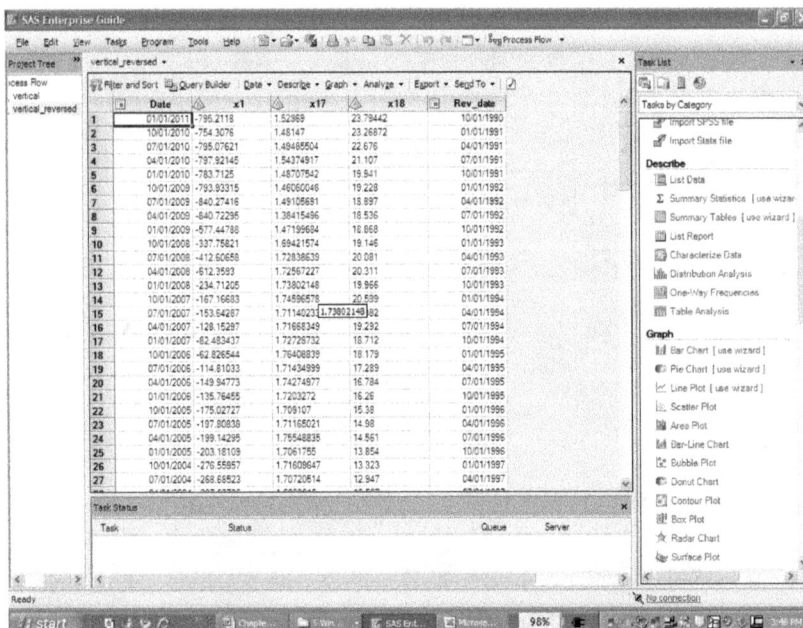

This data set is used in SAS Forecast Studio to produce forecasts. (See Figure 6.34.) The data is re-sorted to produce the backcasted data set.

Figure 6.34: SAS Forecast Studio Backcasting Example

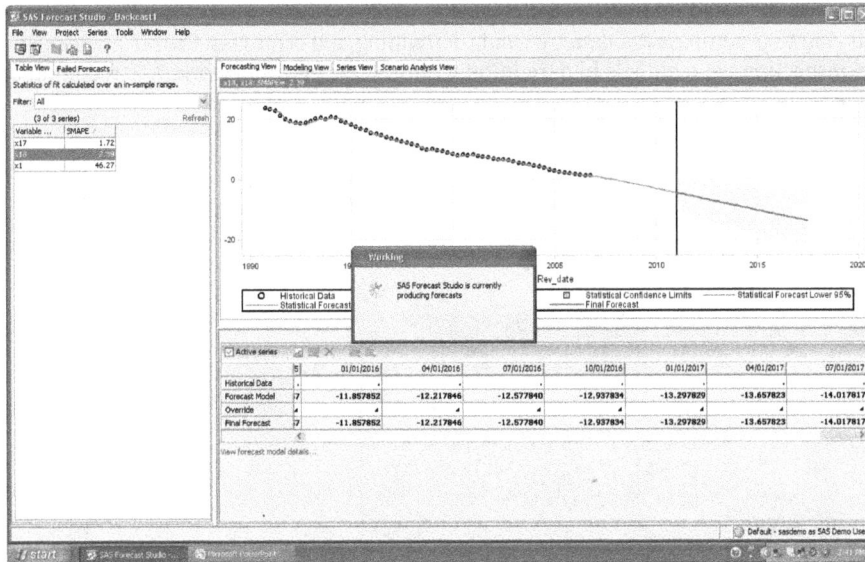

6.6 Outliers

As in any other data analysis endeavor, there are potential outliers in time series data. One of the simplest ways to look for outliers is to graph the data. This can be accomplished by using the SAS Enterprise Guide Line Plot wizard or using raw SAS code such as PROC GPLOT. Other SAS platforms are designed for high-volume time series data mining. One of these platforms has an exploratory graphics node that does large volumes of time series graphs quite effectively. Figure 6.8 shows the Time Series in SAS Enterprise Miner.

Figure 6.35: High Volume Graphing Capability for Identifying Outliers

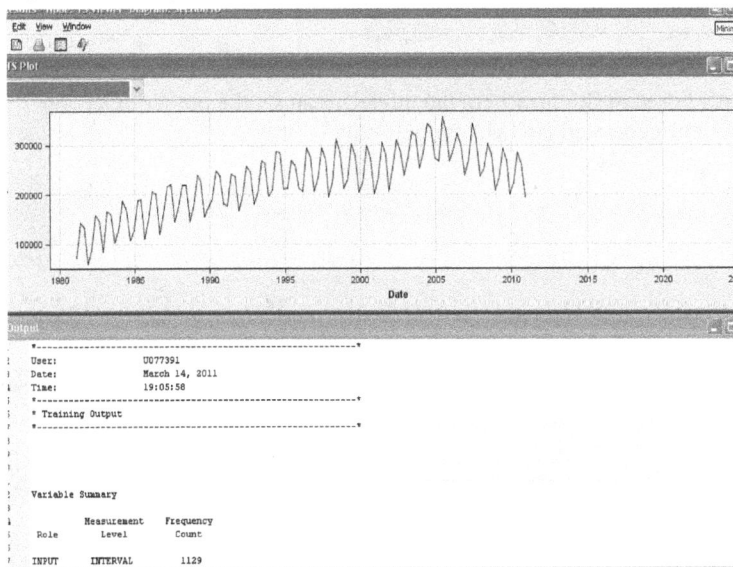

Graphical inspection can only lead you to the proposed outliers. Special events might have induced the supposed outlier, and it might not be an outlier at all. An example of a special event could be a quarter end, a year end, or even a seasonal price reduction to induce a large spike in sales. In these cases, an outlier might be identified. But, the way it is handled in a model is as an event, not as an outlier.

Care has to be taken when identifying outliers because of the fact that time series data is serially correlated. Most traditional outlier tests use the standard random, independent, identical distribution random variables assumption. Rather than trying to preprocess the data by hand (detrending and removing the serial correlation or seasonal effects) and then inspecting the residuals in a traditional outlier testing framework, it is recommended that the modeler use outlier testing in SAS Forecast Studio. For reference, PROC UCM has an OUTLIER statement.

The following example shows two fictitious outliers in a very seasonal time series. The plot itself indicates the potential outliers. (See Figure 6.36.)

Figure 6.36: Outlier Example in SAS Forecast Studio

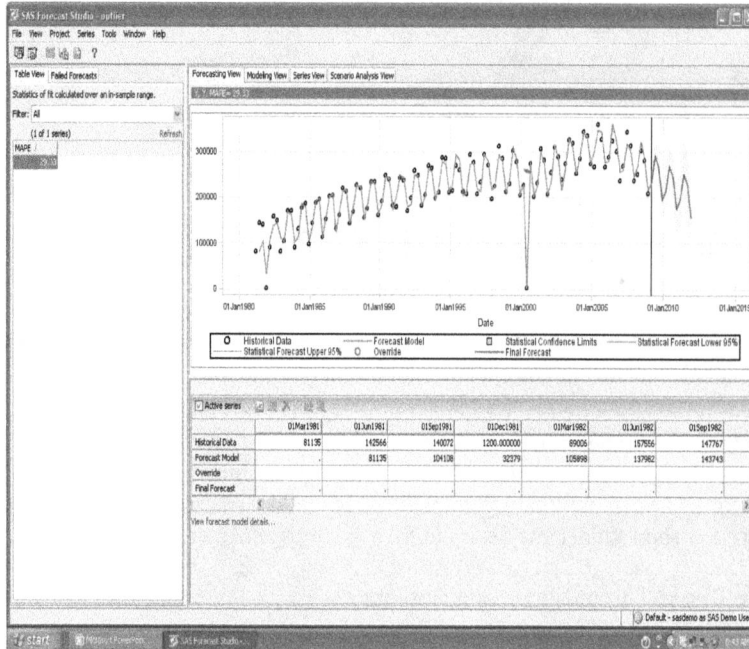

Figure 6.37 shows the dialog box in SAS Forecast Studio where the modeler can set the outlier detection options. These options are limited to system-generated ARIMA models only.

Figure 6.37: Forecasting Settings Dialog Box in SAS Forecast Studio

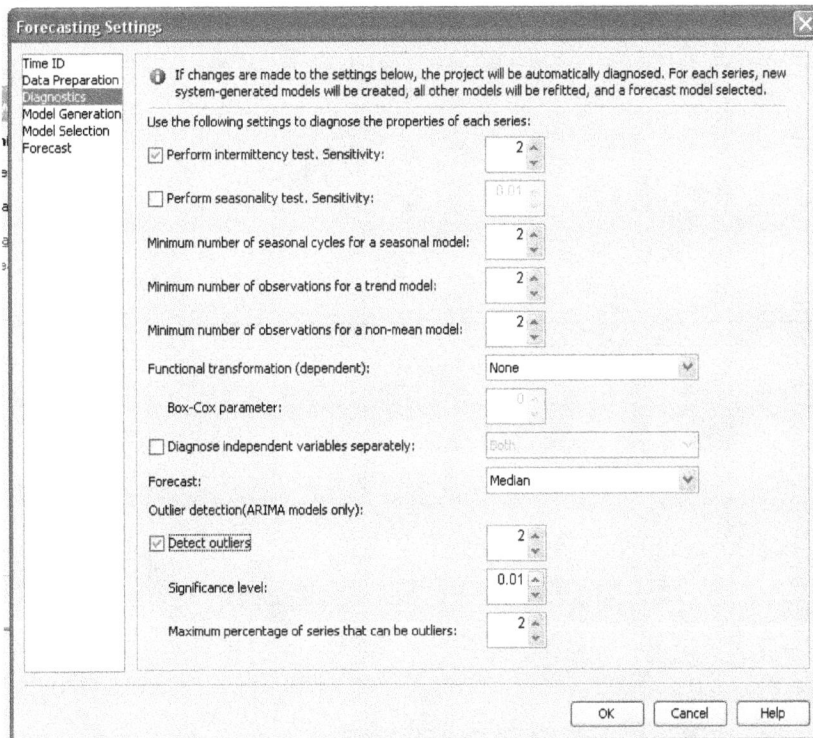

First, click the **Detect outliers** check box. Next, choose the number of outliers that are allowed. Then, choose the significance level for the outlier test, and select the maximum percentage of the series that is allowed to be classified as outliers. Once outliers have been detected by SAS Forecast Studio, they will show up in the model as input variables. This approach removes the variability caused by the outliers and allows the remaining structure in the model to be estimated properly.

Figure 6.38 shows the outliers in SAS Forecast Studio. Figure 6.39 shows the parameter estimates for the outliers.

Figure 6.38: SAS Forecast Studio Model Plot with Outliers Estimated

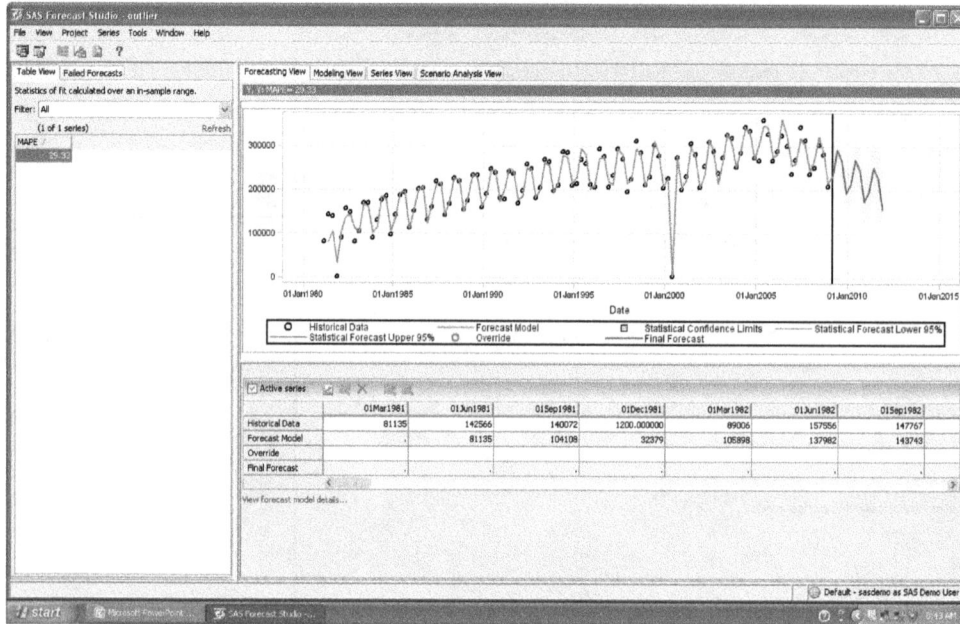

Figure 6.39: Outlier Parameter Estimates

Parameter Estimates

Component	Parameter	Estimate	Standard Error	t Value	Approx Pr > |t|
Y	MA1_1	0.62603	0.07829	8.00	<.0001
Y	AR1_4	0.62510	0.09489	6.59	<.0001
Y	AR1_8	0.36010	0.09728	3.70	0.0003
AO01DEC1981D	SCALE	-85178.9	10512.1	-8.10	<.0001
AO01JUN2000D	SCALE	-298588.0	10318.1	-28.94	<.0001

Once the outliers are identified, other approaches can be used if the outlier variable is not wanted. Because the proposed outliers are verified, they can simply be removed from the data set, and the model can be run without them. Figure 6.40 shows the SAS Forecast Studio model after it was run with the outliers set to missing.

Figure 6.40: Modeling with Identified Outliers Set to Missing

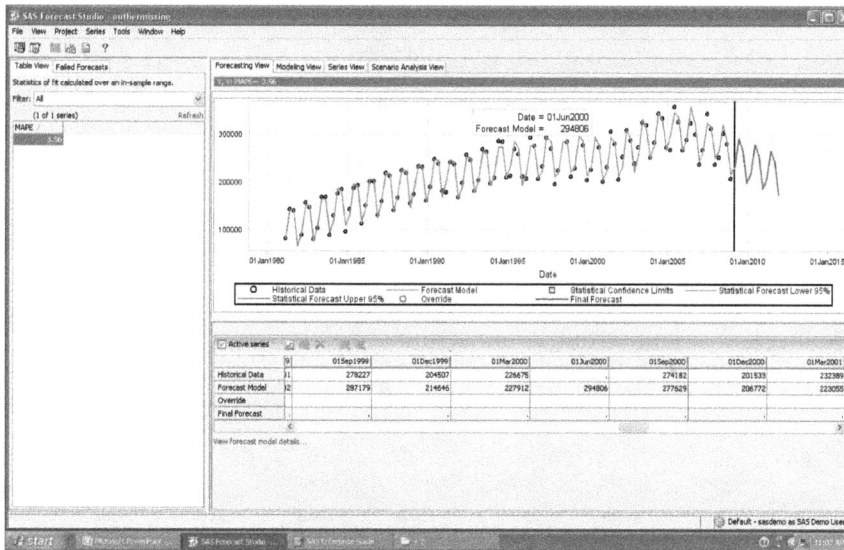

Another approach is to rerun the model. However, this time, do not turn on outlier detection. (See Figure 6.41.)

Figure 6.41: Running the Model without Outlier Detection

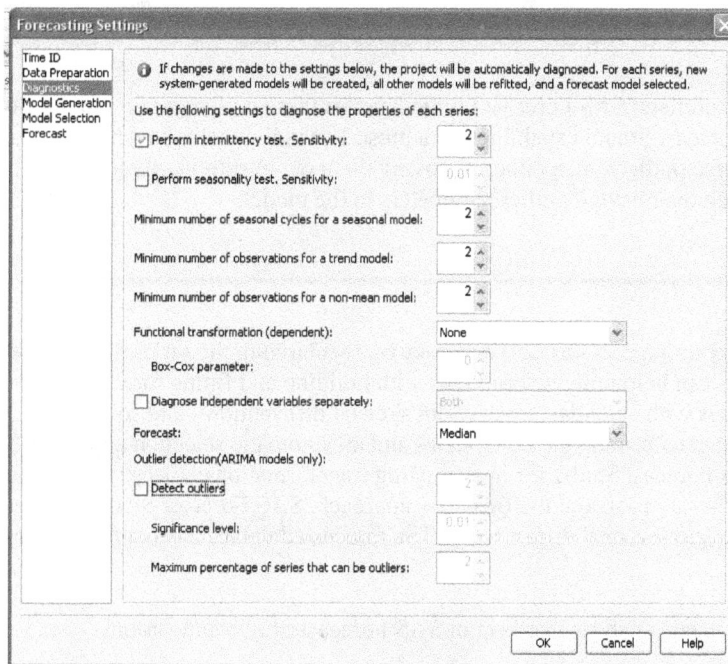

In this scenario, the "model" data can be used to estimate the missing data. Figure 6.42 shows the model plotted without outlier detection on and identifies the estimated value for the outlier.

Figure 6.42: Estimated Value for Outlier

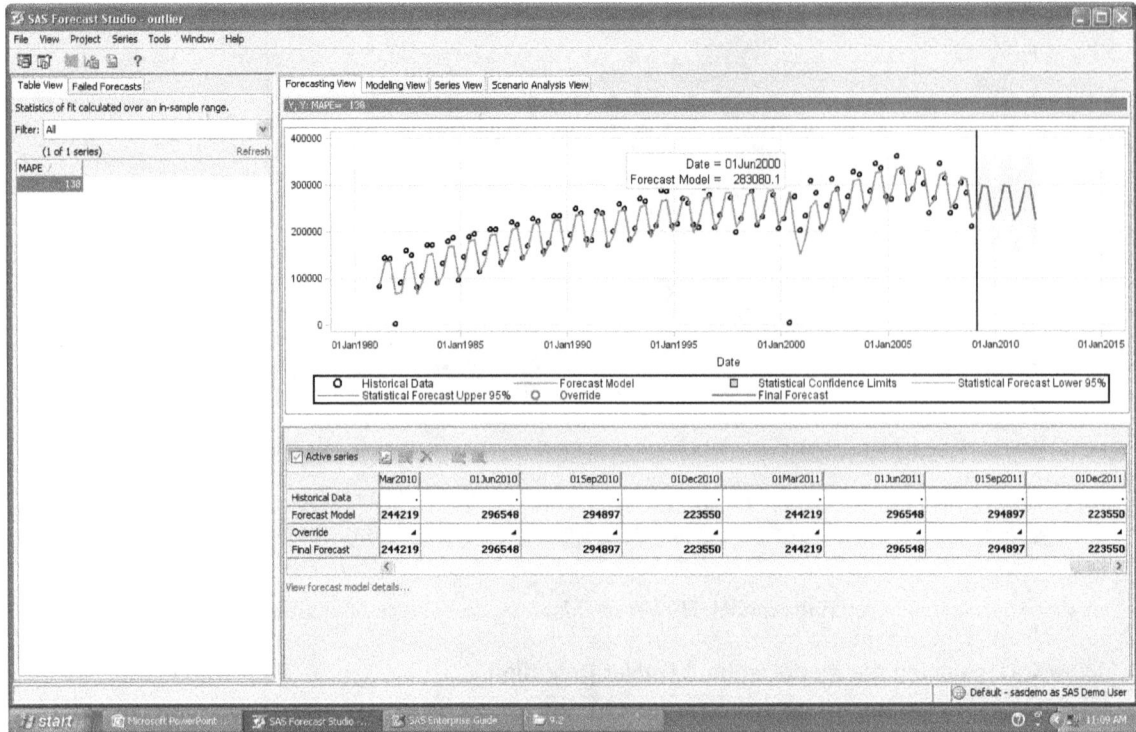

In summary, time series data can have outliers. SAS Forecast Studio handles these outliers easily. In the case of hierarchical forecasting, using SAS Forecast Studio capabilities is a must. For individual model building, using SAS Forecast Studio to first diagnose the outliers, then either removing them or substituting forecasted values before re-forecasting might be preferred over having outlier parameters in the model.

6.7 Transformations

The purpose of transformations is to try to improve forecast accuracy by regularizing the series that will be used in the forecast model. Transformations can help mitigate problems with building and fitting forecasting models, problems associated with outliers, issues with candidate inputs with skewed distributions, and so on. SAS Forecast Studio has both a window for transforming the input series and an automatic selection approach. There is also an exploratory platform in SAS Forecast Studio for investigating transformation choices. The primary pure transformations are log, logistic, square root, and the Box-Cox approach. SAS Forecast Studio enables you to investigate both simple differencing and seasonal differencing. The forecasted values and confidence limits are presented in the original units.

The following figures show the Series View analysis platform in SAS Forecast Studio and various transformations.

Figure 6.43: Series View Analysis Platform (with No Transformations Applied)

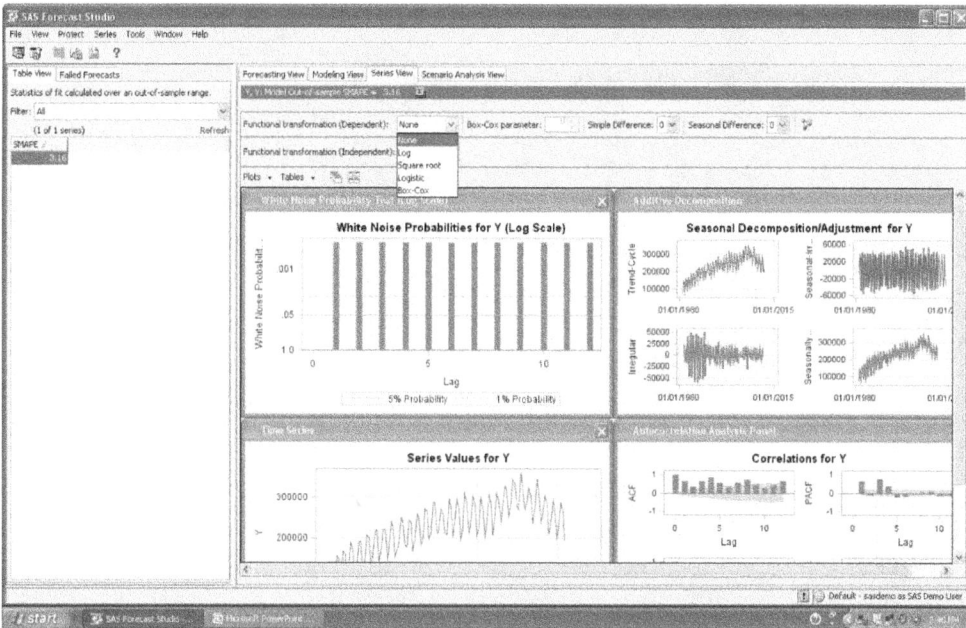

Figure 6.44: Log Transformation Applied

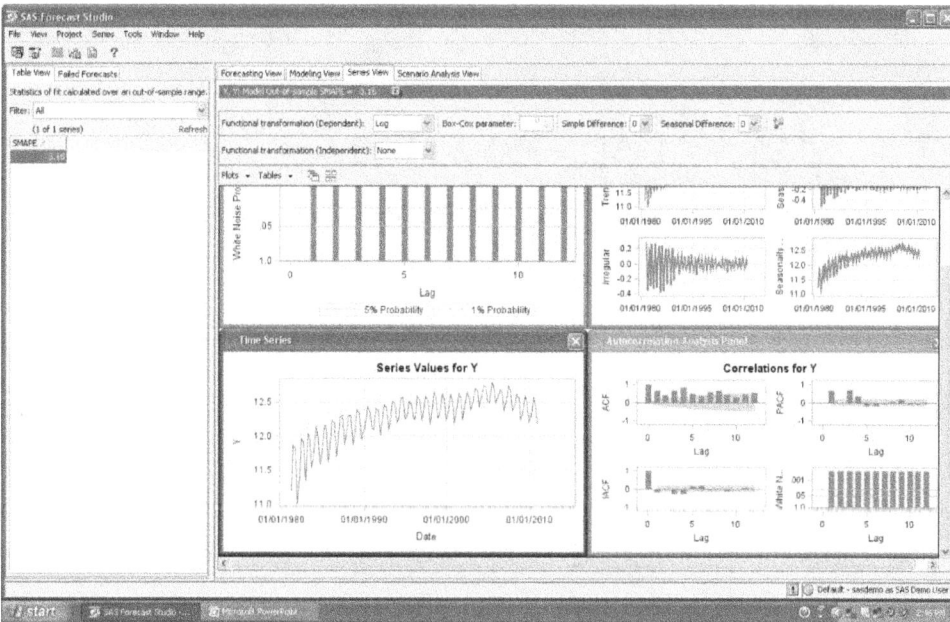

Figure 6.45: First Difference Transformation Applied

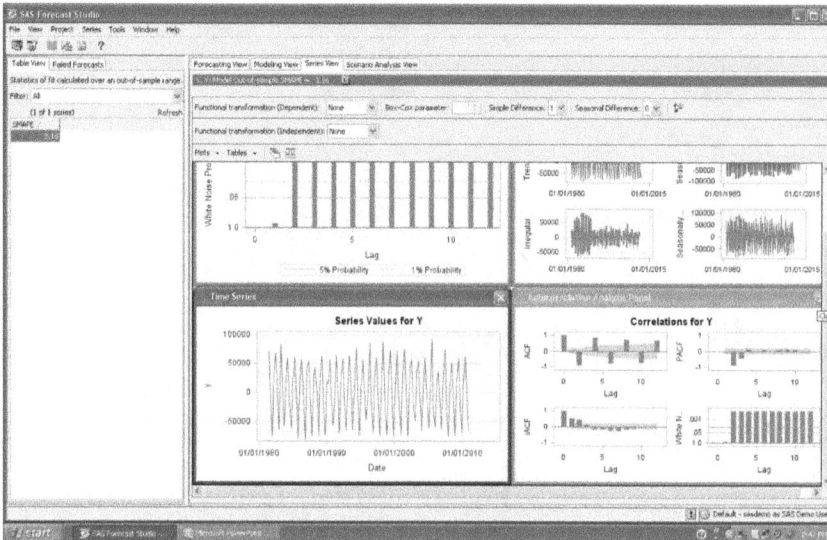

Figure 6.46: Seasonal First Difference Applied

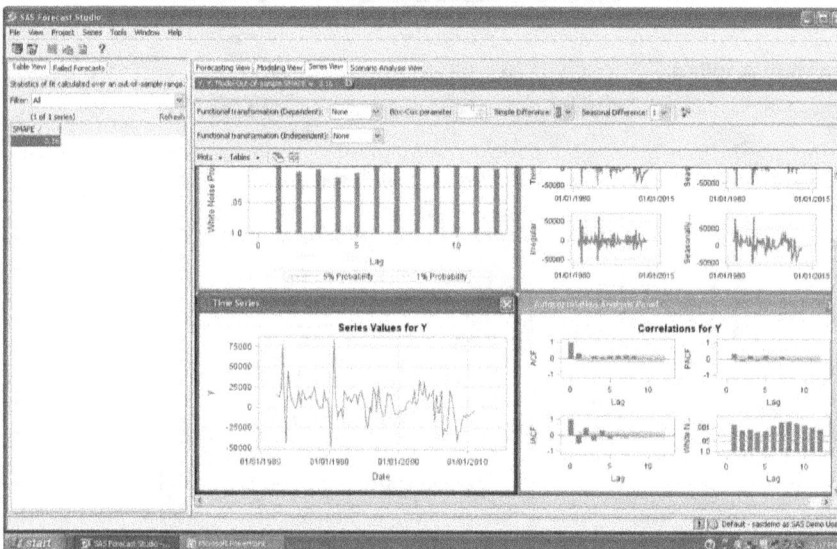

6.8 Summary

Preparing time series data for forecasting is a critical step in the modeling process. Often, time series data starts as transaction data and has to be "accumulated" to make a time series data set. Not all time series data is available in the same frequency. As a result, the data might have to be contracted or expanded. Once all of the data is on the same time frequency, it has to be merged. Once it is merged, there can be missing data (leading, trailing, or embedded). There are various methods for imputing missing data. Sometimes, time series data has outliers that can be handled through various methods. Before modeling, the Y or X variable might need to be transformed and SAS Forecast Studio has automated capabilities to do this for you.

---

[1] Exceptions to this are when you are doing momentum and mixed interval forecasting. See Chapter 10.

[2] A "manual" example of accumulating or converting a transaction data set into a time series data set using the TIMESERIES procedure is provided in Chapter 11.

Chapter 7: A Practitioner's Guide of DMM Methods for Forecasting

7.1 Overview

As mentioned in Chapter 1, a time series database might often have hundreds, if not thousands, of hypothesized Xs for a given set of Ys. Just as with data mining for transactional data, a specific feature or variable selection step is needed. As discussed in Chapter 2, this step of the process exists to help reduce the number of variables for consideration in modeling. There are two aspects of this phase of analysis. First, there is a variable reduction step, and then there is a variable selection step. In variable reduction, various approaches are used to reduce the number of Xs being considered and, at the same time, there are approaches to explain or characterize the key sources of variability in the Xs regardless of the Ys. This variable reduction step is considered a non-supervised method because the Ys are not being considered. Second, there is a variable selection step. This step includes supervised methodologies because the Ys are being considered.

In traditional data mining literature about transaction or non-time series data, many researchers have proposed numerous approaches for variable reduction and selection on transaction data (Koller 1996, Guyon 2003, and so on.). Some of these methods are described below. Given the serial correlation in time series data, methods that specifically consider serial correlation are considered as well and have been found to be more effective by the authors. Thus, for both the variable reduction and variable selection steps in the process, both traditional data mining and time series approaches are presented.

One of the key problems with using transactional variable reduction and variable selection approaches on time series data is that in order to overcome the method not being time series-based, the modeler has to include lags of the Xs in the problem. This dramatically increases the size of the variable selection problem. For example, if there are 1000 Xs in the problem and the data is quarterly, the modeler would have to add a minimum of 4 lags for each X in the variable reduction or selection phase of the project. As a result, there are now 4000 variables to be processed.

7.2 Methods for Variable Reduction

Traditional Data Mining

The most common unsupervised variable reduction method in transactional data mining is principal components analysis (PCA). In this approach, various matrix manipulation techniques are used to derive a new set of variables. Upon reduction, these variables still explain a substantial amount of variability in the Xs. This is known as the Eigenvector–Eigenvalue problem where the original Xs are converted to a set of Zs through a simple scalar multiplier known as the Eigenvector. The hope is that the first few sets of principal components explain 80% or more of the original variability in the Xs. Specific output from the analysis does indicate how much of the variability is explained and what variables have the highest weight or contribution to the new Zs. Most often, modelers use a PCA approach based on the correlation matrix of the Xs as input, and then pick an additional rotation of the resulting factors. This additional rotation (VARMAX is one of the most popular) helps interpret the resulting factors in terms of the original Xs. Another beneficial property of PCA is that the new Zs are orthogonal (having 0 correlation) to one another, which helps tremendously in the model building phase of the project.

The issue with this approach is that it is somewhat misleading in terms of its real variable reduction potential. Remember that the new Zs are a linear combination of the original Xs. Depending on how the Eigen values work out, there is always a loss of information when using a reduced set of Zs or Xs in each Z. Using a reduced set of either or both is common practice.

Another approach to variable reduction is to pick weights to apply to a set of Xs and to develop a new V that is a combination of the Xs based on weights that the analysts pick themselves. In the case of demand forecasting, these weights might simply be the penetration a business has into a set of market segments. Thus, the weights represent the fraction of demand coming from each segment. Forecasting demand generally includes supply, demand, price, capacity, macro, micro, and industry-specific variables. As a result, the indexed demand V is a consolidation of Xs in the demand portion of the problem. Though commonly used by business leaders, the approach of choosing weights in an ad hoc fashion is where the approach can be misleading.

Another approach for variable reduction is variable clustering. The traditional approach to cluster analysis of transactional data involves operating on the rows of the data set and trying to group rows that are similar to one another. The *k*-means and hierarchical clustering methods are common approaches. The objective of cluster analysis is to group rows so that the variance within a cluster is minimized and the variances across the clusters are maximized. In the end, the clusters are differentiated as much as possible. In the case of variable clustering, SAS has a procedure called VARCLUS that operates on the columns of the data set (the variables), rather than on the rows.

PROC VARCLUS is a hierarchical method. Each column can belong to only a certain hierarchy as you choose different numbers of clusters as the solution. The PROC VARCLUS approach actually uses both PCA and hierarchical clustering because there is a linear combination of the variables in the cluster associated with each cluster. PROC VARCLUS tries to maximize the variance that is explained by the cluster components, summed over all of the clusters. In PROC VARCLUS, the cluster components are oblique, not orthogonal, because each cluster component is computed from a different set of variables, not from all of the other cluster components. Because of this fact, the first principal component of one cluster might be correlated with the first principal component of another cluster.

A large set of variables can often be replaced by the set of cluster components with minimal loss of information. That is what variable reduction is all about. As with the standard approach for principal components, a reduced number of cluster components do not generally explain as much variance as the same number of principal components in the full set of variables. The beauty of the approach is that the cluster components are usually easier to interpret than the principal components.

The user must first specify how many clusters are needed to obtain output from the analysis. There are two factors that influence this decision. First, how many variables does the problem start with? Second, how many variables can be used as input to the eventual modeling platform? Normally, about 20% of the variables are set as the number of clusters, which does not exceed 50 or so. Once the number of clusters is chosen, the procedure is run. (Examples are below.) The output of the procedure indicates which variables are in which clusters, what the R square of each variable is and what the cluster centroid is. As a result, you can determine which variable best represents the cluster centroid. This is basically variable reduction because the modeler chooses the variable with the highest R square and with the centroid of the cluster to represent that cluster. When the variable metadata (a short description of each variable) is sorted by cluster, often the theme of the cluster is revealed. To see how cohesive the clusters are, look for variables at the bottom of the list once it is sorted by the R square within each cluster. If there is a step function drop in the R square for a small number of variables at the bottom of the list, this might indicate that more clusters are needed.

There are two ways to use PROC VARCLUS for variable reduction. The first is to completely rely on it to reduce the variable set before doing any variable selection. Then, only perform the variable selection step on the reduced set of variables. The second is to blend variable reduction and variable selection. In this case, the variable reduction step is an independent step in the process and provides a separate set of variables to include in the modeling step with variable selection. Both of these are useful.

Time Series Approach

Because time series data have the added complexity that the data are correlated in time, variable reduction and selection both need to take this complexity into consideration. Leonard and Wolfe (2002) and Leonard and Lee (2008) introduce via SAS an approach for analyzing and measuring the similarity of multiple time series variables. Unlike traditional time series modeling for relating Y (target) to X (input) as detailed in Chapter 8, similarity analysis leverages the fact that the data is ordered. A similarity measure can take various forms, but it is essentially a metric that measures the distance between the X and Y sequences and keeps ordering in mind.

Squared Error
$$d\left(\vec{x}_i, \vec{y}_j\right) = \sum_{k=1}^{K} \left(x_{k,i} - y_{k,j}\right)^2$$

Absolute difference
$$d\left(\vec{x}_i, \vec{y}_j\right) = \sum_{k=1}^{K} \left|x_{k,i} - y_{k,j}\right|$$

This ordering allows the target series to be slid to the left or to the right in time relative to the input sequence. While sliding, the similarity metric being used is calculated, and the slide that minimizes this metric is recorded. Besides sliding the time sequence, the time dimension relative to the target and input sequence can be expanded and compressed. This process is considered time warping and is discussed by Sankoff and Kruskal (1983). Sliding and time warping allow the target and input sequence to be aligned properly to provide a minimization using the distance metrics (such as squared or absolute deviation [weighted or relative] as shown above) of the chosen measure of similarity. This analysis can be done on the Xs themselves. The results from the analysis can produce a matrix of minimized similarity metrics, measuring the similarity amongst all pairs of Xs in the time series data set.

For variable reduction, as with row based PCA, either correlations or the variance-covariance matrix can be used as input for PROC VARCLUS. To use PROC VARCLUS on time series data, the similarity matrix among the Xs(or X-X similarity) should be used as input. The following example illustrates the similarity approach for variable reduction in time series data. The SAS code for conducting similarity analysis and variable clustering is given in Program 7.1.

Program 7.1

```
proc similarity data=lib1.yandxexample outsum=lib1.SIMMATRIX ;
id date interval=quarterly;
target chained_price_index_consumer_pu--
vacancy_rate_of_rental_housing_u/sdif=(1,4) normalize=absolute
measure=mabsdev
```

```
expand=(localabs=12)
compress=(localabs=12);
run;
proc varclus data=lib1.simmatrix maxc=6
outstat=lib1.outstat_full noprint;
var chained_price_index_consumer_pu--vacancy_rate_of_rental_housing_u;
   run;
```

In this code sequence, both the similarity analysis and variable clustering take place. The input data set should have both the Ys and the Xs in it, along with a proper date variable as in a traditional time series database. Note that the date variable is in the ID statement, and the interval is identified as quarterly. In the case of X-X similarity, the TARGET command identifies this list. Using the shorthand syntax -- for variables listed in a continuum in the SAS data set, the complete list of X variables is included. In this example, the data is being seasonally differenced and normally should be (s(1,4)). The data is being normalized using the absolute approach. The similarity metric is MABSDEV (mean absolute difference). The time warping being assessed is up to 12 time periods, and it is done using the local absolute approach (LOCALABS). The similarity matrix is being sent to lib1.simmatrix for use in PROC VARCLUS.

In PROC VARCLUS using the 20% rule, six clusters are being chosen because there are only 29 Xs in this data set. The VAR list in PROC VARCLUS should be the same as the target list in PROC SIMILARITY. Figure 7.1 shows the input data set for running PROC SIMILARITY.

Figure 7.1: Input Data for PROC SIMILARITY Code

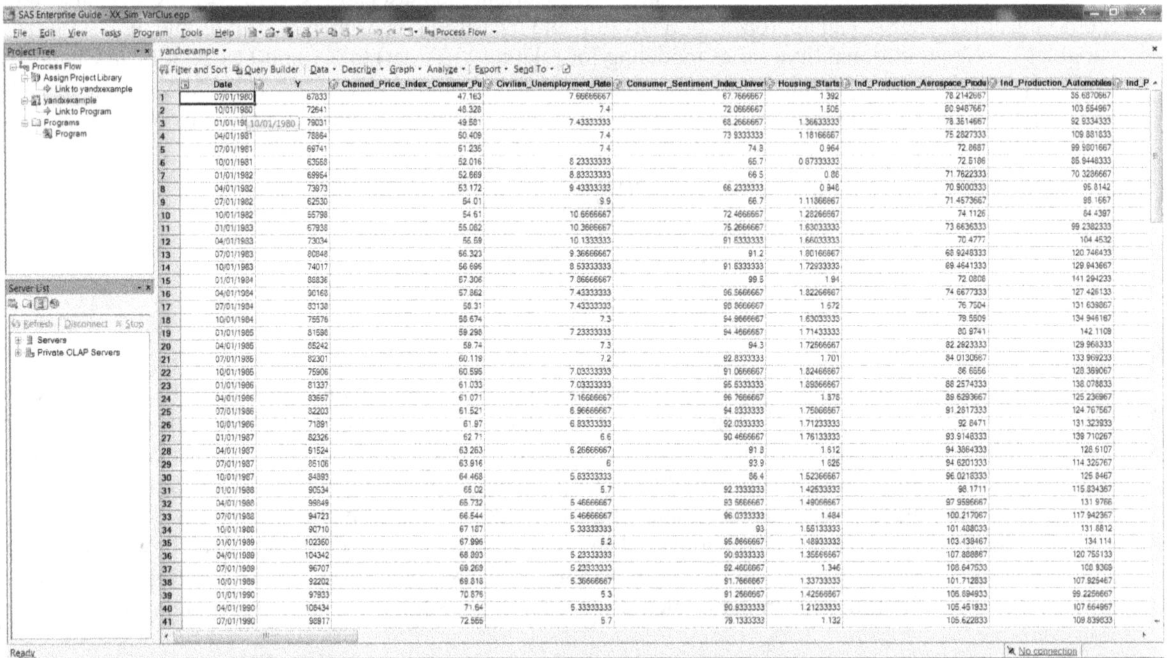

Figure 7.2 shows the SAS Enterprise Guide flow for conducting the analysis before post-processing the output data set.

Figure 7.2: SAS Enterprise Guide Flow for PROC SIMILARITY and PROC VARCLUS Analysis

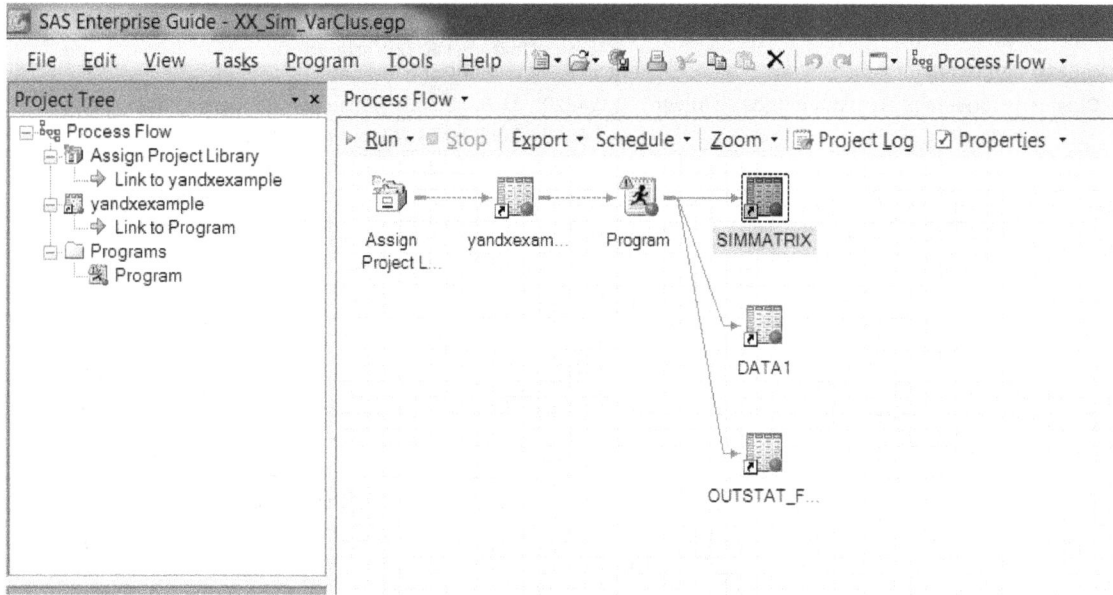

Figure 7.3 shows the PROC SIMILARITY matrix as output by PROC SIMILARITY for input into PROC VARCLUS.

Figure 7.3: PROC SIMILARITY Matrix for Input into PROC VARCLUS

PROC VARCLUS outputs all of the cluster solutions from cluster 2 through the number of clusters specified in the code (in this case, 25). This feature can be turned off. Figure 7.4 shows the output for the two-cluster solution. It is generally recommended that the NOPRINT option be used, and that the output be post-processed to reduce it to the appropriate number of clusters. In Figure 7.4, the variables are sorted by cluster, but not by the R square within a cluster. This can also be accomplished with post-processing. In this case, Ind Production Durable Goods has the highest R square with its own cluster, 0.9036. It would be the X chosen to represent this cluster. In cluster 2, the X chosen would be Ind Production Chemicals with an R square of 0.9142. Note that in both cases there are numerous Xs within each cluster with an R square of less than 0.2. This could indicate the need for more clusters.

Figure 7.4: Cluster Output for 22 Clusters

2 Clusters		R-squared with			
Cluster	Variable	Own Cluster	Next Closest	1-R**2 Ratio	Variable Label
Cluster 1	Consumer_Sentiment_Index_Univer	0.4820	0.0240	0.5307	Consumer_Sentiment_Index_Univer
	Housing_Starts	0.4616	0.0092	0.5434	Housing_Starts
	Ind_Production_Aerospace_Produ	0.1935	0.0003	0.8068	Ind_Production_Aerospace_Produ
	Ind_Production_Automobiles	0.6687	0.0027	0.3322	Ind_Production_Automobiles
	Ind_Production_Beverages	0.4688	0.2620	0.7198	Ind_Production_Beverages
	Ind_Production_Cement	0.6079	0.0005	0.3923	Ind_Production_Cement
	Ind_Production_Computers_And_E	0.6253	0.0590	0.3982	Ind_Production_Computers_And_E
	Ind_Production_Concrete_And_Ce	0.8186	0.0307	0.1871	Ind_Production_Concrete_And_Ce
	Ind_Production_Durable_Goods	0.9036	0.1539	0.1139	Ind_Production_Durable_Goods
	Ind_Production_Electrical_Equi	0.7427	0.3492	0.3954	Ind_Production_Electrical_Equi
	Ind_Production_Fabricated_Meta	0.8890	0.1578	0.1317	Ind_Production_Fabricated_Meta
	Ind_Production_Food	0.1230	0.0651	0.9380	Ind_Production_Food
	Ind_Production_Machinery	0.7067	0.3426	0.4461	Ind_Production_Machinery
	Ind_Production_Manufacturing_	0.8721	0.2534	0.1713	Ind_Production_Manufacturing_
	Ind_Production_Motor_Vehicles_	0.7186	0.0019	0.2819	Ind_Production_Motor_Vehicles_
	Ind_Production_Paints_And_Misc	0.5750	0.0283	0.4374	Ind_Production_Paints_And_Misc
	Ind_Production_Total_Ind_Prod	0.8949	0.2451	0.1392	Ind_Production_Total_Ind_Prod
	Real_Consumer_Spending_On_Durabl	0.8214	0.1256	0.2042	Real_Consumer_Spending_On_Durabl
	Real_Gross_Domestic_Product	0.8602	0.2671	0.1908	Real_Gross_Domestic_Product
	Vacancy_Rate_Of_Rental_Housing_U	0.3290	0.0471	0.7041	Vacancy_Rate_Of_Rental_Housing_U
Cluster 2	Chained_Price_Index_Consumer_Pu	0.5196	0.0416	0.5013	Chained_Price_Index_Consumer_Pu
	Civilian_Unemployment_Rate	0.0545	0.0129	0.9579	Civilian_Unemployment_Rate
	Ind_Production_Chemicals	0.9142	0.1421	0.1001	Ind_Production_Chemicals
	Ind_Production_Household_Appli	0.1872	0.1351	0.9397	Ind_Production_Household_Appli
	Ind_Production_Nondurable_Good	0.7412	0.4709	0.4891	Ind_Production_Nondurable_Good
	Ind_Production_Oil_And_Gas_Ext	0.1860	0.0014	0.8151	Ind_Production_Oil_And_Gas_Ext
	Ind_Production_Resins_And_Synt	0.6194	0.2745	0.5246	Ind_Production_Resins_And_Synt
	Ind_Production_Ship_And_Boat_B	0.5791	0.0807	0.4578	Ind_Production_Ship_And_Boat_B
	Real_Consumer_Spending_On_Nondur	0.7392	0.1514	0.3073	Real_Consumer_Spending_On_Nondur

Figure 7.5 contains the 6-cluster solution. It is evident that the clusters are more cohesive because there is no longer an R square less than .2. Cluster 1 looks like a large expenditure of durables, cluster 2 looks like nondurables, cluster 3 looks like smaller durables, cluster 4 looks like employment and government (laggers), cluster 5 looks like housing and automotive, and cluster 6, food, is by itself and is the fastest moving non-durable. Similar themes are present in many analyses of this nature. So, in this case, the number of variables can be reduced from 30 to 6. In reality, the modeling platform (here, SAS Forecast Server) can easily handle 30 Xs. Therefore, neither variable reduction nor variable selection would have been conducted except for example purposes only.

Figure 7.5: 6-Cluster Solution

6 Clusters		R-squared with			
		Own Cluster	Next Closest	1-R**2 Ratio	Variable Label
Cluster	Variable				
Cluster 1	Ind_Production_Beverages	0.5212	0.3897	0.7845	Ind_Production_Beverages
	Ind_Production_Computers_And_E	0.6531	0.4780	0.6646	Ind_Production_Computers_And_E
	Ind_Production_Durable_Goods	0.9400	0.6641	0.1787	Ind_Production_Durable_Goods
	Ind_Production_Electrical_Equi	0.8261	0.5023	0.3495	Ind_Production_Electrical_Equi
	Ind_Production_Fabricated_Meta	0.8949	0.6643	0.3130	Ind_Production_Fabricated_Meta
	Ind_Production_Machinery	0.7842	0.4751	0.4111	Ind_Production_Machinery
	Ind_Production_Manufacturing_	0.9624	0.5677	0.0869	Ind_Production_Manufacturing_
	Ind_Production_Total_Ind_Prod	0.9666	0.6063	0.0849	Ind_Production_Total_Ind_Prod
	Real_Consumer_Spending_On_Durabl	0.8193	0.6604	0.5322	Real_Consumer_Spending_On_Durabl
	Real_Gross_Domestic_Product	0.9090	0.6077	0.2320	Real_Gross_Domestic_Product
Cluster 2	Ind_Production_Chemicals	0.8885	0.3354	0.1677	Ind_Production_Chemicals
	Ind_Production_Nondurable_Good	0.8572	0.6193	0.3752	Ind_Production_Nondurable_Good
	Ind_Production_Resins_And_Synt	0.7770	0.3477	0.3418	Ind_Production_Resins_And_Synt
	Ind_Production_Ship_And_Boat_B	0.6309	0.1250	0.4218	Ind_Production_Ship_And_Boat_B
	Real_Consumer_Spending_On_Nondur	0.7114	0.3417	0.4385	Real_Consumer_Spending_On_Nondur
Cluster 3	Chained_Price_Index_Consumer_Pu	0.6295	0.3050	0.5331	Chained_Price_Index_Consumer_Pu
	Ind_Production_Household_Appli	0.8066	0.3992	0.3219	Ind_Production_Household_Appli
	Ind_Production_Oil_And_Gas_Ext	0.5148	0.0850	0.5303	Ind_Production_Oil_And_Gas_Ext
Cluster 4	Civilian_Unemployment_Rate	0.7646	0.0398	0.2451	Civilian_Unemployment_Rate
	Ind_Production_Aerospace_Produ	0.7646	0.1632	0.2813	Ind_Production_Aerospace_Produ
Cluster 5	Consumer_Sentiment_Index_Univer	0.5991	0.3497	0.6165	Consumer_Sentiment_Index_Univer
	Housing_Starts	0.5583	0.3459	0.6753	Housing_Starts
	Ind_Production_Automobiles	0.8374	0.4869	0.3168	Ind_Production_Automobiles
	Ind_Production_Cement	0.7669	0.4307	0.4094	Ind_Production_Cement
	Ind_Production_Concrete_And_Ce	0.8192	0.6953	0.5935	Ind_Production_Concrete_And_Ce
	Ind_Production_Motor_Vehicles_	0.8298	0.5581	0.3851	Ind_Production_Motor_Vehicles_
	Ind_Production_Paints_And_Misc	0.6083	0.4605	0.7261	Ind_Production_Paints_And_Misc
	Vacancy_Rate_Of_Rental_Housing_U	0.4143	0.2364	0.7671	Vacancy_Rate_Of_Rental_Housing_U
Cluster 6	Ind_Production_Food	1.0000	0.1299	0.0000	Ind_Production_Food

Other useful output from the procedure includes the inter-cluster correlation and the variation explained tables (Figures 7.6 and 7.7). In the inter-cluster correlation table, clusters 1 and 2 are most similar to one another, and clusters 4 and 6 are the least similar to one another. As for the variation explained table, Figure 7.7 shows that the 6 clusters represent 76% of the total variation in the Xs.

Figure 7.6: Inter-Cluster Correlation Table

Inter-Cluster Correlations						
Cluster	1	2	3	4	5	6
1	1.00000	0.63758	0.05769	−0.26158	0.80759	0.30420
2	0.63758	1.00000	−0.43103	0.00502	0.26198	0.36036
3	0.05769	−0.43103	1.00000	−0.22119	0.55334	0.19886
4	−0.26158	0.00502	−0.22119	1.00000	−0.30868	0.05371
5	0.80759	0.26198	0.55334	−0.30868	1.00000	0.32915
6	0.30420	0.36036	0.19886	0.05371	0.32915	1.00000

Figure 7.7: Variation Explained Table

Number of Clusters	Total Variation Explained by Clusters	Proportion of Variation Explained by Clusters	Minimum Proportion Explained by a Cluster	Maximum Second Eigenvalue in a Cluster	Minimum R-squared for a Variable	Maximum 1-R**2 Ratio for a Variable
1	14.258543	0.4917	0.4917	5.111798	0.0001	
2	17.303114	0.5967	0.5045	1.672127	0.0545	0.9579
3	18.781675	0.6476	0.4297	1.487669	0.2276	0.8616
4	20.031430	0.6907	0.5747	1.287440	0.2276	0.8616
5	21.222382	0.7318	0.6503	1.010351	0.2127	0.8829
6	22.055153	0.7605	0.6503	0.804086	0.4143	0.7845

After post-processing the output from PROC VARCLUS (as in Figure 7.8), the table in Figure 7.9 can be obtained.

Figure 7.8: Post-Processing PROC VARCLUS Output in SAS Enterprise Guide Flow

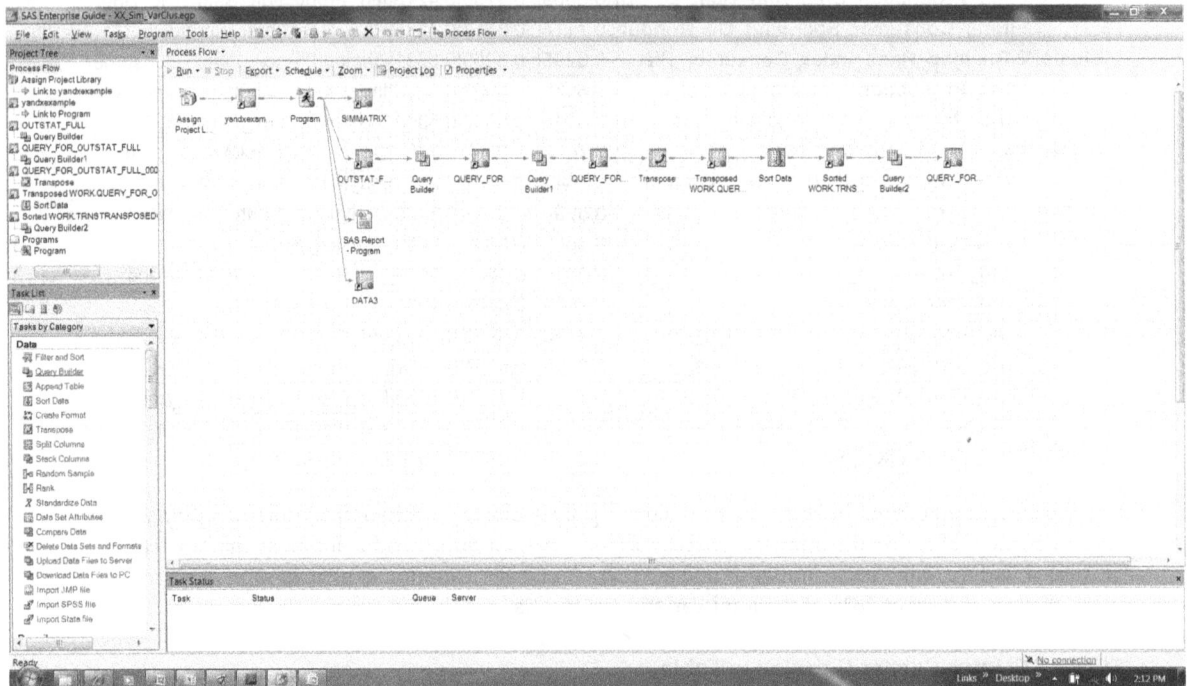

Figure 7.9: Final Output from PROC VARCLUS

	Source	Label	Cluster	Rsquare
1	Ind_Production_Total_Ind_Prod	Ind_Production_Total_Ind_Prod	1	0.96858822
2	Ind_Production_Manufacturing_	Ind_Production_Manufacturing_	1	0.96241896
3	Ind_Production_Durable_Goods	Ind_Production_Durable_Goods	1	0.93996816
4	Real_Gross_Domestic_Product	Real_Gross_Domestic_Product	1	0.90898074
5	Ind_Production_Fabricated_Meta	Ind_Production_Fabricated_Meta	1	0.89481191
6	Ind_Production_Electrical_Equi	Ind_Production_Electrical_Equi	1	0.82607912
7	Real_Consumer_Spending_On_Durabl	Real_Consumer_Spending_On_Cursoi	1	0.81927303
8	Ind_Production_Machinery	Ind_Production_Machinery	1	0.78424811
9	Ind_Production_Computers_And_E	Ind_Production_Computers_And_E	1	0.66309437
10	Ind_Production_Beverages	Ind_Production_Beverages	1	0.62123306
11	Ind_Production_Chemicals	Ind_Production_Chemicals	2	0.88853992
12	Ind_Production_Nondurable_Good	Ind_Production_Nondurable_Good	2	0.86716876
13	Ind_Production_Resins_And_Synt	Ind_Production_Resins_And_Synt	2	0.7770226
14	Real_Consumer_Spending_On_Nondur	Real_Consumer_Spending_On_Nondur	2	0.71136099
15	Ind_Production_Ship_And_Boat_B	Ind_Production_Ship_And_Boat_B	2	0.63090634
16	Ind_Production_Household_Appli	Ind_Production_Household_Appli	3	0.80561171
17	Chained_Price_Index_Consumer_Pu	Chained_Price_Index_Consumer_Pu	3	0.02940834
18	Ind_Production_Oil_And_Gas_Ext	Ind_Production_Oil_And_Gas_Ext	3	0.51477427
19	Civilian_Unemployment_Rate	Civilian_Unemployment_Rate	4	0.76483294
20	Ind_Production_Aerospace_Produ	Ind_Production_Aerospace_Produ	4	0.76483294
21	Ind_Production_Automobiles	Ind_Production_Automobiles	5	0.8374294
22	Ind_Production_Motor_Vehicles_	Ind_Production_Motor_Vehicles_	5	0.82903719
23	Ind_Production_Concrete_And_Ce	Ind_Production_Concrete_And_Ce	5	0.81919402
24	Ind_Production_Cement	Ind_Production_Cement	5	0.76693288
25	Ind_Production_Paints_And_Misc	Ind_Production_Paints_And_Misc	5	0.60026628
26	Consumer_Sentiment_Index_Univer	Consumer_Sentiment_Index_Univer	5	0.59906482
27	Housing_Starts	Housing_Starts	5	0.55827822
28	Vacancy_Rate_Of_Rental_Housing_U	Vacancy_Rate_Of_Rental_Housing_U	5	0.414258
29	Ind_Production_Food	Ind_Production_Food	6	1

In any modeling endeavor, the estimation part of the modeling process is limited by the number of observations available for modeling. But, assuming a long enough time series, it has been our collective experience that SAS Forecast Server can handle about a hundred variables easily. That is, SAS Forecast Server is able to diagnose possible functional forms for various Xs, estimate the parameters for those forms, and forecast. Therefore, for problems having hundreds of Xs or even thousands, variable reduction is critical.

7.3 Methods for Variable Selection

Variable selection is differentiated from variable reduction in that all variable selection approaches are supervised approaches because they consider the relationship between the Y and the X. In traditional transactional data mining problems, various variable selection methods are commonly used on the same problem. These various methods are used in conjunction with one another, not by themselves. This is because this early stage of the modeling process is exploratory in nature. There are no modeling methods that can handle thousands of variables simultaneously. Even after or with variable reduction, there is a need for a variable selection step in the process.

When modeling transactional data, it is a common practice to consider interaction terms between two independent variables. The reason this is an important consideration in transactional problems is because there is a debate over when to consider interactions (from the outset or after the first round of variable selection takes place). For a problem with 100 Xs, this would be another 100, taken two at a time, number of interactions. This calculates to be $(100!)/(100-1)!*2! = 100*99/2 = 50*99 = 4950$ additional Xs! Fortunately, this is not a common practice in time series problems.

The objective of this section is to give you a condensed overview of the variety of methods for variable selection. It is assumed that the number of potential economic drivers has been significantly reduced during the previous phase. The section is divided into two parts. The first part includes the traditional transactional data mining approaches, such as stepwise regression, decision trees, partial least square (PLS), and a new approach based on genetic programming. The second part includes time series approaches, such as co-integration, similarity, cross-correlation functions, time series stepwise regression, and PROC VARCLUS.

Traditional Data Mining

Recently, variable selection has become the focus of much research and development in data mining where applications with tens or hundreds of thousands of variables are available. (For a comprehensive survey, see Guyon and Elisseeff 2003.) The objective of variable selection methods is to reduce the input set to a subset that includes only relevant inputs to the target variable. Unfortunately, the criteria for relevancy do not guarantee an optimal subset selection (that is, the selected subset is generally suboptimal [Guyon and Elisseeff 2003]). A by-product of variable selection is variable ranking, indicating the relative importance of the variables. Ranking could be used in heuristic-based variable selection.

Variable selection methods that apply to transactional data are briefly described in this subsection. It is important to introduce additional lags in the Xs in order to compensate for the time series nature of the data at hand. This can often dramatically increase the number of variables to be considered. Approaches that are specific to time series data are described in the next subsection.

Data Preparation for Variable Selection

To apply the standard transactional data statistical and data mining methods for variable selection, some data preparation of the time series is needed. Because time series data are inherently correlated with time, it is necessary to reduce this correlation by making the data stationary. It is important to remove the seasonal and cyclical effects. Because the transactional variable selection methods do not consider the time series nature of the data, it is crucial to include the lags for each input. The key methods for preparing the data for applying transaction-based statistical and data mining approaches for variable selection are briefly presented below.

Stationarity

A stationary time series is one whose statistical properties such as mean, variance, and autocorrelation are all constant over time. One of the key reasons for trying to make a time series stationary is to be able to obtain meaningful sample statistics such as mean, variance, and correlation with other variables. These statistics are useful as descriptors of future behavior only if the time series is stationary. For example, if the series has a consistently increasing trend over time, the sample mean and variance will grow with the size of the sample, and they will always underestimate the mean and variance in future periods. If the mean and variance of a series are not well defined, then neither are its correlations with other variables (that is, it is statistically incorrect to apply transactional methods for variable selection).

In reality, most economic time series are far from stationary, and they exhibit trends, cycles, seasonality, and other nonstationary behavior. If the series has a stable long-run trend and tends to revert to the trend line following a disturbance, it might be possible to make it stationary by de-trending (for example, by fitting a trend line and removing it before fitting a model). This type of series is considered trend-stationary. However, sometimes even de-trending is not sufficient to make the series stationary. In that case, it might be necessary to transform it into a series of period-to-period or season-to-season differences. If the mean, variance, and autocorrelation of the original series are not constant in time even after de-trending, perhaps the differences or deltas in the statistics in the series between periods or seasons will be constant. This type of series is considered difference-stationary.

Differencing

The first difference of a time series is the series of changes from one period to the next. For example, suppose $Y(t)$ is a monthly time series. If $Y(t)$ denotes the value of the time series Y at period t, then the first difference of Y at period t is equal to $Y(t)-Y(t-1)$. One possible way to calculate this difference in SAS is by using the option DIF=(1) from the TIMESERIES procedure.

In the case of seasonal data, it is recommended that non-stationarity be reduced by using the seasonal difference (that is, the difference between an observation and the corresponding observation a year ago). For example, the first seasonal difference for the monthly data $Y(t)$ with an annual period is equal to $Y(t)-Y(t-12)$. It is possible to calculate this difference by either using the DIF=(1,12) or SDIF=(1) option of the TIMESERIES procedure.

When both seasonal and first differences are applied, it is recommended that the seasonal difference be done first because there is a chance the resulting time series is stationary, and there is no need for another first difference (Makridakis et al. 1998). If stationarity cannot be accomplished by a first difference, it is possible to apply a second-order regular and seasonal difference on the data.

De-Trending

A trend in a time series is a slow, gradual change in a property of the series over the whole interval under investigation. A trend is sometimes loosely defined as a long-term change in the mean, but it can also refer to a change in another statistical property. Identifying a trend in a time series is subjective because a trend cannot be unequivocally distinguished from low-frequency fluctuations. What looks like a trend in a short-time series segment often proves to be a low-frequency fluctuation (perhaps part of a cycle) in the longer series.

De-trending is the statistical or mathematical operation of removing a trend from the series. Many alternative methods are available for de-trending. A simple linear trend in the mean can be removed by subtracting a least squares fit straight line. More complicated trends might require different procedures. In the order of popularity, the most popular approaches for de-trending are first differencing (see previous section), curve fitting, digital filtering, and piecewise polynomials.

Curve Fitting—If a time series changes in level gradually over time, it makes sense to represent the trend by a simple function of time itself. A simple and widely used function of time is the least squares fit straight line, which assumes a linear trend. The straight line might be unrealistic because it restricts the functional form of the trend. Other functions of t (for example, quadratic) might be better, depending on the type of data.

Digital Filtering—Another procedure for handling a trend is to describe the trend as a linearly filtered version of the original series. The original series is converted to a smooth trend line by weighting the individual observations.

Piecewise Polynomials—An alternative to fitting a curve to the entire time series (curve fitting) is to fit polynomials of time to different parts of the time series. Polynomials used in this way are called piecewise polynomials.

A possible way to apply a trend analysis of a time series in SAS is by using the TREND option of the TIMESERIES procedure. An example of applying the TREND option to an industrial time series with a seasonal pattern (shown in Figure 7.10) is given in Figure 7.11. The increasing trend component is not based on a linear or other simple function of time, and it includes several piecewise polynomials.

Figure 7.10: Time Plot of the Monthly Industrial Production of Business Y from January 1997 to January 2011

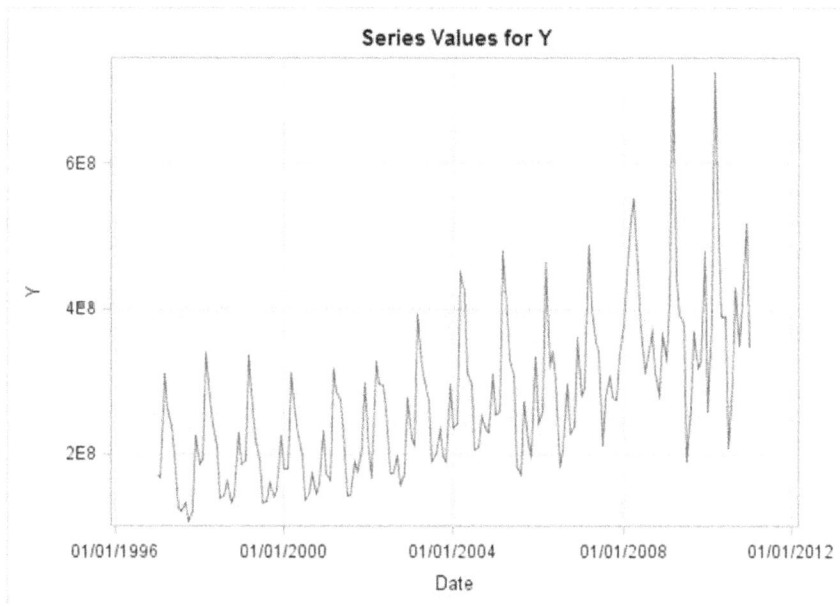

Figure 7.11: Time Plot of the Original Data and the Trend Component for the Monthly Industrial Production of Business Y

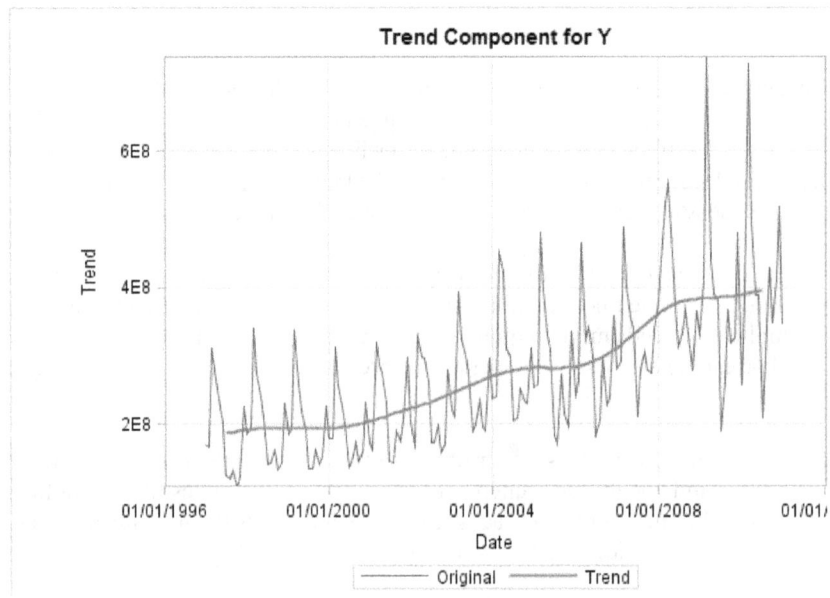

De-Seasonalize

De-seasonalization is the statistical or mathematical operation of removing seasonal patterns from the series. Usually, the removal of the seasonal patterns is combined with the decomposition of the time series from cyclical and irregular components. The major distinction between a seasonal and a cyclical pattern is that the former is of a constant length and appears in regular periods, and the latter varies in length. A possible way to apply seasonal decomposition of a time series in SAS is by using the DECOMP statement with the TIMESERIES procedure. The following components are available:

TC—trend component

TCC—trend-cycle component

SC—seasonal component

CC—cycle component

IC—irregular component

TCS—trend-cycle-seasonal component

The seasonal component of the monthly industrial production time series is shown in Figure 7.12. The corresponding cycle component is shown in Figure 7.13, and the trend-cycle-seasonal component is shown in Figure 7.14.

Figure 7.12: Time Plot of the Seasonal Component for the Monthly Industrial Production of Business Y

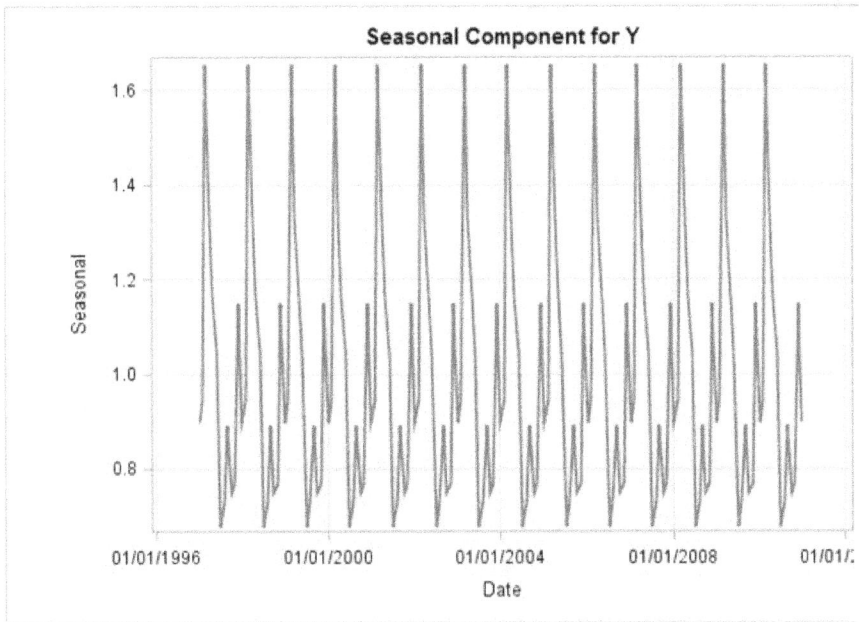

Seasonal Component for Y

Figure 7.13: Time Plot of the Cycle Component for the Monthly Industrial Production of Business Y

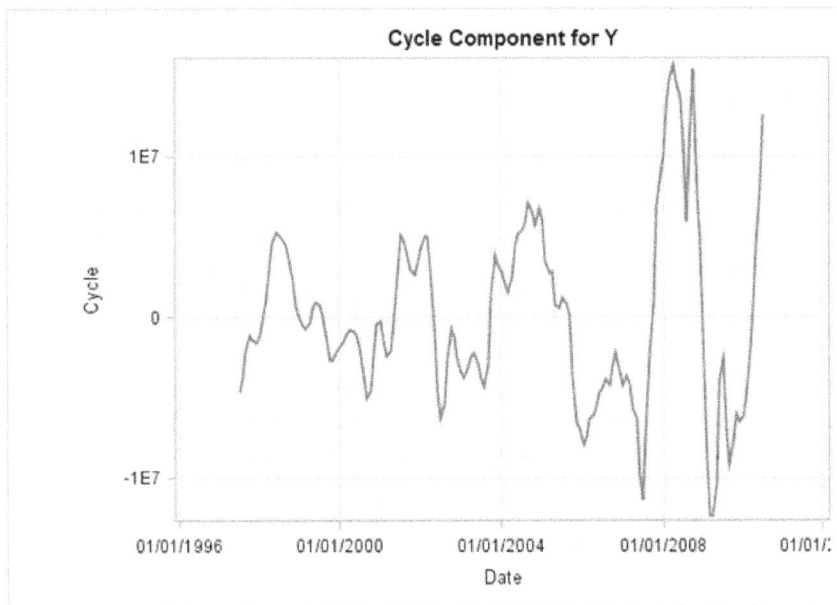

Cycle Component for Y

Figure 7.14: Time Plot of the Trend-Cycle-Seasonal Component for the Monthly Industrial Production of Business Y

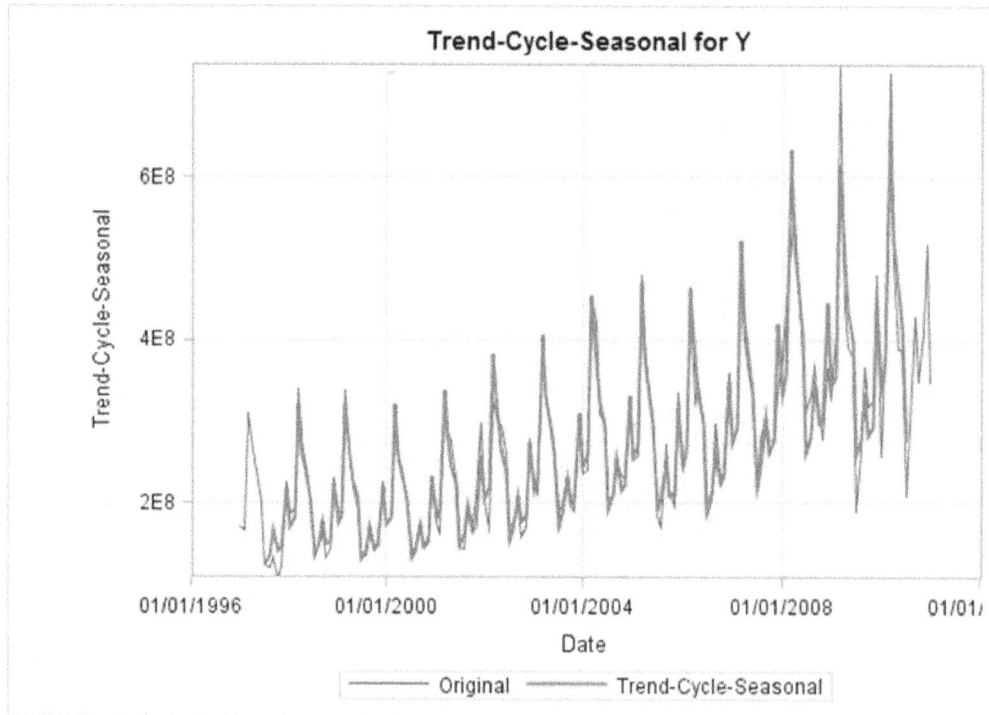

Seasonal Adjustment by X11and X12 Procedures

One of the main objectives for decomposition is to estimate seasonal effects that can be used to create and present seasonally adjusted values. A seasonally adjusted value removes the seasonal effect from a series so that trends can be seen more clearly. The most commonly used seasonal adjustment packages are in the X11 family. X11 was developed by the U.S. Bureau of the Census and began operation in the United States in 1965.[1] It was soon adopted by many statistical agencies around the world, and it has been integrated into a number of commercially available software applications including SAS. The X11 method involves applying symmetric moving averages to a time series to estimate the trend, seasonal, and irregular components. However, at the end of the series, there is insufficient data available to use symmetric weights, known as the "endpoint" problem. Consequently, either asymmetric weights are used or the series must be extrapolated. This limitation has been reduced by the X11ARIMA method, developed by Statistics Canada in 1980. Essentially, the use of ARIMA modeling on the original series helps reduce the revisions in the seasonally adjusted series so that the effect of the endpoint problem is minimized.

However, the X11ARIMA method still has the following limitations:

- There are no user-defined regressors for special situations.
- ARIMA modeling is not robust against outliers.
- Seasonal adjustment is not robust against level shifts.

In the late 1990s, the U.S. Census Bureau released X12ARIMA. It uses regARIMA models (regression models with ARIMA errors) to enable the user to extend the series with forecasts and pre-adjust the series for outlier and calendar effects before seasonal adjustment takes place. The advantages of X12ARIMA are as follows:

- wide variety of seasonal and trend filter options
- extensive time series modeling and model selection capabilities
- linear regression models with ARIMA errors (regARIMA models)

- automatic model selection options
- user-defined regression variables
- suite of modeling and seasonal adjustment diagnostics, including spectral diagnostics
- out of sample forecast error model selection diagnostics

Both X11 and X12 procedures are available in SAS. An example of using PROC X11 for the seasonal adjustment of the monthly industrial production time series follows:

```
proc x11 data=Seasonalexample;
monthly date=date;
var Y;
tables b1 d11;
output out=out b1=series d10=d10 d11=d11
d12=d12 d13=d13;
run;
```

The original data are in table B1, table D10 contains the final seasonal factors, table D11 contains the final seasonally adjusted series, table D12 contains the final trend cycle, and table D13 contains the final irregular series. The plot of the original and seasonally adjusted time series is shown in Figure 7.15.

Figure 7.15: Time Plot of the Original and Seasonally Adjusted Data for the Monthly Industrial Production of Business Y

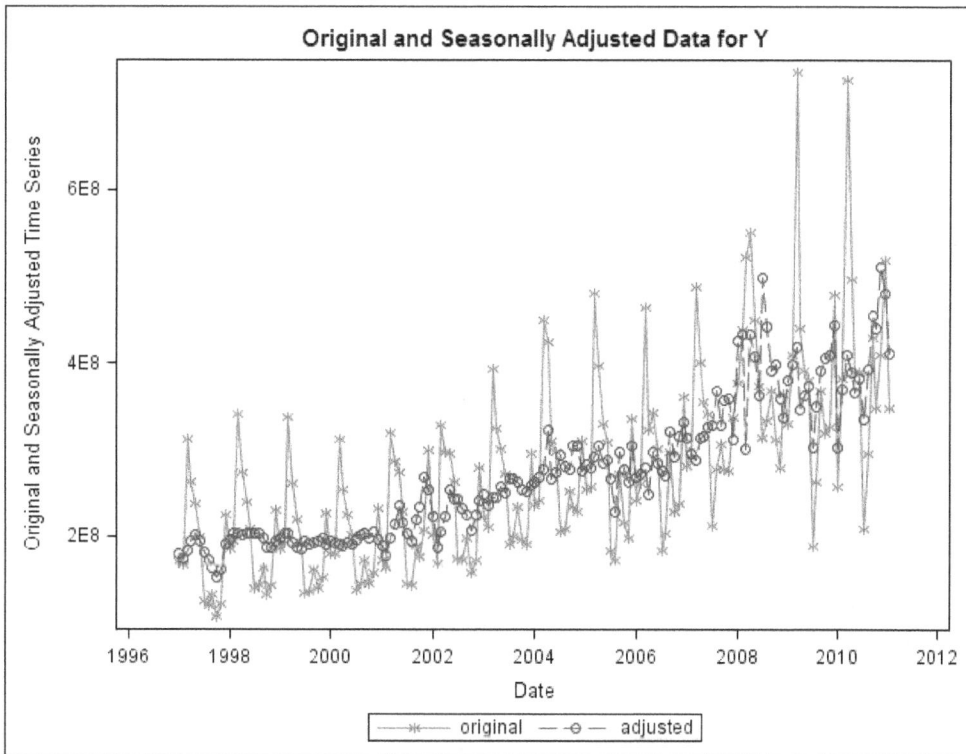

Seasonality Detection by PROC SPECTRA

Another method to detect seasonality in the time series is spectral analysis. A straightforward way to determine whether a series has a periodic component is to plot the periodogram or spectral density of the series onto the period or the frequency. A significant ordinate value at the period or frequency indicates the numerical value of that periodic element. In SAS, the procedure that can be used for spectral analysis is PROC SPECTRA, which outputs estimates of the spectral and cross-spectral densities of multiple time series. These estimates are produced using a finite Fourier transform, which decomposes the series into a sum of sine and cosine waves of varying amplitudes and wavelengths. In a general sense, PROC SPECTRA regresses the time series under analysis onto the sine and cosine variates for frequencies varying from 0 to ∞ by small increments. The plotted periodogram or spectral density function is the sum of squares of the regression model associated with each frequency. Periodicity is determined by a high value for the ordinate of the periodogram or spectral density.

An example of invoking PROC SPECTRA for the spectral analysis of the monthly industrial production time series follows:

```
proc spectra data= Seasonalexample
out=b p s adjmean whitetest;
var Y;
weights 1 2 3 4 3 2 1;
run;
```

The VAR statement specifies the variables to be analyzed (in this case, Y). The P option computes the periodogram of the series. The S option computes the estimate of the spectral density of the series. The ADJMEAN option changes the frequency value to zero and avoids the need to exclude that observation from the plots. The WHITETEST option performs Fisher's test for white noise. The WEIGHTS statement is required if the S option is used. The WEIGHTS statement specifies a triangular spectral window for smoothing the periodogram to produce the spectral density estimate. The results from the procedure follow.

```
                 The SPECTRA Procedure

           Test for White Noise for Variable Y

                   M        =         84
                   Max(P(*))     6.034E17
                   Sum(P(*))     2.002E18


           Fisher's Kappa: M*MAX(P(*))/SUM(P(*))
                   Kappa    25.31259
           Bartlett's Kolmogorov-Smirnov Statistic:
          Maximum absolute difference of the standardized
          partial sums of the periodogram and the CDF of a
                  uniform(0,1) random variable.

     Test Statistic                          0.503284
     Approximate P-Value                       <.0001
```

The Fisher's Kappa test for the white noise statistic of 25.31 is larger than the 5% critical value of 7.2, so the null hypothesis that the industrial production time series is a white noise is rejected. In the same way, the Bartlett's Kolmogorov-Smirnov statistic of 0.5 has a small p-value of < 0.001 (in other words, the hypothesis that the spectrum is based on white noise is rejected).

The spectral density estimate by period plot is shown in Figure 7.16. The spectral analysis confirms the strong 12-month cycle, shown by the reference line. Two-month and six-month sub-cycles also have to be taken into account.

Figure 7.16: Plot of the Spectral Density Estimate by Period for the Monthly Industrial Production of Business Y Data

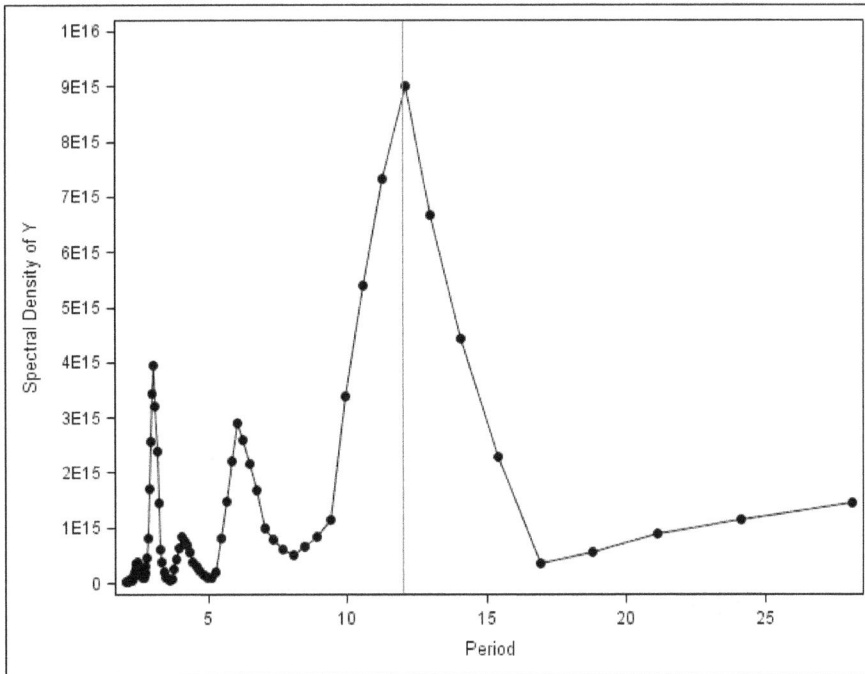

Lags

An important requirement for the variable selection of potential economic drivers is considering the dynamics in the relationships. A possible way to represent the dynamic relationships is by introducing lagged variables into the analysis data set. Unfortunately, this increases significantly the number of variables. The selection of the number of lags is a trade-off between fully representing the dynamics and the increased dimensionality of the search algorithms for variable selection. From a practical perspective, adding between three and six lags is a well-tested rule. An example of a data set, based on adding a first difference and four lags variable to each input, is shown in Table 7.1.

Table 7.1: Data with Added First-Difference and Four Lags Variable to Each Input

Date	x4	Diff_x4	Diff_x4_lag1	Diff_x4_lag2	Diff_x4_lag3	Diff_x4_lag4
Jan-2005	281.88					
Feb-2005	323.56	41.68				
Mar-2005	387.75	64.19	41.68			
Apr-2005	342.66	-45.09	64.19	41.68		
May-2005	256.25	-86.41	-45.09	64.19	41.68	
Jun-2005	257.60	1.35	-86.41	-45.09	64.19	41.68
Jul-2005	292.17	34.57	1.35	-86.41	-45.09	64.19

Example for Variable Selection

The key methods for variable selection are illustrated with a real industrial data set of 45 potential economic drivers to the volume of product Y1. These inputs are a result of the variable reduction analysis done on the original set of 211 variables. The data are monthly data from January 2005 to December 2008. The following variable selection methods are explored in SAS Enterprise Miner: Pearson product-moment correlation coefficient, stepwise regression, the SAS Enterprise Miner variable selection algorithm, partial least squares, and decision trees. The corresponding SAS Enterprise Miner diagram is shown in Figure 7.17.

Figure 7.17: SAS Enterprise Miner Diagram for Variable Selection

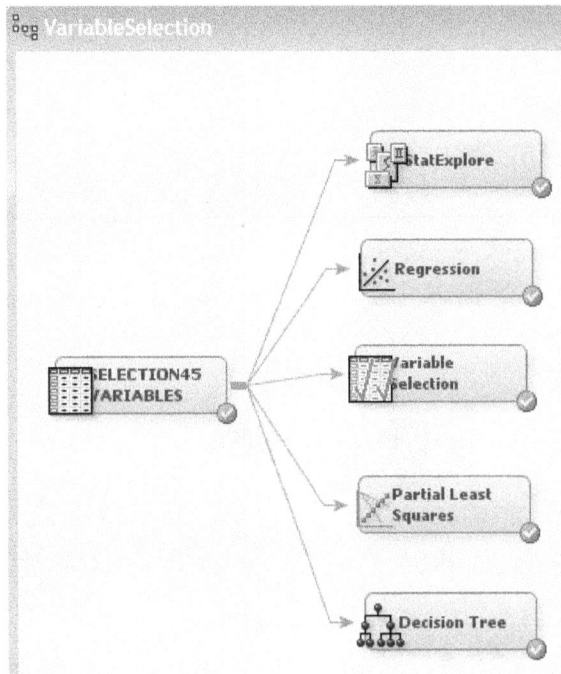

A brief description of these methods and the implementation details of their corresponding SAS Enterprise Miner nodes follow.

Variable Selection Based on Pearson Product-Moment Correlation Coefficient

The most obvious method for variable selection is based on the simple correlation of the potential economic drivers to the target variable. The most well-known measure of correlation is the Pearson product-moment correlation, which quantifies the strength and direction of the relationship between two variables. The calculated correlation coefficient is in a range from –1.0 to +1.0. This enables you to get an idea of the strength of relationship or the strength of linear relationship between the variables. The closer the coefficient is to +1.0 or –1.0, the greater the strength of the linear relationship.

One possible way to rank variables based on their Pearson product-moment correlation is by using the StatExplore node of SAS Enterprise Miner. The selected node options are shown in Figure 7.18, where the critical choice is the selection of the correlation statistics. In Figure 7.18, the standard Pearson Correlations option has been selected.

Figure 7.18 Selected Options for EM StatExplore Node

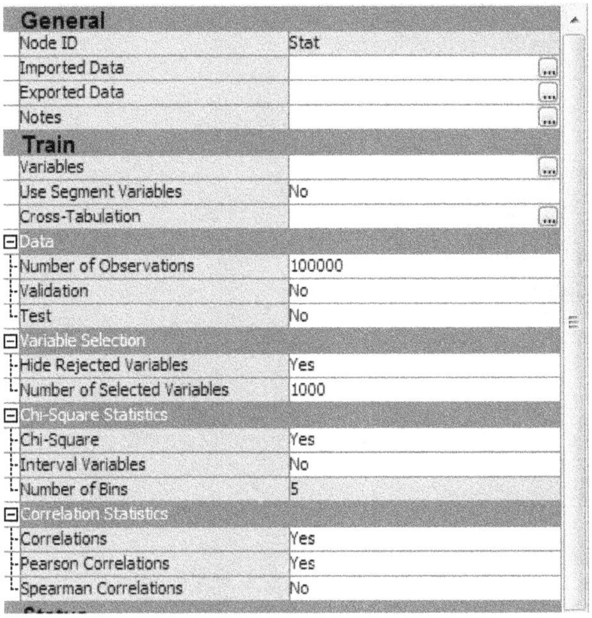

General	
Node ID	Stat
Imported Data	
Exported Data	
Notes	
Train	
Variables	
Use Segment Variables	No
Cross-Tabulation	
⊟ Data	
⋮ Number of Observations	100000
⋮ Validation	No
⋮ Test	No
⊟ Variable Selection	
⋮ Hide Rejected Variables	Yes
⋮ Number of Selected Variables	1000
⊟ Chi-Square Statistics	
⋮ Chi-Square	Yes
⋮ Interval Variables	No
⋮ Number of Bins	5
⊟ Correlation Statistics	
⋮ Correlations	Yes
⋮ Pearson Correlations	Yes
⋮ Spearman Correlations	No

The inputs, sorted in decreasing order of their correlation coefficient, are shown in Figure 7.19. This input selection includes 18 to 19 highly correlated variables with a correlation coefficient > 0.8 or < -0.8. The list of the top-correlated inputs with correlation coefficients > 0.9 or < -0.9, which can be chosen as key Xs, follows.

```
Correlation Statistics
(maximum 500 observations printed)


Data Role=TRAIN Type=PEARSON Target=Y1


Input      Correlation

 X44         0.94129
 X5          0.91768
 X26         0.91605
 X41         0.90763
 X37         0.90596



...


 X35        -0.93007
 X38        -0.93322
```

Figure 7.19: Sorted Inputs Based on the Decreasing Pearson Product-Moment Correlation Coefficient

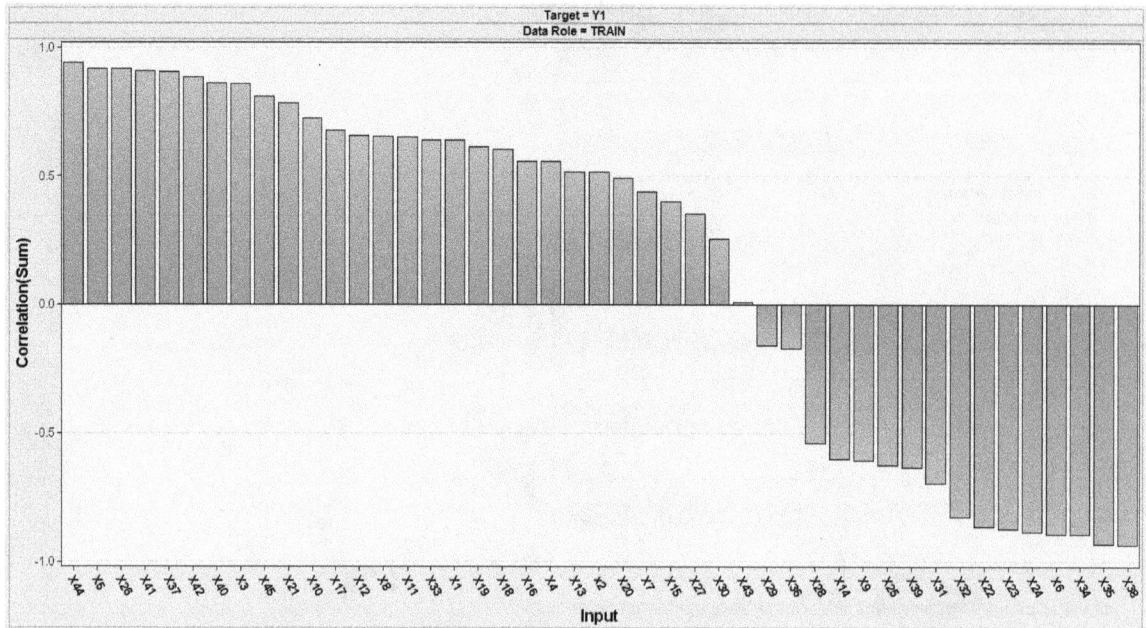

Variable Selection Based on Stepwise Regression

The most-used methods for variable selections are based on multivariate regression modeling techniques. The key techniques are forward selection, backward elimination, and stepwise regression.

Forward Selection—This method begins with the assumption that there are no inputs in the model. The procedure begins by fitting the intercept-based model, and then the number of inputs into the model is increased one by one. Only those inputs related to the target variable above a predefined level of statistical significance are added in the model. To focus on the very significant effects on the target variable, it is recommended that this entry-significance level be relatively low (for example, 0.01). In principle, once an input variable is integrated in the model, it is never removed. The forward selection process stops when all remaining inputs are statistically insignificant.

Backward Elimination—This method begins with the assumption that all possible inputs are part of the model. The procedure begins with removing the input with the least statistical significance relative to the others. A detail of importance is that the level of statistical significance (named stay level) might differ from the model entry level in forward selection. The backward selection process stops when all remaining inputs are statistically insignificant relative to the stay significance threshold.

Stepwise Regression—This method is a modified version of the forward selection method. In each step of the stepwise method, the statistical significance of all previously entered inputs is reassessed. In this process, it is possible that an input added in an earlier step might become statistically insignificant and will be dropped from the model. This procedure requires two significance levels: an entry significance level to allow the input to be kept in the models, and a stay significance level to allow the algorithm to remove inputs already selected in the regression model. The stepwise regression process stops when all remaining inputs are statistically insignificant relative to the stay significance level. One of the advantages of this method is that the final model is selected based on a larger set of potential model evaluations than the other methods.

The limitations of stepwise regression (such as generating different variable selection from the same data set, non-optimal solutions, and issues with multicollinearity) are discussed in Chapter 4. Some of these limitations are illustrated with an example of applying stepwise regression in SAS Enterprise Miner.

The regression-based methods for variable selection are part of the SAS Enterprise Miner Regression node. The selection options of this node are shown in Figure 7.20. The method selection option is in the Model Selection section. The stepwise selection model shown in Figure 7.20 invokes the stepwise regression method. In this case, the Use Selection Defaults option is selected, which sets both entry and stay levels to 0.05. It is recommended that the default option for large data sets not be used and that a lower significance level be selected that will more strongly filter the potential inputs (Matignon 2007).

Figure 7.20: Selected Options for the SAS Enterprise Miner Regression Node

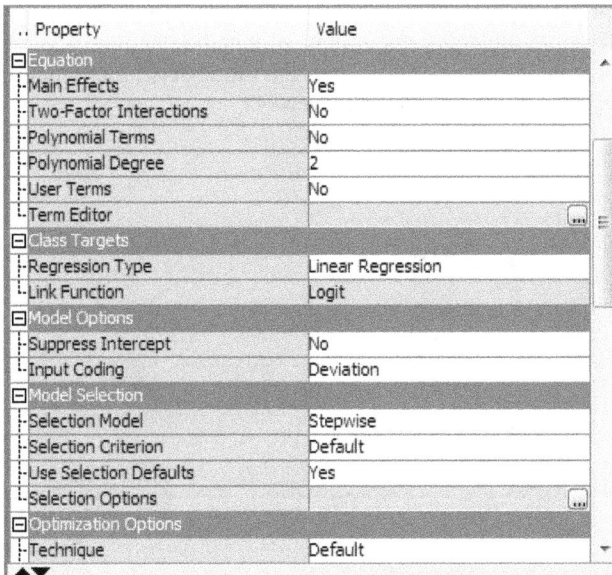

.. Property	Value
☐ Equation	
├ Main Effects	Yes
├ Two-Factor Interactions	No
├ Polynomial Terms	No
├ Polynomial Degree	2
├ User Terms	No
└ Term Editor	
☐ Class Targets	
├ Regression Type	Linear Regression
└ Link Function	Logit
☐ Model Options	
├ Suppress Intercept	No
└ Input Coding	Deviation
☐ Model Selection	
├ Selection Model	Stepwise
├ Selection Criterion	Default
├ Use Selection Defaults	Yes
└ Selection Options	
☐ Optimization Options	
├ Technique	Default

The key results from the stepwise regression, extracted from the output report, follow.

```
           Analysis of Maximum Likelihood Estimates

                            Standard
Parameter    DF    Estimate    Error    t Value    Pr > |t|

Intercept    1      0.0572     0.0464     1.23      0.2238
X1           1     0.000891   0.000394    2.26      0.0287
X44          1      0.00197   0.000757    2.60      0.0127
X5           1     0.000223   0.000043    5.14      <.0001


NOTE: No (additional) effects met the 0.05 significance level for entry into the
model.


               Summary of Stepwise Selection

            Effect           Number
    Step    Entered    DF      In      F Value    Pr > F

     1      X44         1       1      357.60     <.0001
     2      X5          1       2       20.57     <.0001
     3      X1          1       3        5.12     0.0287


The selected model is the model trained in the last step (Step 3). It consists of
the following effects:

Intercept  X1  X44  X5
```

The stepwise selection process included three steps, and the final statistically significant inputs are below the 0.05 significance level for entry. The selected inputs are x1, x44, and x5. Although the selection of the highly correlated inputs x44 and x5 is not a surprise, input x1 is not in the highly correlated inputs list with a correlation coefficient > 0.9 or < -0.9. (See the previous section.)

The results from the forward selection with default parameters are identical. However, the results from the backward elimination (given below) are very different. It took 40 steps of input elimination until no additional effects met the 0.01 significance level for removal from the model. Only one from the selected four inputs (x19, x4, x43, and x44) is similar to the forward and stepwise selection results (input x44).

```
              Analysis of Maximum Likelihood Estimates

                                    Standard
Parameter     DF     Estimate        Error      t Value    Pr > |t|

Intercept      1       1.1805        0.2999       3.94       0.0003
X19            1       0.00243       0.000578     4.20       0.0001
X4             1      -0.00199       0.000457    -4.36      <.0001
X43            1      -0.0129        0.00296     -4.37      <.0001
X44            1       0.00519       0.000303    17.16      <.0001


NOTE: No (additional) effects met the 0.01 significance level for removal from
the model.


                  Summary of Backward Elimination

              Effect              Number
     Step      Removed      DF      In      F Value    Pr > F

       1       X27          1       43       0.00      0.9750
       2       X16          1       42       0.00      0.9549
..
      39       X13          1        5       1.97      0.1678
      40       X20          1        4       4.03      0.0510


The selected model is the model trained in the last step (Step 40). It consists
of the following effects:

Intercept  X19  X4  X43  X44
```

Variable Selection Based on the SAS Enterprise Miner Variable Selection Node

Another option for variable selection in SAS Enterprise Miner is to use the Variable Selection node itself. The node is based on a two-step process for selecting numerical (interval) variables. In the first step, the R^2 value between each potential input and the target variable is calculated by a simple linear regression. Then, the inputs with $R^2 <$ Minimum R square (a predefined cutoff value) are removed. In the second step, a forward stepwise regression is performed on the remaining inputs. In contrast to the stepwise regression procedure in the Regression node, the criterion for selection or rejection is not the statistical significance of candidate models, but their calculated R^2. The procedure begins with the input with the highest R^2. At each consecutive step, a new input that provides the largest increase in the total R^2 is added. Those inputs, which have a negligible effect for model R^2 improvement (defined by another threshold Stop R-Square), are removed from the model and assigned as rejected. The variable selection consists of the remaining list of variables with a model role input. The results from applying the SAS Enterprise Miner Variable Selection node for variable selection from the 45 potential inputs are shown below. The selected options are shown in Figure 7.21.

Figure 7.21: Selected Options for SAS Enterprise Miner Variable Selection Node

.. Property	Value
Notes	
Train	
Variables	
Max Class Level	100
Max Missing Percentage	50
Target Model	R-Square
Manual Selector	
Rejects Unused Input	Yes
Bypass Options	
Variable	None
Role	Input
Chi-Square Options	
Number of Bins	50
Maximum Pass Number	6
Minimum Chi-Square	3.84
R-Square Options	
Maximum Variable Number	3000
Minimum R-Square	0.8
Stop R-Square	5.0E-4
Use AOV16 Variables	No
Use Group Variables	Yes
Use Interactions	No
Use SPDE Library	Yes
Print Option	Default
Score	

The available options for numerical or interval variables are in the R-Square Options area (the Chi-Square Options area is for nominal or ordinal variables). Two of these options—Minimum R-Square and Stop R-Square—are very important for the analysis. The Minimum R-Square option is the cutoff value for assigning a rejected model role to an input variable if it is below this threshold. Setting a high Minimum R-Square value tends to exclude more input variables and suggests more parsimonious models. Comparing to the Pearson product-moment correlation distribution where important correlations were defined as > 0.9 or < -0.9, the value for Minimum R-Square is selected as 0.8 (approximately equal to $\sqrt{0.9}$). Stop R-Square specifies the termination criterion for the forward stepwise regression when the corresponding incrementally added inputs have no statistically significant effect in the model. The default value for Stop R-Square of 0.005 has been selected. A summary of the variable selection procedure, extracted from the output section of the results, follows.

```
Summary of Variable Selection from Training
----------------------------------------------------------
Target Name : Y1
Target Level : INTERVAL
Variable Selection Model : R-Square
Total number of inputs : 45

Total number of used inputs : 45
Total number of unused inputs : 0
Total number of rejected inputs : 39
```

Only six from the original 45 inputs have been selected. The remaining 39 inputs have been rejected due to a small R square value. The list of the selected inputs, which includes inputs x5, x26, x35, x38, x41, and x44, is shown in Figure 7.22. The variable effects plot, shown in Figure 7.23, illustrates the individual effect of adding, sequentially, each significant input in the model. The decreasing order input ranking is based on the sequential R-square values. Based on Figure 7.23, the effect of x26 and x35 to the model is negligible, even if the variables have been selected as inputs.

Figure 7.22: Variable Selection Table of SAS Enterprise Miner Variable Selection Node

Variable Name	ROLE	LEVEL	TYPE	Variable Label	Reasons for Rejection ▲
X26	INPUT	INTERVAL	N	X26	
X35	INPUT	INTERVAL	N	X35	
X38	INPUT	INTERVAL	N	X38	
X41	INPUT	INTERVAL	N	X41	
X44	INPUT	INTERVAL	N	X44	
X5	INPUT	INTERVAL	N	X5	
X1	REJECTED	INTERVAL	N	X1	Varsel:Small R-square value
X10	REJECTED	INTERVAL	N	X10	Varsel:Small R-square value
X11	REJECTED	INTERVAL	N	X11	Varsel:Small R-square value
X12	REJECTED	INTERVAL	N	X12	Varsel:Small R-square value
X13	REJECTED	INTERVAL	N	X13	Varsel:Small R-square value
X14	REJECTED	INTERVAL	N	X14	Varsel:Small R-square value
X15	REJECTED	INTERVAL	N	X15	Varsel:Small R-square value
X16	REJECTED	INTERVAL	N	X16	Varsel:Small R-square value
X17	REJECTED	INTERVAL	N	X17	Varsel:Small R-square value
X18	REJECTED	INTERVAL	N	X18	Varsel:Small R-square value
X19	REJECTED	INTERVAL	N	X19	Varsel:Small R-square value
X20	REJECTED	INTERVAL	N	X20	Varsel:Small R-square value
X21	REJECTED	INTERVAL	N	X21	Varsel:Small R-square value
X22	REJECTED	INTERVAL	N	X22	Varsel:Small R-square value

Figure 7.23: Variable Effects in the Selected Model Plot

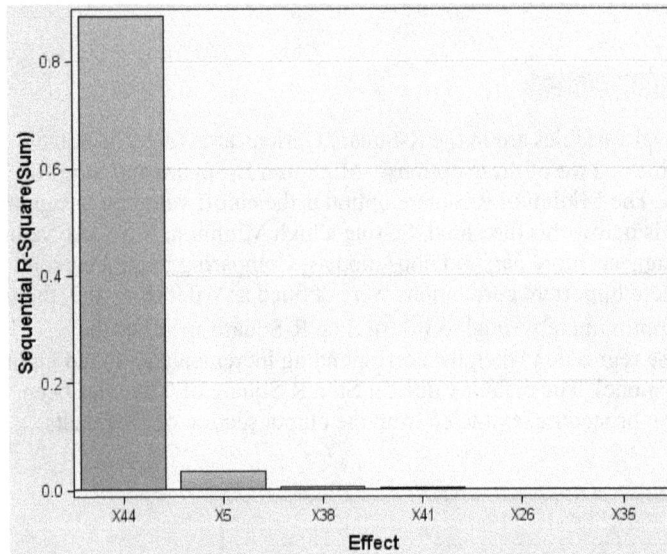

Variable Selection Based on the SAS Enterprise Miner Partial Least Squares Node

The key application area for partial least squares (PLS) is building predictive models when the data set includes many highly correlated inputs. The first step in PLS is to reduce the dimensionality of the system by defining the so-called latent vectors that account for most of the variation in both the inputs and the output or target. Usually, the number of latent vectors is much smaller than the number of original inputs. However, each latent vector is a linear combination of the original inputs with weights proportional to the variance contribution. The second step in PLS is to use these latent vectors for modeling the target. First, the PLS algorithm uses input-based latent vectors (also referred to as X-scores) to predict the Y-scores, and then the predicted Y-scores are used to construct the final predictions for the target.

PLS can be used for variable selection in two ways. The first way is by ranking the original variables based on their regression coefficients in the predicted target equation. The second way is by using the Variable Importance in the Projection (VIP) metric, which represents the relative importance of each original input to the latent vectors. The rule is that if an input has a relatively small coefficient (in absolute value) and the VIP value is below some threshold, then it is removed. The selected options for the SAS Enterprise Miner PLS node, applied to the 45 inputs data set, are shown in Figure 7.24.

Figure 7.24: Selected Options for SAS Enterprise Miner Partial Least Squares Node

., Property	Value
Notes	
Train	
Variables	
⊟Modeling Techniques	
├Regression Model	PLS
├PLS Algorithm	NIPALS
├Maximum Iteration	200
└Epsilon	1.0E-12
⊟Number of Factors	
├Default	Yes
└Number of Factors	15
⊟Cross Validation	
├CV Method	None
└CV N Parameter	7
⊟Random CV Options	
├Number of Iterations	10
├Default No. of Test Obs.	Yes
├No. of Test Obs.	100
├Default Random Seed	Yes
└Random Seed	1234
Score	
⊟Variable Selection	
├Variable Selection Criterion	Either
├Para. Est. Cutoff	0.2
├VIP Cutoff	0.7
├Export Selected Variables	No
└Hide Rejected Variables	Yes

The key options of interest are in the Variable Selection section. It is recommended that the Either option for the Variable Selection Criterion be selected because the selection is based on the inputs with both high regression coefficients (absolute standardized parameter estimates) and high VIP. The recommended cutoff values for the absolute standardized parameter estimates (Para. Est. Cutoff) is 0.1–0.2 (the last value is selected based on the distribution in Figure 7.25). The recommended cutoff values for the VIP metric (VIP Cutoff) is 0.7 – 0.8 (the first value is selected based on the distribution in Figure 7.26). As it is shown in both distribution plots, there are differences in the original inputs impact to the target variable, measured by the regression coefficients and VIP. The final list of selected inputs, based on both criteria, is shown in Figure 7.27. The list includes eight variables: x6, x24, x35, x37, x38, x41, x42, and x44.

Figure 7.25: Absolute Standardized Parameter Estimation Plot

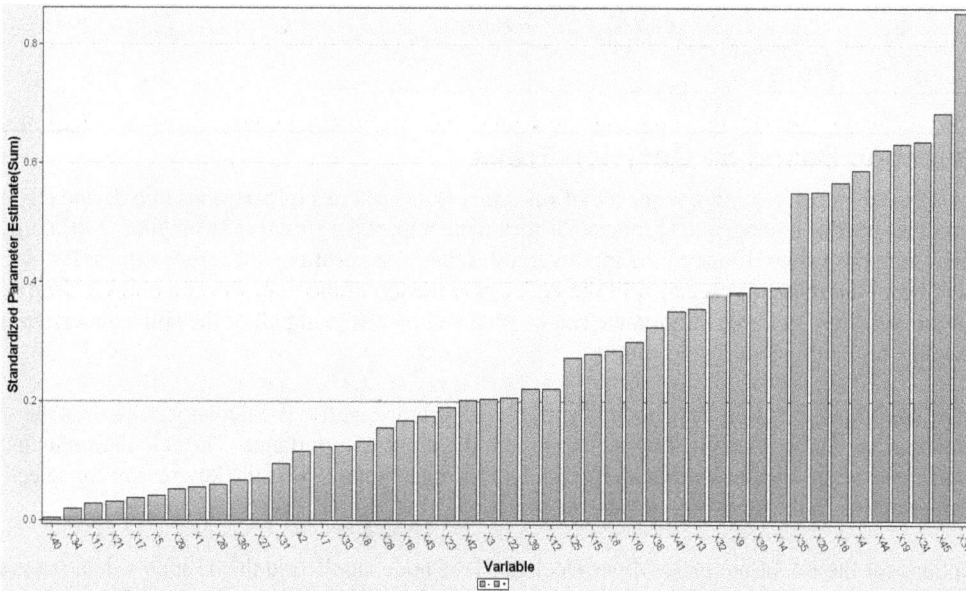

Figure 7.26: Variable Importance in the Projection (VIP) Plot

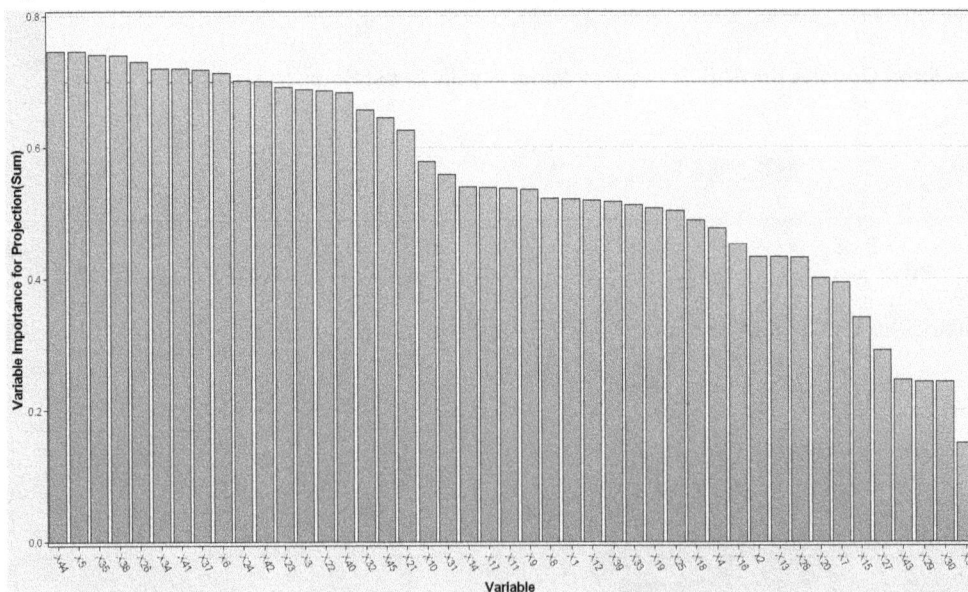

Figure 7.27: Variable Selection by the SAS Enterprise Miner PLS Node

Variable	Standardize d Parameter Estimate	Variable Importance for Projection	Rejected by Parameter Estimate	Rejected by VIP	Role ▲
X24	-0.63679	0.701191	No	No	Input
X35	-0.55114	0.740883	No	No	Input
X37	-0.85392	0.718341	No	No	Input
X38	0.32553	0.740289	No	No	Input
X41	-0.35255	0.719731	No	No	Input
X42	-0.20120	0.700582	No	No	Input
X44	0.62316	0.74664	No	No	Input
X6	0.28594	0.713568	No	No	Input
X1	-0.05485	0.521416	Yes	Yes	Rejected
X10	-0.30032	0.578163	No	Yes	Rejected
X11	0.02750	0.537266	Yes	Yes	Rejected
X12	0.22137	0.519643	No	Yes	Rejected
X13	0.35598	0.432858	No	Yes	Rejected
X14	-0.39173	0.539854	No	Yes	Rejected

Variable Selection Based on Decision Trees

Decision trees are based on performing a sequence of rules that forms a series of partitions that divide the target values into a small number of homogenous groups that formulate a tree-like structure (Matignon 2007). Each split is performed from the values of one of the input variables that best partitions the target values. By examining which input variables are used to split the nodes near the top of the tree, you can quickly determine the most important variables. Variable importance can be obtained by analyzing all of the splits generated by each variable and the selection of surrogate splitters.

In the SAS Enterprise Miner Decision Tree node, a variable importance metric is calculated, based on the normalized amount of variability explained by the input with the highest importance. The calculation includes not only inputs selected as split variables, but also considers surrogate inputs. It is the key reason for selecting the Surrogate Rules option in the node box in variable selection.

The selected options for the SAS Enterprise Miner Decision Tree node, applied to the 45 inputs data set, are shown in Figure 7.28. The important options are indicated with arrows. Defining a proper Significance Level specifies the threshold value for the worthiness of a candidate splitting rule. The default value is 0.2. Specifying the Number of Surrogate Rules as 1 in the Node section enables surrogate splits to be included in the variable selection. It is also recommended that you specify the Observation Based Importance and Variable Selection options as Yes, as shown in Figure 7.28.

Figure 7.28: Selected Options for SAS Enterprise Miner Decision Tree Node

Train	
Variables	
Interactive	
Use Frozen Tree	No
Use Multiple Targets	No
Splitting Rule	
Interval Criterion	ProbF
Nominal Criterion	ProbChisq
Ordinal Criterion	Entropy
Significance Level	0.2 ←
Missing Values	Use in search
Use Input Once	No
Maximum Branch	2
Maximum Depth	6
Minimum Categorical Size	5
Node	
Leaf Size	5
Number of Rules	5
Number of Surrogate Rules	1 ←
Split Size	
Split Search	
Exhaustive	5000
Node Sample	20000
Subtree	
Method	Largest
Number of Leaves	1
Assessment Measure	Decision
Assessment Fraction	0.25
Cross Validation	
Perform Cross Validation	No
Number of Subsets	10
Number of Repeats	1
Seed	12345
Observation Based Importance	
Observation Based Importance	Yes ←
Number Single Var Importance	5
P-Value Adjustment	
Bonferroni Adjustment	Yes
Time of Kass Adjustment	Before
Inputs	No
Number of Inputs	1
Split Adjustment	Yes
Output Variables	
Leaf Variable	Yes
Performance	Disk
Score	
Variable Selection	Yes ←
Leaf Role	Segment

The results of interest are in the Variable Importance table of the output. The list includes the inputs above the critical importance threshold of 0.05. An interesting fact is that the second ranked variable, x21, with almost the same variability as the most important input, x23, does not appear in a tree, but in the surrogate tree. The list of selected inputs by the SAS Enterprise Miner Decision Tree node includes eight variables: x1, x10, x14, x21, x22, x23, x37, and x40.

```
Variable Importance

Obs    NAME    LABEL    NRULES    NSURROGATES    IMPORTANCE

 1     X23     X23        2            0           1.00000
 2     X21     X21        0            1           0.98185
 3     X37     X37        0            1           0.29903
 4     X40     X40        1            0           0.29903
 5     X1      X1         2            0           0.29223
 6     X14     X14        0            1           0.26895
 7     X10     X10        0            1           0.12435
 8     X22     X22        0            1           0.09269
```

Variable Selection Based on Genetic Programming

Genetic programming (GP) is an algorithm that mimics the evolutionary process observed in nature to "breed" equations (formulas) to predict an output from a given set of input data. This evolutionary process is based on the rule of survival of the fittest. Those equations that have a higher fit (better prediction of the Y) get to produce offspring for a next generation. Offspring (children) can either be generated by mutation (a part of the equation is changed) or by crossover (two-parent equation mix).

Another key concept in genetic programming algorithms is the "gene pool." Gene pools are the building blocks for equations. The gene pool consists of the basic arithmetic operators (plus, minus, multiply, and divide) and other transforms (such as square root, power, logarithm, exponent, sine, and cosine). The gene pool is further expanded by the set of input variables (x1,x2,...,xn) and constants. At the beginning of the genetic programming model generation process, a starting population is randomly generated from the gene pool. Then, the fitness (performance) of the population of potential equations is determined. They are ranked based on their performance and complexity. The highest ranked equations get a higher probability to produce offspring. This process is repeated until the maximum number of generations is reached. At the end of model generation, hundreds or even thousands of possible solutions (some better than others) exist. The best equations need to be selected. More details for the genetic programming algorithm are in a classic book by John Koza entitled *Genetic Programming: On the Programming of Computers by Means of Natural Selection* (1992). The current state of the art is summarized in *A Field Guide to Genetic Programming* (Poli et al. 2008).

One of the unique features of GP is its built-in mechanism to select variables related to the target variable during the simulated evolution process and to gradually ignore variables that are not. In this way, variable selection based on evolutionary computation principles is being used. Inputs ranking is based on the normalized frequency of its use in equations during the simulated evolution. An advantage of this type of variable selection is that it implicitly includes nonlinear relationships between the target and the inputs. Details for the principles of GP-based variable selection can be found in the article "Variable Selection in Industrial Data Sets Using Pareto Genetic Programming" (Smits et al. 2006). An example of using GP for variable selection of the 45 inputs is given below.

Unfortunately, the GP algorithm is not yet available in standard SAS products. A specialized MATLAB toolbox, GPTools, has been used to run the example. The reader can use the commercially available product DataModeler of Evolved Analytics with similar functionality (www.evolved-analytics.com).

To take into account the stochastic nature of evolutionary computation and to generate consistent variable selection, 20 different GP processes with fitness maximal correlation have been run. Each GP process is initialized by a random set of equations, which continues for 625 generations and includes a population of 200 equations. The functional set from which the GP models are generated is shown in Figure 7.29.

Figure 7.29: Functional Set for Generating GP Models

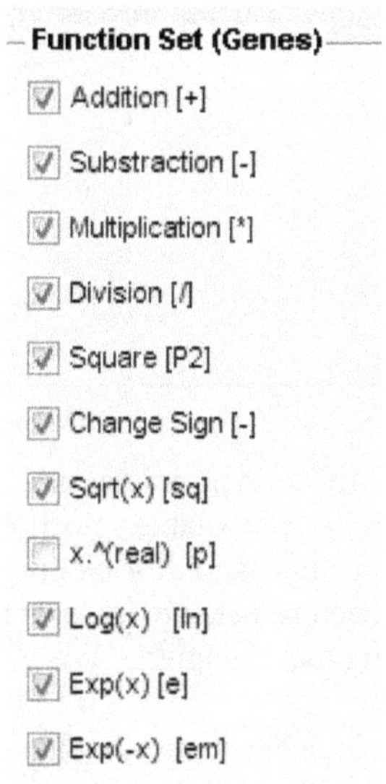

── **Function Set (Genes)** ──

☑ Addition [+]

☑ Substraction [-]

☑ Multiplication [*]

☑ Division [/]

☑ Square [P2]

☑ Change Sign [-]

☑ Sqrt(x) [sq]

☐ x.^(real) [p]

☑ Log(x) [ln]

☑ Exp(x) [e]

☑ Exp(-x) [em]

There are 834 equations with unique structures that have been generated from the 20 GP processes. The variable ranking, based on the normalized frequency of participation of the corresponding variable in GP-generated equations, is shown in Figure 7.30. On top of the ranking is x5, which has been selected in 24% of GP-generated equations, followed by x44 with 20%, and x45 with 14%. Unfortunately, there is no theoretically derived method to define a threshold for GP variable selection based on statistical significance. One practical solution is to define this threshold as a 5% normalized frequency limit (Normalized Sensitivities in Figure 7.30). It is based on the assumption that 95% of any linear or nonlinear equation generated by GP will include the selected set of inputs. The threshold is shown as a thick line in Figure 7.30. The selected variables are x1, x5, x28, x38 (selected even if it is slightly below the threshold, but with a big difference from the next in the rank x31), x43, x44, and x45.

Figure 7.30: Variable Ranking Based on Normalized Frequency of Participation in GP-Generated Functions

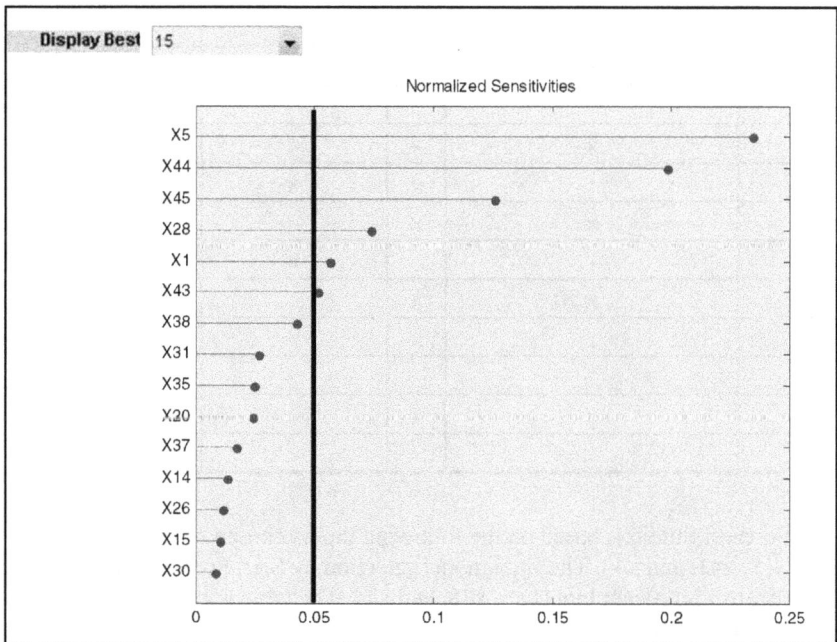

A side effect of GP-based variable selection is a list of nonlinear transforms that have shown high fitness during the simulated evolutions. Often, these transforms give insight for potential relationships. The top three transforms in the 45 inputs follow:

$$\text{Transform1} = \sqrt{x5 * x45}$$
$$\text{Transform2} = x44*x38^2$$
$$\text{Transform3} = \sqrt{x1 + \log(\frac{x28^3}{x38})}$$

Comparison of Data Mining Variable Selection Results

Because the statistical and data mining methods for variable selection are based on different principles and include specific thresholds, the results differ. To increase the consistency and reliability of variable selection, it is strongly recommended that several methods be used and results be compared. One suggestion for picking variables for time series selection and forecasting is to include all inputs that have been selected by at least two variable selection methods. This approach is illustrated below with an example using the 45 inputs.

First, the results from all explored methods are summarized in Table 7.2. The table includes inputs that have been selected by at least one of the methods.

Table 7.2: Selected Variables from All Explored Methods

Input	Correlation	Stepwise	EM variable selection	PLS	Decision tree	GP
x1		X			X	X
x5	X	X	X			X
x6				X		
x10					X	
x14					X	
x21					X	
x22					X	
x23					X	
x24				X		
x26	X		X			
x28						X
x35	X		X	X		
x37	X			X	X	
x38	X		X	X		X
x40					X	
x41	X		X	X		
x42				X		
x43						X
x44	X	X	X	X		X
x45						X

The final list to be used in SAS Forecast Studio is based on the following inputs chosen by at least two methods: x1, x5, x26, x35, x37, x38, x41, and x44. The final model selection by SAS Forecast Studio starting with all 45 Xs included the statistically significant inputs x5, x38, and x44. The table with the model parameter estimates and statistical significance of the model components is shown in Figure 7.31. The final model selection is not a surprise because these three inputs have been selected by most of the methods.

Figure 7.31: Final Variable Selection for the Time Series Multivariate Model Generated by SAS Forecast Studio

Component	Parameter	Estimate	Standard Error	t Value	Approx Pr > |t|
Y1	CONSTANT	-1.02197	0.39786	-2.57	0.0138
Y1	MA1_1	0.55722	0.129	4.32	<.0001
X5	SCALE	0.0001728	0.00001843	9.38	<.0001
X38	SCALE	0.0071179	0.0027593	2.58	0.0134
X44	SCALE	0.0063828	0.001239	5.15	<.0001

The forecasting plot of the multivariate model with acceptable performance measured by a MAPE of 3.40 and a holdout MAPE of 4.26 for a six-month period is shown in Figure 7.32.

Figure 7.32: Forecasting Plot of the Time Series Multivariate Model Generated by SAS Forecast Studio

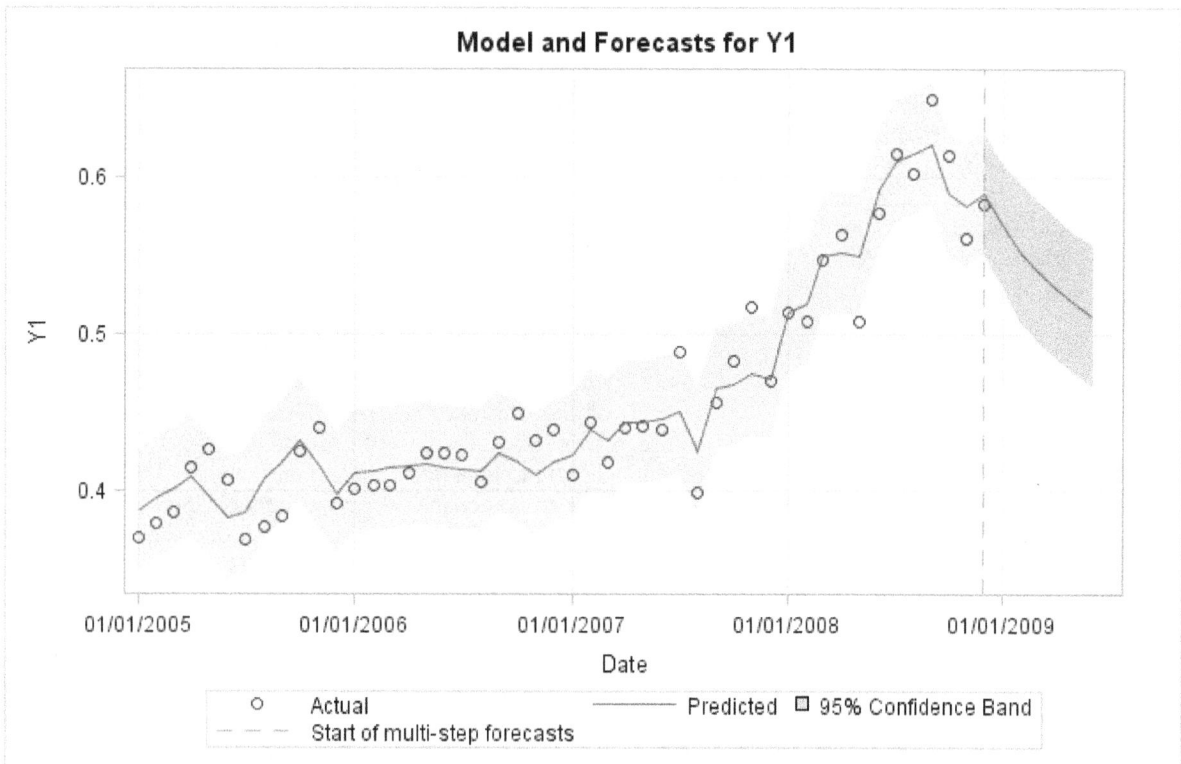

As a reminder, it is important to expand the list of Xs being considered while using transaction-based variable selection methods with additional lags because the transaction-based systems are not time series–based approaches. The following section shows examples of time series-based approaches for variable selection.

7.4 Time Series Approach

The same variable selection considerations apply to time series data. But, the approaches are different because of the serial correlation among variables. Literature concerning variable selection for time series data is currently not that extensive. We have found three different methods to be of practical use. Each one measures a different aspect of the relationship between Ys and Xs in time series data.

Introduced in section 7.2.2, the first method is similarity. In the same way that X-X similarity was developed, the same similarity metric can be used on Y versus X.

Second, there is the co-integration test, which is a test of the economic theory that two variables move together in the long run. (See Engle and Granger 2001 for more information.) The traditional approach to measuring the relationship between Y and X is to make each series stationary (generally by taking first differences), and then to see whether they are related using a regression approach. This differencing might result in a loss of information about the long-run relationship. Differencing has been shown to be a harsh method for rendering a series stationary. Co-integration takes a different approach. First, the following simple OLS regression model (called the co-integrating regression) is used, where x is the independent variable, y is the dependent variable, t is time, α and β are coefficients, and ε is the residual.

$$x_t = \alpha + \beta\, y_t + \varepsilon_t$$

The actual test statistic used to see whether the residuals of the model are stationary is either the Dickey-Fuller test or the Durbin-Watson test. In the implementation examples below, the Dickey-Fuller test is used.

A common approach used in time series modeling to show the relationship between Ys and Xs is called the cross-correlation function (CCF). The traditional regression approach (or what is called the poor man's approach to time series modeling) assumes that the effects are contemporaneous or coincident in time. The first way to avoid this assumption in the poor man's time series modeling is to introduce lags as additional effects. This still does not fully incorporate the potential complexity of effects over time. The only way to do this properly is to build full ARIMAX or transfer function models (Box and Jenkins 1968 and Pankratz 1991). Models of this nature are described in Chapters 8 through 12.

To build the case for using the CCF as one of the key approaches for variable selection, an overview of the concept of a transfer function is given below. Figure 7.3.3 shows the complex nature of multiple (non-contemporaneous) effects of one X on one Y. A common mistake made by modelers not familiar with time series data is to look at simple correlation or regression models between the Y and the X. Because of the added dimension of time, each of the variables might be related to time rather than to one another. Thus, a best practice is to make sure that the time series of interest are stationary. The most common approach for obtaining a stationary time series is to take the first difference. The use of the CCF is normally conducted on stationary series.

When transitioning from data mining with transaction-based data to time series-based data, the first thing to understand is that the effect of an X on a Y can be spread over time. If a Y changes instantaneously in the same time period as the X changed, this is known as a contemporaneous effect. But, this might not be the only time period that the change in X affects the Y. This is one of the most common mistakes novice time series modelers make—the assumption that the effect of an X on a Y is contemporaneous. Another common mistake is to look at varying lag structures, but to include only one particular lag's effect on the Y.

More often than not, the effect of an X on the Y looks more like the first histogram in Figure 7.33. That is, there are multiple time period effects of a change in an X on the Y. Taking this a step further, if the line (Input X_t) shows where three different Xs change, and then the three histograms below the line represent each of the effects on the Y, then the final effects on Y_t are represented in the output Y_t line.

Figure 7.33: Transfer Function Concept

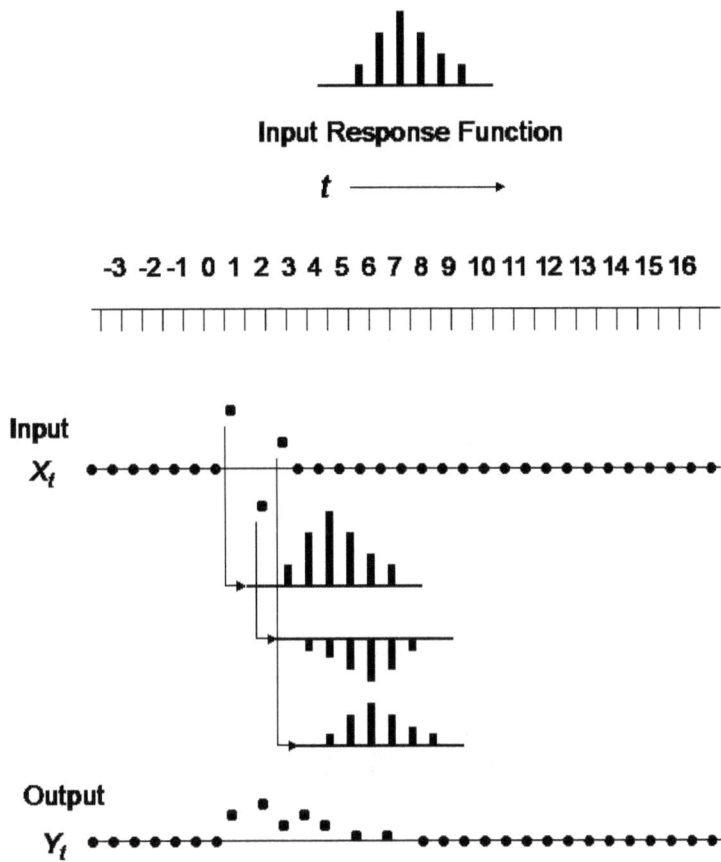

Input Response Function

t ⟶

-3 -2 -1 0 1 2 3 4 5 6 7 8 9 10 11 12 13 14 15 16

Input X_t

Output Y_t

By definition, the CCF is the Pearson product-moment correlation between the differenced Y and X for a particular lag. Figure 7.34 represents this type of CCF. There is a One Standard Error line and a Two Standard Errors line in the graph for interpreting real correlations.

Figure 7.34: Example of a CCF

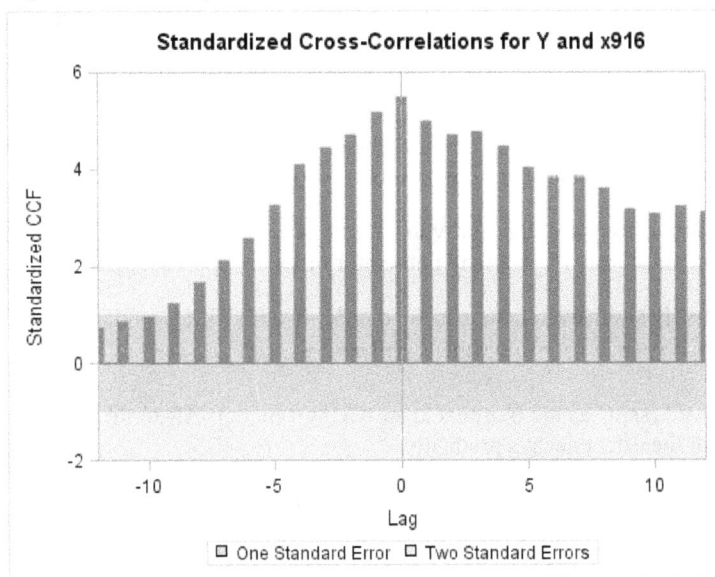

Generating a CCF in SAS Forecast Studio is easy. Figure 7.35 shows the click sequence. Once a project is started and the Ys and Xs have been identified, go to the **Series View** tab, select **Plot**, and select **Standardized Cross-Correlation Function**.

Figure 7.35: Running CCFs in SAS Forecast Studio

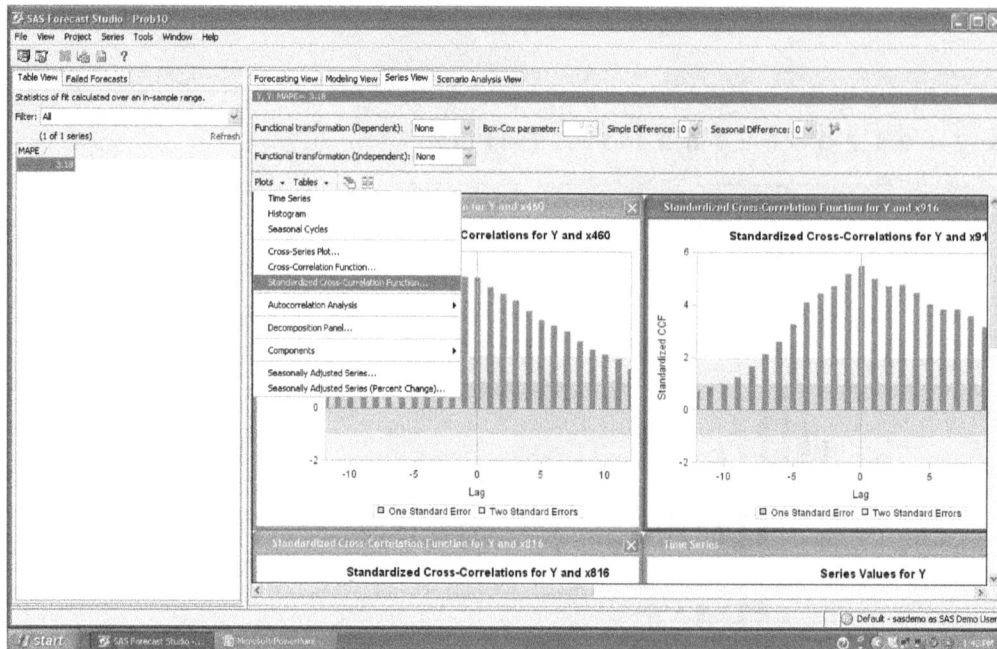

Though Chapters 8–11 more fully describe transfer function models, this discussion helps discern why the CCF is useful for variable selection. Think of a transfer function as interpreting the effects of a one-time pulse in the input variable on the dependent or forecast variable.

The transfer function has potentially three parts:

- Scale

  The scale variable provides the contemporaneous (unless there is a lag) effect (direction and magnitude) of the pulse on the Y variable.

- Numerator

  The parameter associated with a numerator of order 1 provides the effect of the pulse the interval after the pulse occurs.

  The parameter associated with a numerator of order 2 provides the effect of the pulse two intervals after the pulse occurs. The effect of the pulse lasts for as many numerator orders as specified, plus one.

  Numerator order effects are relatively short-lived (same as the q order in an ARIMA model) MA term.

- Denominator

  Denominator orders are longer-lived (same as the p orders in an ARIMA model) AR term; the pulse has a scale effect (same as above), but the effect decays gradually.

  The parameter associated with the denominator order determines the decay in the effect of the pulse or the trajectory back to the steady state mean or trend line.

The CCF is used to determine the order of the transfer function. To interpret the CCF relative to a transfer function form, the following rules are useful.

There are some things to note about the rules below. Usually, lags are defined as significant spikes following the first significant spike in the CCF, contemporaneous or not. The shift is the number of lags after lag 0 that the first significant spike shows up. (See Chapter 9 for details.)

- Rule 1

    Ignore all to the left (negative) of zero on the X axis. This would be a StateSpace model (having feedback). An approach for handling feedback effects is presented in Chapter 11. This is not discussed in this section.

    For remaining rules, spikes have to be larger than +/– 2 Sigma (the second line on the CCF chart).

- Rule 2

    Determine the contemporaneous term (lag).

    If there is a large (significant) spike at 0, then there is no lag.

    Count the number of non-significant spikes to the right to the first significant spike. That gives you the lag order.

- Rule 3

    For the autoregressive term (denominator), we are looking at the decay after the initial jump or spikes. If it is low, then the denominator order is 1. If it is persistent or oscillating, then it is 2.

- Rule 4

    For the moving average term (numerator), we are looking to see whether the spikes are building first before a decay. If there is a build (two successive increasing spikes), then it is order 2. If there are three successive increasing spikes, then it is order 3.

The CCF not only identifies which Xs might be of use in the eventual model, but, it also shows what the best shift and lags might be.

There is some debate concerning whether to pre-whiten the Y and X before calculating the CCF. We would agree that in the final modeling stages, this is probably a best practice. But, in the early exploratory stages of variable selection, the raw first differenced CCFs will suffice because in this stage, the objective is to see which potential Xs might be of interest.

To show each of these measures in the context of an example, the data set used to introduce similarity is analyzed again. Because the use of all three measures together has been found to be very useful as a major variable reduction and variable selection process, a SAS Enterprise Guide flow was developed to do all three of these variable selection methods simultaneously. Figures 7.36 and 7.37 capture the essence of the SAS Enterprise Guide flow. Each of the three analyses is shown separately below.

Figure 7.36: First Half of the SAS Enterprise Guide Flow for Similarity, Co-Integration, and CCF

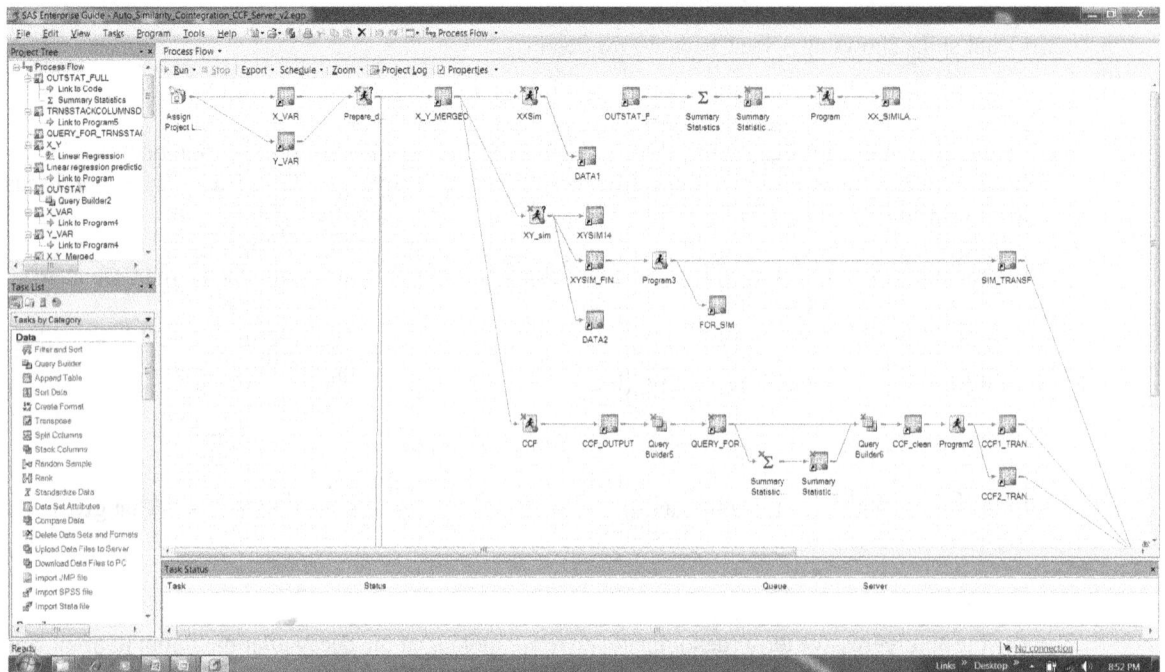

Figure 7.37: Second Half of the SAS Enterprise Guide Flow for Similarity, Co-Integration, and CCF

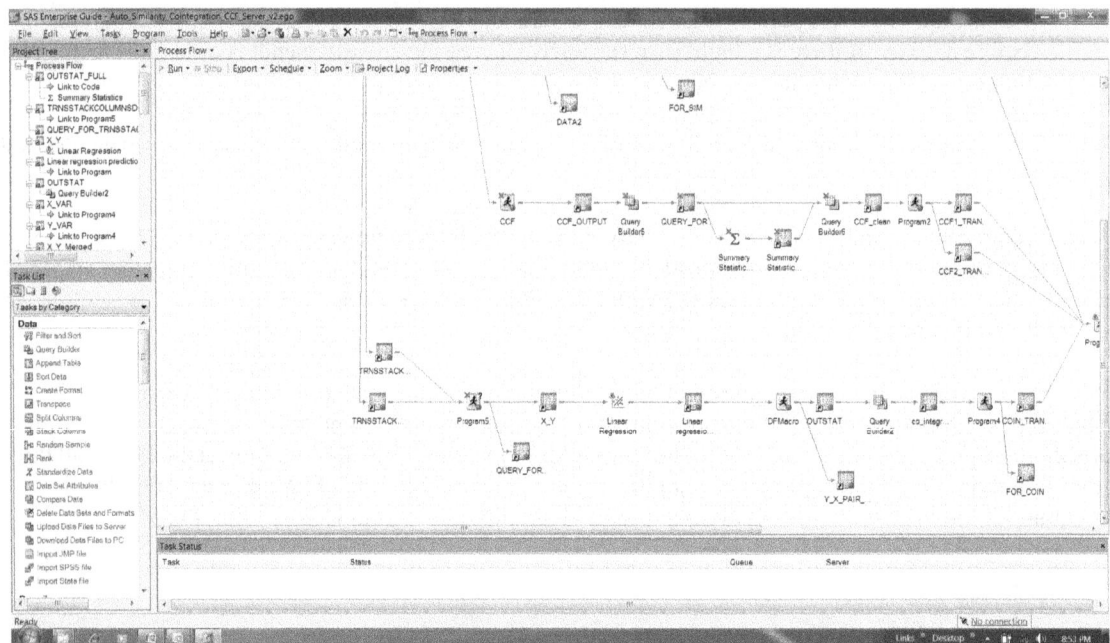

For the X-Y similarity, the following SAS code is embedded in the SAS Enterprise Guide flow above (the XY_Sim code node).

```
proc similarity data=lib1.yandxexample outsum=lib1.xysim14;
id date interval=&=quarterly;
input chained_price_index_consumer_pu--vacancy_rate_of_rental_housing_u /
scale=standard normalize=standard;
target y/ measure=mabsdev normalize=standard
expand=(localabs=12)compress=(localabs=12);
run;
```

For the CCF, the following SAS code is embedded in the SAS Enterprise Guide flow above.

```
proc arima data=lib1.yanxexmaple;
identify var=y crosscorr=( chained_price_index_consumer_pu--
vacancy_rate_of_rental_housing_u) nlag=10
outcov=CCF_output noprint;
```

For the co-integration, the data first has to be set up in a slightly different format to do all the combinations of Y and X variables. This format leverages the BY capability in Base SAS. Once the data is in this vertical format, the X becomes the Y and the Y becomes the X, and then a simple regression model is run. The trick is that the residuals are where the actual test is conducted. A Dickey-Fuller test is conducted to see whether the residuals are stationary. The null hypothesis is that the residuals are not stationary. As a result, we are looking to reject the null hypothesis (in other words, we are looking for small p-values (significance levels)). In Figure 7.38, the Linear model interface in SAS Enterprise Guide shows how to set up the co-integration test.

Figure 7.38: Setting Up the Co-Integration Model

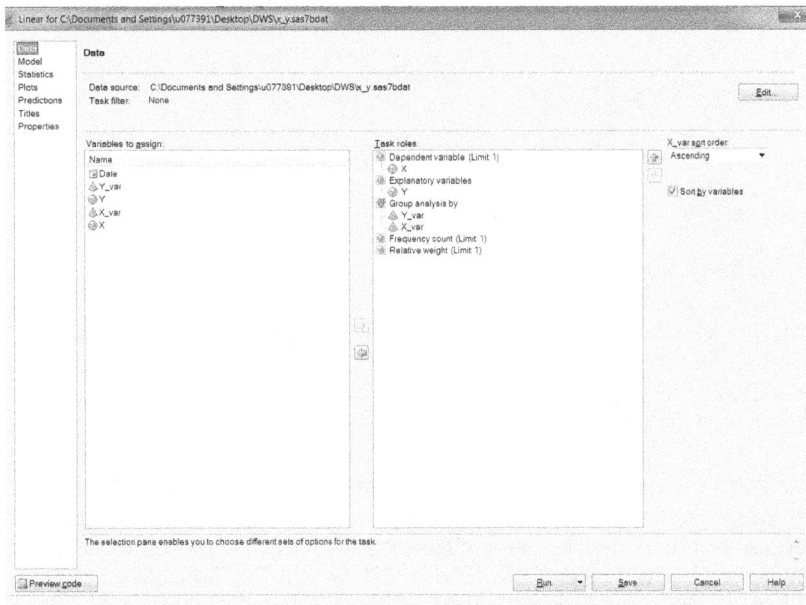

The key is to output the residuals, as shown in Figure 7.39.

Figure 7.39: Saving the Residuals for the Co-Integration Test

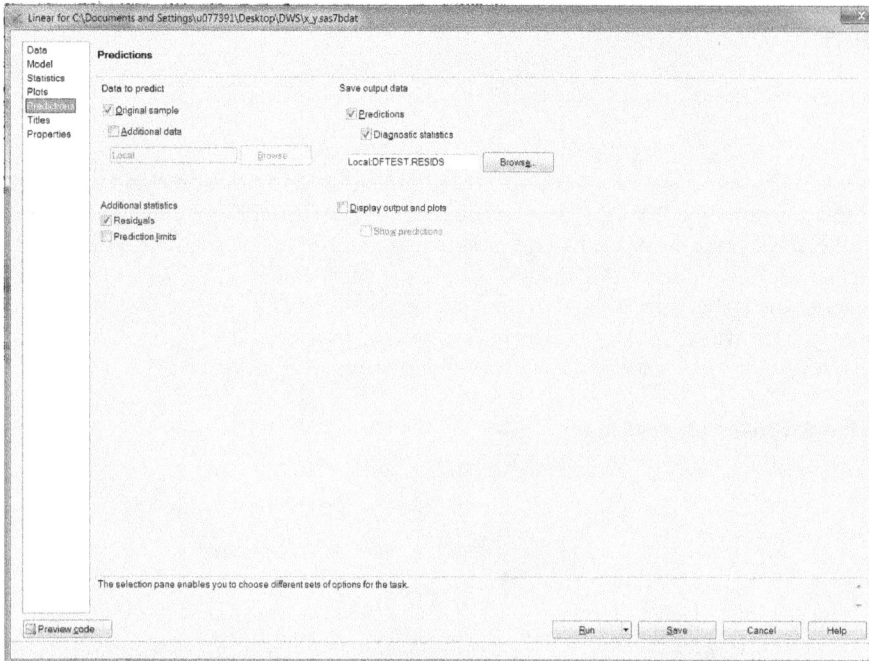

Using the x and y example data, the following output was generated.

One additional approach for variables selection is time series-based stepwise regression. This approach is similar to a blended transactional and time series approach, and it is readily available in SAS Forecast Studio. This analysis is available in the additional Stored Process Reports in SAS Forecast Studio. It enables the modeler to introduce various lags in both the dependent and independent variables, and to introduce simple or seasonal differencing in both the Y and the X. There are various stepwise regression options. In this case, the objective is to determine the potential Xs to pass onto the final modeling step. We have found that allowing lags only in the Xs, and generally allowing only about four lags (one year for quarterly data) is as much a limitation on the algorithm as anything because it is limited by the ratio of data to parameters. Simple differencing is used, but not seasonal differencing. We wait for SAS Forecast Server to model the seasonal structure in the modeling step. Stepwise selection is used because of the nature of the output provided, which is a reduced set of Xs and their associated lags.

Figure 7.40: Picking the X Variables

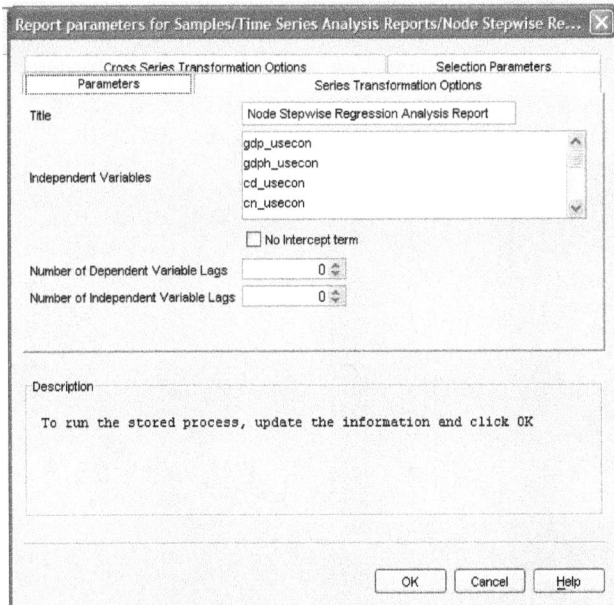

Figure 7.41: Picking the Y Variable Transformations

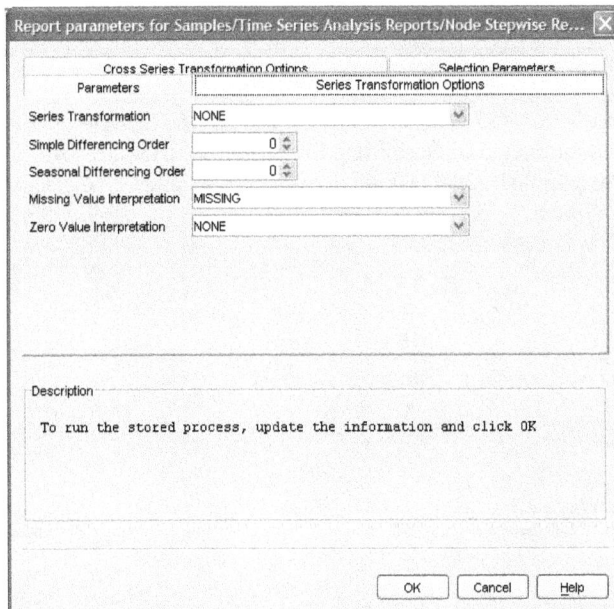

Figure 7.42: Picking the Stepwise Regression Options

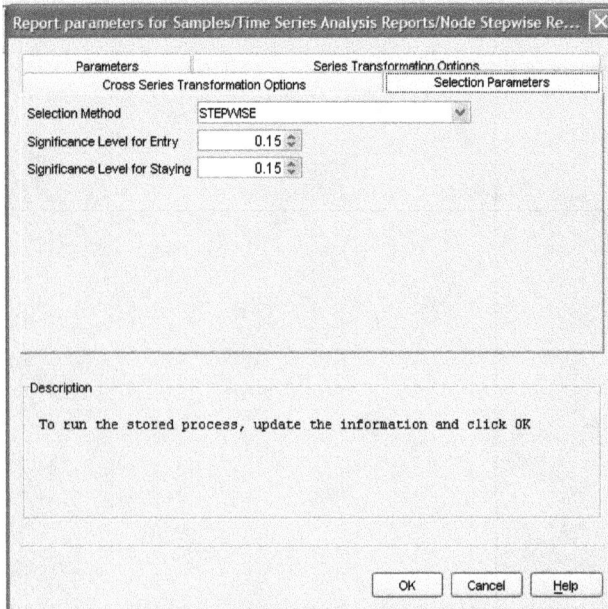

7.5 Summary

Using time series-based approaches for variable reduction and variable selection enables the modeler to attack much larger time series data mining problems. A combination of Base SAS, SAS Enterprise Guide, SAS/STAT, SAS Enterprise Miner, and SAS Forecast Studio accomplishes the task in an expedient manner. Using rigorous time series statistical approaches for variable reduction and variable selection increases the probability that the modeler will find a high-quality exogenous variable forecasting model.

---

[1] Details about all X11 and X12 related methods are given in www.census.gov/srd/www/sapaper/.

Chapter 8: Model Building: ARMA Models

Introduction

This chapter is intended to be a model building guide for the forecasting practitioner. The applied, data-driven focus of the book continues here. The aim is to provide useful tools for extracting and extrapolating relevant information from the data at hand in a robust and straightforward way. Emphasis is placed on developing your intuition regarding the techniques presented and on the process of model building associated with time series data. Technical details are provided as supplementary resources or references.

More specifically, the focus is on implementing univariate or one-series-at-a-time forecasting models. The primary forecasting tool presented is the rational polynomial transfer function. This modeling approach is also known as the ARIMAX, or ARIMA models with eXogenous variables. The transfer function has proven to be a workhorse in a wide range of forecasting and inference-based time series applications. As we will see, it also provides a powerful and flexible framework that easily accommodates dynamic relationships (for example, lags, shifts, and persistence between inputs and the dependent variable).

The rational polynomial transfer function model also provides other practical advantages. The forecaster might face a situation where data on important independent variables are either unavailable or unreliable. This often manifests itself in the time series model's residuals. The error series of the model might exhibit systematic variation or correlation. Transfer function specifications can be easily extended and refined with Autoregressive (AR) or Moving Average terms (MA) in order to diminish correlation in the model's residuals. That is, this model can be used to build more of the systematic variation in the data into the "explained" part of the model.

The focus on a one-series-at-a-time approach and the transfer function framework is not intended to indoctrinate readers. However, these methods have been proven to be robust, flexible, and, with the concepts and tools presented here, fairly straightforward to implement. There are other forecasting approaches and several other families of models to choose from. Some of these are summarized in this chapter.

Multivariate (in the dependent or Y variable) modeling approaches are becoming more widely used by forecasting practitioners. Other families of forecast models include Exponential Smoothing Models (ESM), Unobserved Components Models (UCM) and Neural Network models. Concepts and some details associated with alternative models and approaches are presented in Chapter 11. The purpose is to give you a basis for making informed decisions when comparing alternative approaches and models to the primary tools presented here.

Regardless of the approach and model family chosen, there is a specific methodology to building models on time series data. The process involves the systematic application of sound techniques, but there is also an art to model building. You might initially be dismayed to find out that the diagnostics and techniques covered here and in other texts do not usually lead the analyst to *the* forecast model. You usually have to validate and choose between several candidate model specifications. The modeling process consists of deciding on likely candidate specifications (also known as model identification), model estimation, forecast generation, and then returning to identification for further refinement as time permits. The first two sections in this chapter (and Chapter 9) presents the time series model building process. To help you develop your intuition about the process, details about diagnostics, best practices, and the techniques that make them up are provided.

This chapter presents the Autoregressive Moving Average (ARMA) model. Placing this topic first might seem strange, but we will see that there is a close relationship between the ARMA model and the rational polynomial transfer function. The specification of the ARMA model enables us to motivate the specification of the transfer function in a particularly straightforward way.

8.1 ARMA Models

Knowledge of how the use of Autoregressive (AR) models has evolved can be useful in understanding how they work. "Auto" Regression indicates that a variable is modeled as a function of its own (=auto=self) past values. The idea of regressing a variable on its past has been around for a long time, but was not widely used in applications until two events occurred.

First, Box and Jenkins (1972) brought the theory together in a systematic and understandable way. Second, econometricians had a hand in it. In the 1970s, econometricians were fond of creating large systems of structural equations to explain and forecast economic activity. "Structural equations" used here means equations with explanatory variables (for example, national income modeled as a function of government spending, some measure of aggregate savings, and so on). These systems of equations had three hallmark characteristics: 1) they were expensive to maintain and estimate (the number of parameters to be estimated was large and computer memory was expensive), 2) they were complex, and 3) they did not forecast very well. They were fine for testing theory and other inference-based activities, but extrapolation is a different problem.

AR models became more widely used as a forecasting alternative to the structural equations approach. The motivation begins with the following idea; let's assume that we don't know anything about what inputs cause some variable to move around. All we observe about a variable is its own past, and that some portion of the variation in it is systematic. An AR model can be used to capture the systematic part, and then extrapolate it. Nothing could be simpler. AR models provided a less expensive way to create economic forecasts, and the AR forecasts were competitive with the ones the structural equations produced.

However, you might ask: what if you do know something structural about the variable you want to forecast? The price you charge for a product impacts the amount sold, right? The transfer function model provides the best of both the AR and structural approaches. We use a structural or transfer function equation to capture the effect of inputs that are easy to get and measure (for example; the price that we charge, promotional activity, and seasonality). We then let AR or moving average (MA) terms capture the effect of inputs that we do not have access to or that are not well measured. This describes the application of the rational polynomial transfer function or ARIMAX (X stands for eXogenous) model discussed in this chapter. Understanding how to build an appropriate AR model is the first step in implementing it.

8.1.1 AR Models: Concepts and Application

Most forecasting textbooks begin the presentation of AR models with a discussion of stationary and nonstationary processes. In the application discussed here, AR terms are used to augment the regression or transfer function model by refining its residuals. Residuals usually are stationary, so the discussion of nonstationary processes, unit roots, and associated topics are put into Appendix 2. We begin with two questions; what is a stationary AR process, and how is an appropriate model specified to capture the systematic variation in it?[1]

Ideas

The most obvious characteristic of a stationary process is that it has a well-defined mean. The simplest type of stationary process is a white noise process. A white noise process or series varies randomly around its mean.

The example white noise process used here are observations taken over time on weekly sales of white tennis shoes (WTS). Because sales of WTS vary randomly over time, observing that an above average number are sold today does not really give any additional information about what is going to happen in the future. The best guess or forecast for next week's sales is the mean. Another way of thinking about this is that observations on the WTS series are uncorrelated. A white noise specification looks like the following, equation 1.

$$WTS_t = \delta + \varepsilon_t \qquad\qquad 1$$

In equation 1, δ is the average of white tennis shoe sales, and ε_t can be thought of as a random draw at time t from a normal distribution that is centered around zero. More details are provided in the appendices to this chapter.

The next simplest type of stationary process is an AR order one, or AR 1. (See the RTS series plot below.) The AR 1 example series is weekly sales of red tennis shoes (RTS). The order 1 means that sales of RTS in adjacent weeks (lag 1) are correlated. AR 1 indicates that there is a systematic component in the variation in RTS sales. In this case, observing that an above average number of RTS are sold this week does say something about what might happen next week. Because adjacent observations are correlated, there is a better than 50% chance that sales in the next week will be above average as well.[2] Note that this works both ways. A below average sales week would tend to be followed by weeks of below average sales. An AR 1 specification looks like the following (equation 2a and 2b).

$$(RTS_t - \mu) = \varphi_1(RTS_{t-1} - \mu) + \varepsilon_t \qquad\qquad 2a$$
$$\rightarrow RTS_t = \varphi_0 + \varphi_1 RTS_{t-1} + \varepsilon_t \qquad\qquad 2b$$
$$\varphi_0 = \mu(1 - \varphi_1) \rightarrow \mu = \varphi_0 / (1 - \varphi_1) \qquad\qquad 2c$$

Where μ is the mean of weekly red tennis shoe sales, and the AR 1 parameter, φ_1, regulates the correlation between adjacent weeks. Note that the AR 1 parameter does not have a time subscript. This indicates that the correlation "structure" does not change over time.[3]

Although equation 2a is not as compact as 2b, it provides an illustration of what is going on in an AR 1 process. An innovation or "jump" high, away from the mean today (t–1) tends to be followed by another jump high away from the mean next week (t). Equation 2b shows the more common way to write an AR 1 specification. Getting from 2a to 2b involves a little algebra, which is shown in 2c. For readers who are interested in stationarity topics (see Appendix 2), equation 2c shows why a unit root (that is, $\phi_1 = 1$) can be a problem for the mean of the series.

Why is this correlation in the data for RTS and not WTS? It might be a result of RTS being endorsed by a celebrity, and the observed correlation is a result of the celebrity's appearance, or non-appearance, in tabloid headlines. However, it could also be an artifact of production scheduling, a competitor's promotional activity, and so on. The point is that with AR models it is not necessary to know what "causes" it. The approach in this section is to identify the correlation or systematic variation in the data, estimate an AR model that captures it, and then use that model to extrapolate or forecast into the future.

Other orders or lags of autocorrelation are also possible and common in time series data. An example of an AR 2 process could be black tennis shoe sales (BTS, shown in equation 3). In this process, sales in adjacent weeks (lag 1), and in weeks one week apart (lag 2) are correlated. Seasonal tennis shoe sales (STS, shown in equation 4) might exhibit correlation between weeks 52 weeks apart as well as lower order autocorrelation at lag 1. The STS process would be written AR(1 52).

$$BTS_t = \phi_0 + \phi_1 BTS_{t-1} + \phi_2 BTS_{t-2} + \xi_t \qquad\qquad 3$$

$$STS_t = \phi_0 + \phi_1 STS_{t-1} + \phi_{52} STS_{t-52} + \xi_t \qquad\qquad 4$$

A couple of notes on modeling terminology are helpful. The letter p is commonly used to denote the AR order of the model. For example, the AR 2 model can be described as p=2. This notation describes a saturated AR model, and means that the BTS series is autocorrelated at lag 1 *and* lag 2. The STS model is actually a subset AR model, and is usually denoted with parentheses as follows; p=(1, 52). This notation indicates that the STS series is autocorrelated at lags 1 and 52, but not at lags 2, 3, ..., 51. Note that the season cycle length for weekly data is usually assumed to be 52 weeks. The 52-week lag indicates that like weeks move together; the first week of the year this year is like the first week of the year last year, and so on.

Model Identification

Stationary autoregressive series (the uncorrelated series, WTS, can be thought of as AR 0 or p=0 in this context) have been discussed. Natural next questions are: does the series at hand contain systematic components that can be modeled as an autoregressive process, and, if so, what AR model is appropriate? Because observations move together in an autocorrelated series, jumps high (low) away from the average value tend to be followed by more observations above (below) the series average. Autocorrelated data tends to look different than non-correlated data. The correlation manifests itself in clusters or clumps of observations above or below the mean (plots of autocorrelated series are provided below).

However, even an analyst with a particularly keen eye for differentiating correlated versus non-correlated data is still faced with a model building problem, which is how to decide on the appropriate AR specification. The problem is not really different from selecting input variables when building a regression model. Selecting too many AR terms or lags is the equivalent to having irrelevant variables in a regression specification. Having too few AR terms is analogous to omitting relevant inputs. For weekly data, relevant AR lags could range anywhere from 0 to approximately 52. One approach would be to fit all combinations of lags from 0 to 52 and then rely on some fit diagnostic to choose the best one. This is a lot of models to have to fit (1.4 E+80). The diagnostic graphs presented next help in narrowing the field of candidate models considerably.

Note that, in this topic, it is assumed that the only type of systematic variation a series can contain is AR variation. The primary model identification tool that is used here is the ARIMA procedure. PROC ARIMA provides flexible and interactive options that can follow the identification, estimation, and forecasting framework for building univariate forecasting models.[4] However, the cost of flexibility is that the syntax is somewhat less than transparent to new users. Key components and interactivity between statements are discussed as they come up in the process of building models. Generated diagnostic plots are presented in the order in which they are generally used and consist of the following:

- the series plot
- the series white noise probability plot
- the autocorrelation function plot
- the partial autocorrelation function plot

The code below generates the described diagnostics. The ODS GRAPHICS statement turns on the plotting functionality in the ARIMA procedure. The ARIMA procedure statement (PROC ARIMA) denotes the data set that contains the series to be identified, and lists instructions for generating the plots. All plots that are associated with the IDENTIFY statement are produced (a subset is shown below). The IDENTIFY statement lists the variable or series to be identified, sales of red tennis shoes (RTS).

Program 8.1

```
ods graphics on;
proc arima data=work.ar1_5 plots(unpack)=series(all);
   identify var=RTS nlags=12;
quit;

ods graphics off;
```

The plot of the time series RTS is shown below. Although particular details about the process generating the data are difficult to discern, the plot provides useful summary details about the series. Features of interest could include outlier observations, changes in series volatility, trends, and seasonal patterns. Even though none of these features are readily apparent, it looks like RTS sales average around 10 units per week with minimum and maximum observations around 5 and 15 units over the range of the data. There are approximately 240 weeks of observations, or a little less than 4.5 years of data.

Output 8.1

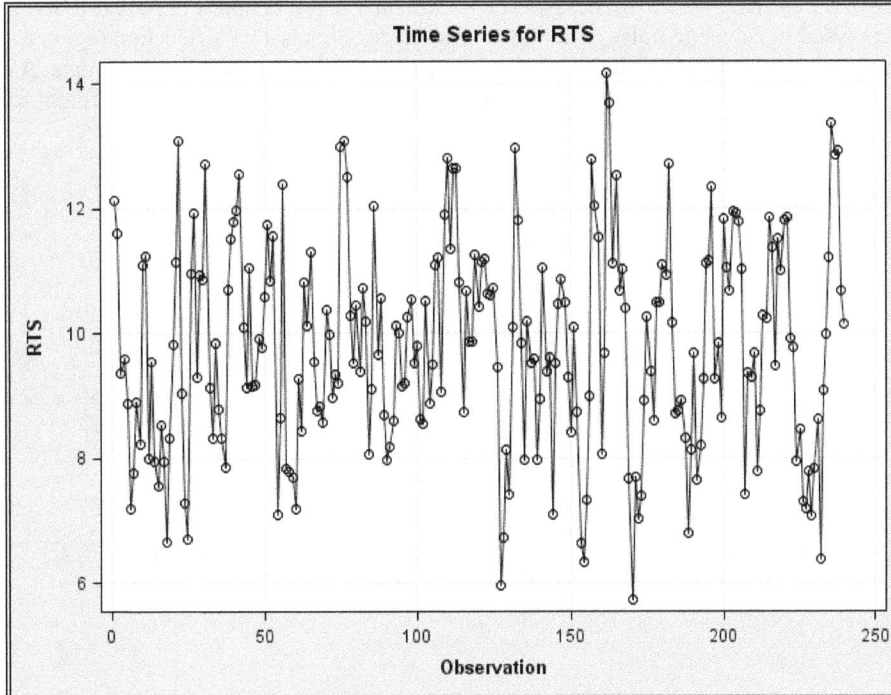

Time Series for RTS

The next diagnostic gives information about whether the series contains any systematic variation, in this case autocorrelation. The autocorrelation test for white noise tests the null hypothesis that the series is uncorrelated, or white noise. The test provides information to help decide whether it is worth spending any time trying to build a model for this series. If the test fails to reject (Pr > .05 as a rule of thumb), then the answer is probably no. This result would indicate that the series is white noise and the mean of the series provides the best guess about what is going to occur in the future. If the test rejects, as is the case below (Pr < .001), then this is evidence that there is some autocorrelation in the series, and spending time building an AR model is warranted.

Output 8.2

	Autocorrelation Check for White Noise									
To Lag	Chi-Square	DF	Pr > ChiSq	Autocorrelations					
6	90.81	6	<.0001	0.540	0.227	0.089	-0.025	-0.057	-0.135
12	118.02	12	<.0001	-0.185	-0.205	-0.144	-0.058	-0.059	-0.070

The test is cumulative. The p value on the first line, "To Lag 6", tests the null hypothesis that there is no autocorrelation at any lag between 1 and 6 in the series. The six numbers listed under Autocorrelations are the estimated autocorrelation at each lag. For example, the first number on the first line under Autocorrelations indicates the following: $RTS_t = \delta + 0.540RTS_{t-1}$, or a lag one AR model with the listed number denoting the estimated coefficient. The second number lists the estimated autocorrelation at lag 2, or an AR 2 model with the first and second numbers listed denoting the estimated coefficients, and so on.

To summarize the results above, there is evidence of some autocorrelation in the series. The null hypothesis of white noise is rejected. The largest (farthest from zero) estimated autocorrelation coefficient occurs at lag 1. Using an AR 1 model is likely the best starting place. Note that only lags of RTS between 1 and 12 were tested.

This was controlled using the NLAGS option in the IDENTIFY statement of the ARIMA procedure. Significant correlation at higher lags is possible. The white noise probability diagnostic also serves a useful role in the estimation step. It is used to validate candidate models' residuals. The next diagnostic plots presented provide information that enables us to narrow down which models that might be appropriate for the data.[5]

The autocorrelation function plot (ACF) is shown below. The information that it contains is similar to the autocorrelation estimates listed in the white noise plot, above. The shaded area is a 95% rejection region for each listed lag. The magnitude of the bars indicates the estimated autocorrelation at each lag in the data. A bar extending beyond the shaded region indicates that the autocorrelation at that lag is significantly different from zero.

Output 8.3

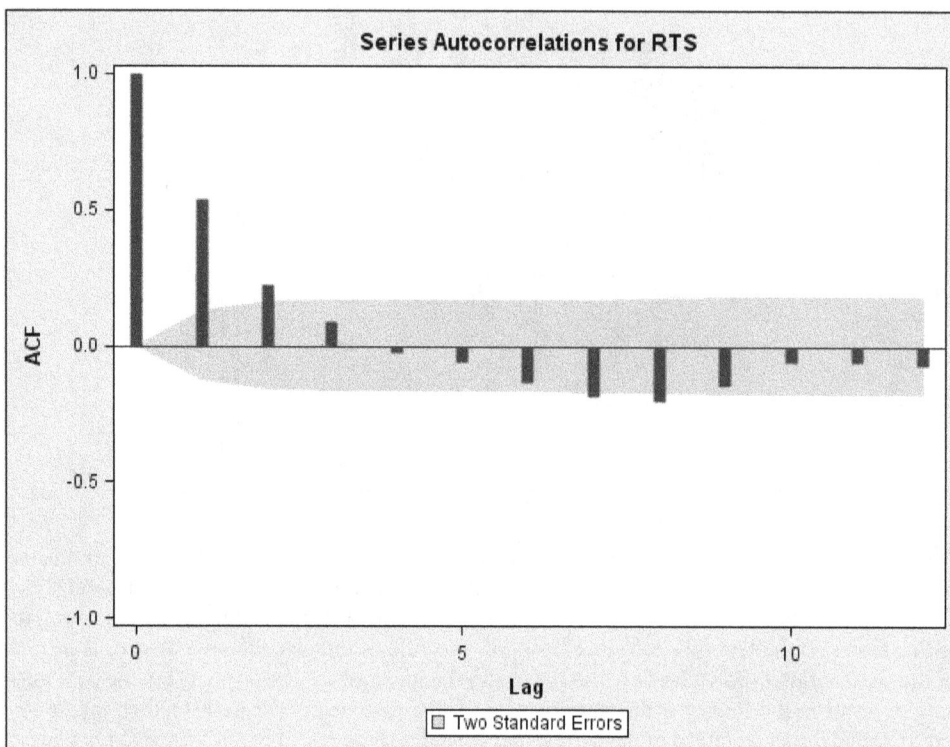

It appears that there is strongly significant (nonzero) autocorrelation at lag one in the data. Marginally significant autocorrelation at lag two and also at some higher lags is indicated. Note that there is a spike at lag 0 with a magnitude of 1. This indicates that any value is perfectly correlated with itself, and is included simply for scaling purposes.

The ACF plot is useful, but it has a major flaw for identifying AR processes—the proximity effect. To understand the proximity effect, consider the following: if the data follow an AR 1 process, then adjacent observations are correlated. Starting with the current value for this process, this week's observation is correlated with last week's observation. Last week's observation is correlated with the week before, and so on. Because of this pattern, it is going to look a lot like this week's observation is correlated with the observation two weeks ago (lag 2), when in fact it is not. For data that follow an AR 1 process, the ACF plot usually shows a significant spike a lag 1, with significant but diminishing spikes at higher lags. The diminishing spikes at lags greater than 1 are spurious and an artifact of the proximity effect. As another example, consider an AR 3 process. In this case, the first three spikes in the ACF would indicate actual correlation in the data, and significant spikes at lags larger than 3 would be spurious.

The partial autocorrelation plot (PACF), shown below, provides a remedy for the proximity effect in identifying AR processes. The PACF works a lot like a partial derivative. Any given bar or spike in the plot—say, the bar at lag 3—can be interpreted as follows; this spike represents the magnitude of correlation at lag 3 in the data *holding constant or accounting for correlation at lags 1 and 2.* The correlation for each lag in the PACF is calculated in a recursive way that generates the partial representation of correlation in the data, and effectively removes the proximity effect.

Output 8.4

Series Partial Autocorrelations for RTS

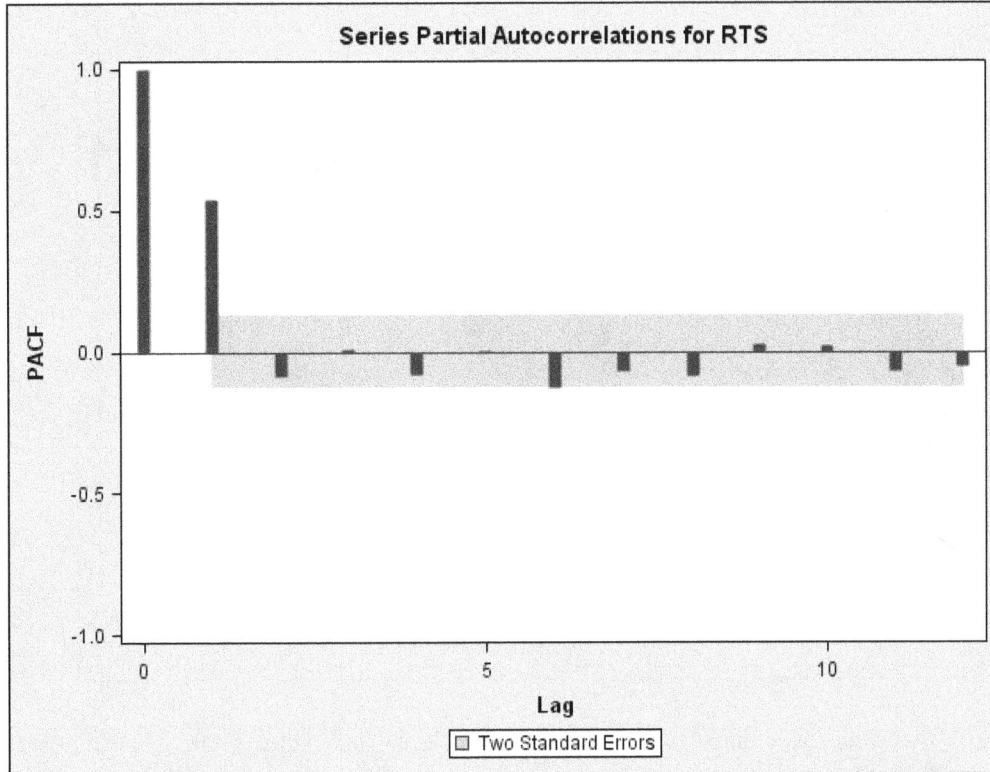

To summarize, the most important plot for identifying AR processes is the PACF.[6] The pattern of bars or spikes contained in the ACF and PACF plots above show the characteristic pattern associated with a stationary AR 1 processes. That is, there is one significant spike at lag 1 in the PACF, and a significant spike a lag one with further significant but rapidly diminishing spikes in the ACF. Significant spikes in the PACF are the primary source of information in identifying AR series.

To develop intuition, a second series needs to be identified. The following PROC ARIMA code generates the diagnostics to identify the yellow tennis shoes (YTS) series.

Program 8.2

```
ods graphics on;

proc arima data=work.ar3 plots(unpack)=series(all);
   identify var=YTS nlags=12;
quit;

ods graphics off;
```

Although the YTS series plot shows no strong evidence of outliers, seasonal patterns, or trends, it does provide summary information about the series.

Output 8.5

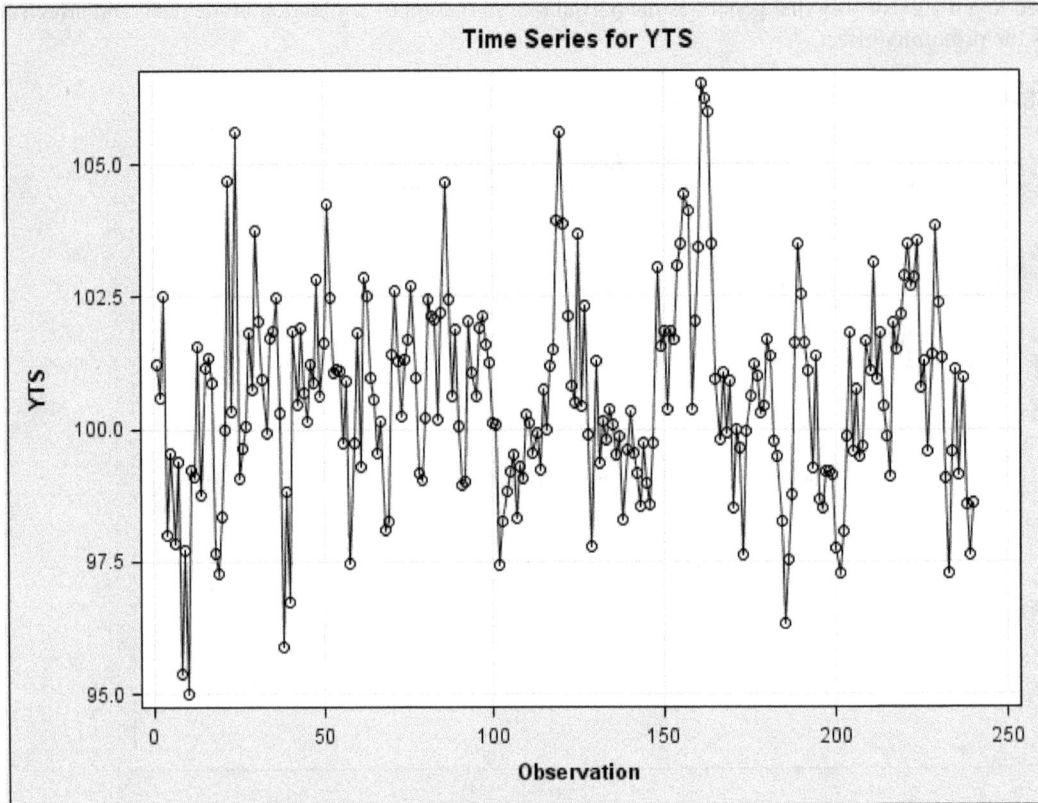

The White Noise Probability table shows evidence that the series is not white noise; some systematic variation seems to be present in the series.

Output 8.6

	Autocorrelation Check for White Noise								
To Lag	Chi-Square	DF	Pr > ChiSq	Autocorrelations					
6	50.63	6	<.0001	0.331	-0.134	0.129	0.232	0.082	0.052
12	61.18	12	<.0001	0.129	0.096	-0.062	-0.086	0.062	0.030

The **ACF** and **PACF** plots for the YTS series are shown next.

Output 8.7

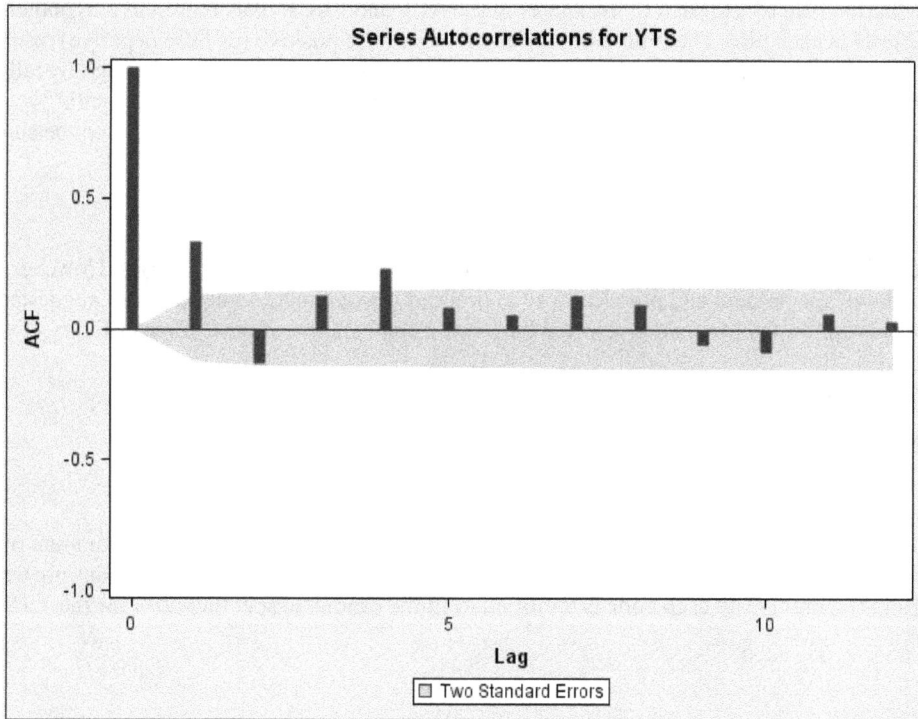

Series Autocorrelations for YTS

Output 8.8

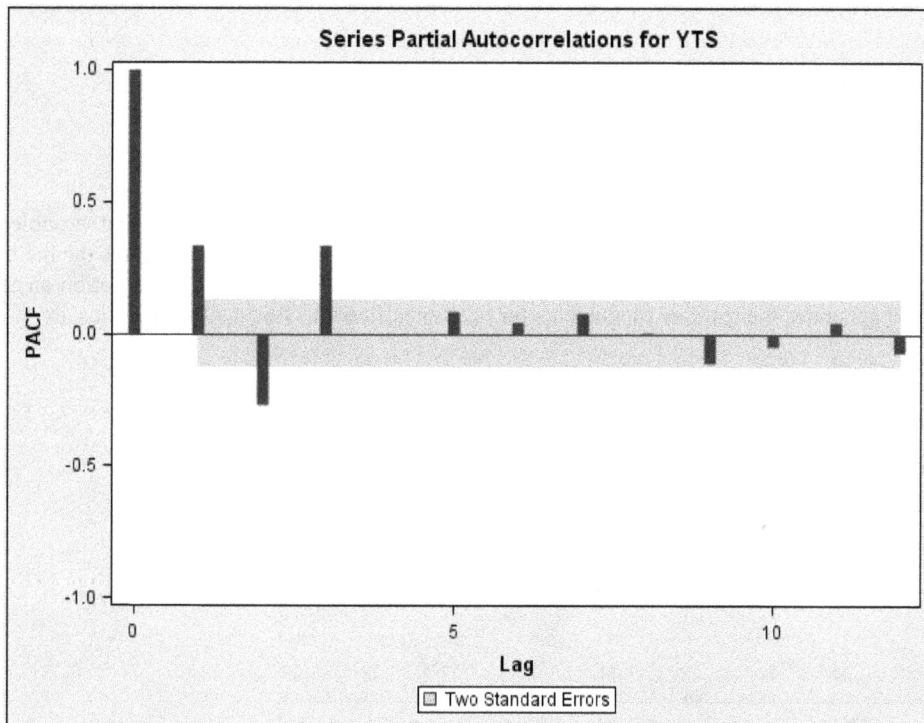

Series Partial Autocorrelations for YTS

The PACF shows strong evidence that the YTS series is an AR 3 (p=3); there are three significant spikes at lags 1, 2 and 3. The ACF is less clear; significant spikes at lags 1 and 4 with marginally insignificant spikes at some other lags. However, the pattern in the ACF does, approximately, fit the rule of thumb for low order stationary AR processes; a pattern of significant but rapidly diminishing spikes as the lag length increases.

Based on the diagnostics presented, an AR 3 model is the likeliest candidate specification for the YTS series. However, a word of caution is in order. Each of the spikes in the ACF and PACF plots represent a hypothesis test, and there are 12 tests in each plot. The probability of observing a false positive (or false negative) result increases with the number of tests done. With 12 tests the probability of a false positive is quite high (recall that the shaded area represents the 95% rejection region). If 20 tests are conducted, the probability of getting a spurious result on any one test is close to one.[7] The important point for our analysis is that we cannot be sure that the AR 3 specification is really the best model for the YTS series.

Model Estimation, Selection, and Forecasting

The AR 3 specification is indicated by the diagnostics as the most appropriate for the YTS series. However, given the discussion above, the prudent and practical way to proceed is to estimate some subset of suggested models. Once the models are fit, the best one is selected based on a combination of these three criteria:

- individual parameter significance within each model
- fit to the data
- analyses of the models' residuals

The range of candidate AR orders to fit is usually regulated by the value of the series at hand, the amount of history available, and the time available for analysis. For the YTS series, the following specifications are tried: AR 2, AR 3, and AR 4. The previously used code is modified to fit the candidate specifications for the YTS series, shown below.

Program 8.3

```
ods graphics on;

proc arima data=work.ar3 plots(unpack)=series(all);
    identify var=YTS nlags=12;
    estimate p=2 method=ml plot;
    estimate p=3 method=ml plot;
    estimate p=4 method=ml plot;
quit;

ods graphics off;
```

A couple of notes on the syntax might be helpful. The IDENTIFY statement defines the dependent variable for all subsequent ESTIMATE statements. The method option in the ESTIMATE statements indicates the method used to estimate parameters in each specification. ML stands for maximum likelihood. The plot option on each ESTIMATE statement generates the residual diagnostics for each specification. Residual diagnostics are discussed below.

The AR 2 Specification

The parameter estimates of the AR 2 specification are shown next.

Output 8.9

Maximum Likelihood Estimation					
Parameter	Estimate	Standard Error	t Value	Approx Pr > |t|	Lag
MU	100.48956	0.12484	804.93	<.0001	0
AR1,1	0.42013	0.06247	6.72	<.0001	1
AR1,2	-0.27293	0.06283	-4.34	<.0001	2

All parameter estimates are significantly different from zero. Recall that AR models do not have explanatory or structural variables, so coefficient interpretation, marginal effects, and so on are relatively unimportant. The interesting result here is that no irrelevant lags seem to be included in the specification.

The AR 2 model's fit diagnostics are shown below.

Output 8.10

Constant Estimate	85.75797
Variance Estimate	2.720848
Std Error Estimate	1.649499
AIC	924.2981
SBC	934.74
Number of Residuals	240

Akaike's Information Criterion (AIC) and the Schwarz Bayesian Criterion (SBC) are the two primary fit diagnostics used in this chapter.[8] (See the ARIMA procedure's details section in SAS/ETS documentation for details.) Here are some rules of thumb to keep in mind about these fit diagnostics:

- They are penalized fit statistics and are designed to prevent overfitting or having too many terms in the specification.
- They do not mean much on their own and are primarily useful for comparing two or more candidate specifications.
- Smaller is better—the perfect AIC or SBC is negative infinity.
- SBC has a harsher penalty than AIC.[9]

The residuals diagnostics are shown next. This is exactly the same White Noise Probability plot discussed in the identification section above, but in this case the AR 2 model's residuals are tested. Recall that the null hypothesis tested in each row is that no autocorrelation exists at any lag between, say, 1 and 6 (first line in the table). The p values in the table indicate that this null hypothesis is rejected at all tested lags. This is a problematic finding for the AR 2 model. A correctly specified model should have white noise residuals. That is, all of the systematic variation in the data is "captured" in the model specification, and all that is "left over" in the residuals is randomness. The results below indicate that the AR 2 model is *inadequately* specified. In this case a relevant term could be missing from the specification.

Output 8.11

To Lag	Chi-Square	DF	Pr > ChiSq	Autocorrelations					
6	31.18	4	<.0001	0.092	-0.095	0.267	0.188	0.025	0.048
12	40.18	10	<.0001	0.108	0.091	-0.038	-0.090	0.067	0.042

The PACF plot on the residuals of the AR 2 model confirms this. It appears that the significant spike at lag 3 is not spurious, and the model should be augmented with at least an additional term at lag 3.

Output 8.12

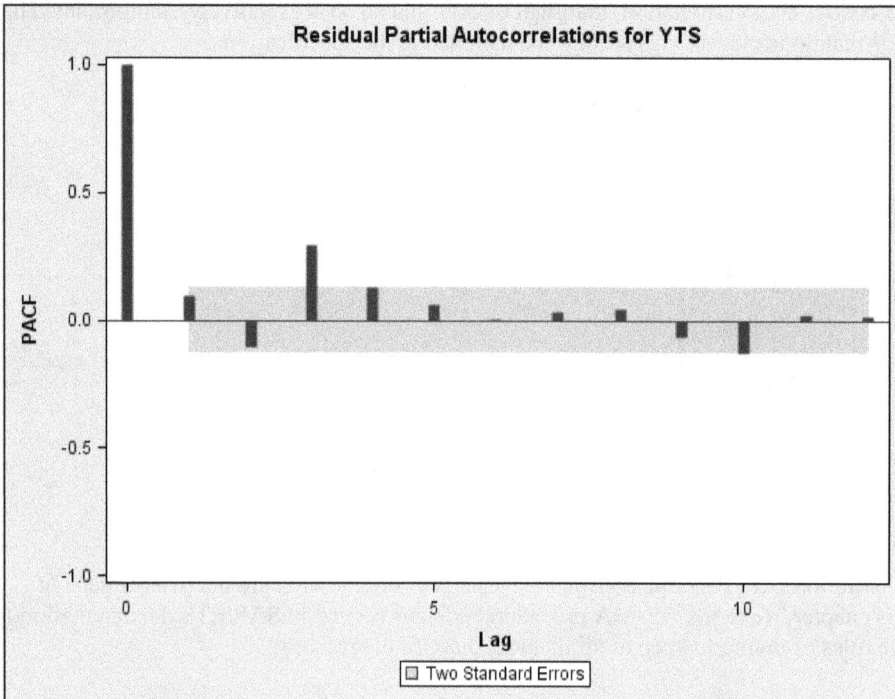

Residual Partial Autocorrelations for YTS

The AR 3 Specification

The parameter estimates show that all terms are significantly different from zero; it appears that no irrelevant terms are included.

Output 8.13

Maximum Likelihood Estimation					
Parameter	Estimate	Standard Error	t Value	Approx Pr > |t|	Lag
MU	100.48531	0.17619	570.32	<.0001	0
AR1,1	0.51210	0.06125	8.36	<.0001	1
AR1,2	-0.41353	0.06460	-6.40	<.0001	2
AR1,3	0.33478	0.06159	5.44	<.0001	3

The fit diagnostics show that the overall fit of the AR 3 model is better than the AR 2 model. This is indicated by the smaller AIC and SBC values shown below.

Output 8.14

Constant Estimate	56.93999
Variance Estimate	2.422129
Std Error Estimate	1.556319
AIC	897.9989
SBC	911.9215
Number of Residuals	240

The residuals diagnostics or white noise test verifies the adequacy of the AR 3 specification; residuals appear to be white noise. That is, all tested lags fail to reject ($Pr > 0.05$), which indicates that all systematic variation in the data seems to have been captured in the model.

Output 8.15

Autocorrelation Check of Residuals									
To Lag	**Chi-Square**	**DF**	**Pr > ChiSq**	**Autocorrelations**					
6	3.63	3	0.3045	0.002	-0.031	0.007	-0.031	0.064	0.092
12	6.77	9	0.6612	0.051	0.026	-0.026	-0.080	0.030	0.035

Note that the white noise test on the residuals is particularly useful for applied analysts. It helps determine when we should stop trying to add terms to the model. Based on the result above, any terms added to the AR 3 model would be merely fitting noise. By definition, this is overfitting.

The AR 3 model is the champion so far, and, based on the presented results, it can be used and easily defended in an analysis of the YTS series. However, because over-specified models occur frequently in analyses, the AR 4 model will be considered for completeness.

The AR 4 specification

The parameter estimates table reveals that the AR 4 term is irrelevant or that the correlation at lag 4 is not significantly different from zero.

Output 8.16

Maximum Likelihood Estimation					
Parameter	**Estimate**	**Standard Error**	**t Value**	**Approx Pr > |t|**	**Lag**
MU	100.48553	0.17582	571.53	<.0001	0
AR1,1	0.51361	0.06518	7.88	<.0001	1
AR1,2	-0.41541	0.07034	-5.91	<.0001	2
AR1,3	0.33706	0.07024	4.80	<.0001	3
AR1,4	-0.0044038	0.06568	-0.07	0.9465	4

The fit diagnostics show that when the lag 4 term enters the model, it does not improve the overall fit enough to overcome the penalty terms. The fit of the model diminishes, relative to the AR 3 specification, based on both AIC and SBC.

Output 8.17

Constant Estimate	57.19085
Variance Estimate	2.432388
Std Error Estimate	1.559611
AIC	899.9944
SBC	917.3976
Number of Residuals	240

As you might have suspected the residuals diagnostic of the AR 4 model show white noise.

Output 8.18

			Autocorrelation Check of Residuals						
To Lag	Chi-Square	DF	Pr > ChiSq			Autocorrelations			
6	3.59	2	0.1662	0.001	-0.030	0.007	-0.029	0.066	0.092
12	6.72	8	0.5674	0.051	0.026	-0.026	-0.079	0.030	0.035

Based on these findings, forecasts for the YTS series will be generated using the fitted AR 3 model. The code is modified to generate the forecast.

Program 8.4

```
ods graphics on;

proc arima data=work.ar3 plots(unpack)=series(all)
         plots=forecast(all);
   identify var=YTS nlags=12;
   *estimate p=2 method=ml plot;
   estimate p=3 method=ml plot;
   *estimate p=4 method=ml plot;
   forecast lead=52 printall out=yts_for;
quit;

ods graphics off;
```

Note that the estimate statements generating the AR 2 and AR 4 models have been commented out. In the FORECAST statement, `lead=52` sets the lead forecast horizon at 52 intervals into the future, and the option `out=yts_for` creates a data set in the work library named yts_for. This data set contains YTS forecasts, actual values, and confidence intervals.

The plot below shows the YTS forecasts based on the AR 3 model. The vertical dashed line indicates where the last observation ends and the future forecasts begin. You might note that the lead forecast converges pretty quickly to the mean or average for the series. This is a characteristic of models for low-order stationary processes. Because the series is stationary, the effect of any shocks or bangs in the data eventually disappears, and the series converges back to the mean.

Output 8.19

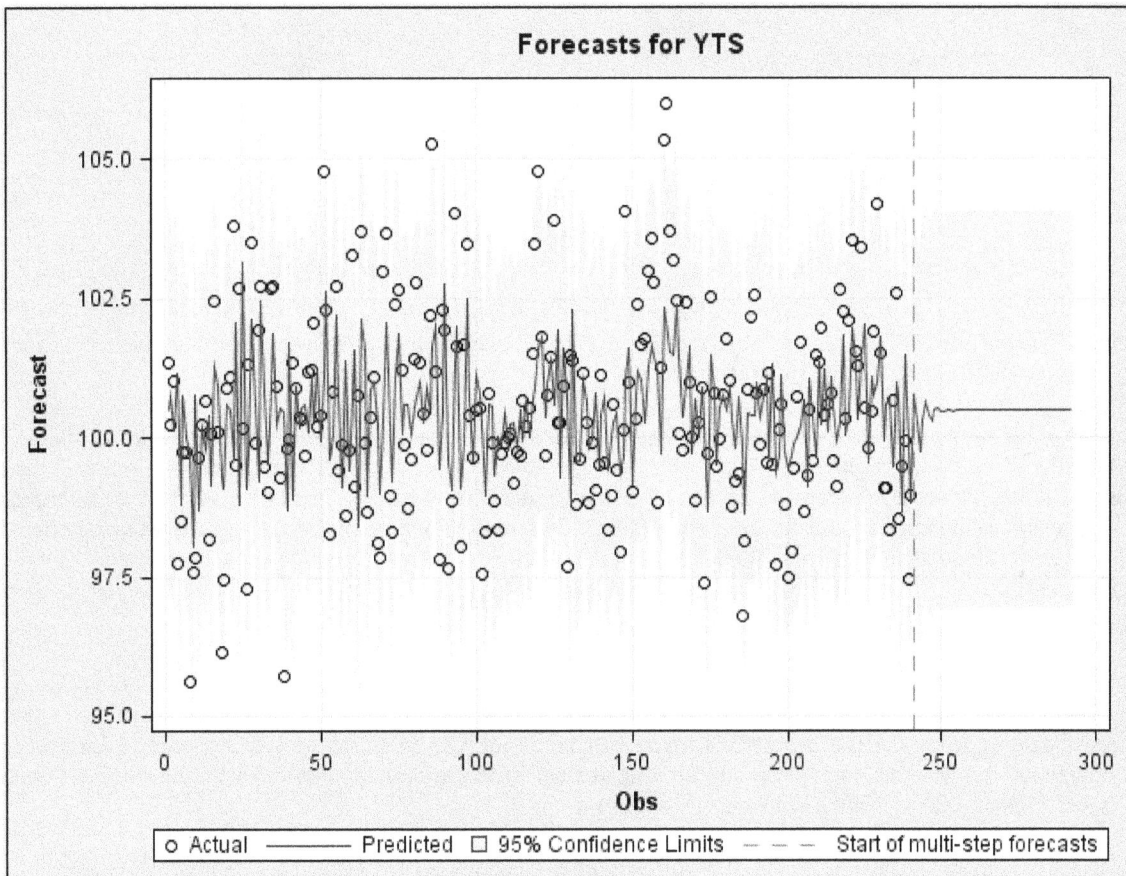

Forecasts for YTS

At this point, you might think that this seems like a lot of work to get to something that is not much better than the mean forecast. Going through this process is helpful, however, because of the following.

- AR models provide a useful introduction to general ARIMAX time series models. Modifications to the AR model will include Moving Average (MA) terms and dynamic regression inputs (X). These added components can now be discussed as straightforward extensions to the ideas already presented.

- The AR model provided an uncluttered basis for presenting the main components in an applied time series analysis. The steps and diagnostics presented comprise the fundamental approach and necessary tools. As we move to the more general ARIMAX model, there are more terms to consider and more candidate models to evaluate. However, the basic tools have been discussed and the modeling framework has been set.

8.1.2 Moving Average Models: Concepts and Application

Ideas

Let's begin our discussion of Moving Average variation with the now familiar AR 1 specification. Reconsider the AR 1 specification for red tennis shoes, RTS. If the equation holds for the current value of RTS, then it must also hold for the lag 1 value for RTS. If we plug this relationship into equation 5 in place of the lagged value of RTS we get equation 6. Doing this over and over results in equation 7.[10]

$$RTS_t = \phi_1 RTS_{t-1} + \varepsilon_t \qquad\qquad 5$$

$$\rightarrow \quad RTS_{t-1} = \phi_1 RTS_{t-2} + \varepsilon_{t-1}$$

$$RTS_t = \phi_1(\phi_1 RTS_{t-2} + \varepsilon_{t-1}) + \varepsilon_t \qquad\qquad 6$$

$$RTS_t = \phi_1^2 RTS_{t-2} + \phi_1\varepsilon_{t-1} + \varepsilon_t \quad \rightarrow \quad \phi_1^2(\phi_1 RTS_{t-3} + \varepsilon_{t-2}) + \phi_1\varepsilon_{t-1} + \varepsilon_t$$

$$RTS_t = \phi_1^3 RTS_{t-3} + \phi_1^2\varepsilon_{t-2} + \phi_1\varepsilon_{t-1} + \varepsilon_t \rightarrow \dots$$

$$RTS_t = \phi_1^n RTS_{t-n} + \phi_1^{n-1}\varepsilon_{t-n-1} + \dots + \phi_1^2\varepsilon_{t-2} + \phi_1\varepsilon_{t-1} + \varepsilon_t \qquad 7$$

In textbooks, equation 7 is often known as the impulse response representation of the process (Hamilton 1994). Recall the properties of the RTS series (also see this chapter's Appendix 1); the coefficient ϕ_1 has to be less than 1 (in absolute value). Otherwise, the effect of disturbances on the series would increase over time and the series would be nonstationary (it would blow up).

Now, think of the RTS series as never varying far from its mean except for a one time "impulse" or shock. Equation 7 shows the response or effect of the shock on the RTS series. The largest effect on RTS is in the time interval that the impulse occurs (in the equations above, the current value or time *t*). The effect then persists but diminishes as the impulse gets further into the past. The length of persistence, or the time it takes the effect to go to zero is regulated by the magnitudes of both the shock and the coefficient. Another implication of this is that the lagged RTS term effectively disappears; a number less than 1 raised to the nth power is very close to zero.

However, the most important thing about the recursion shown above is that it writes the AR process in a different form. The RTS series is represented as a weighted average of error components that are realizations of a white noise process. These error terms are uncorrelated (not AR) by definition (see Appendix 1 for details). Equation 7 is a Moving Average model. Moving average orders are usually denoted with the letter q. Equation 7 is a saturated, order n–1 moving average model, or q = n–1.

How this is useful? Consider equation 7 from a more pragmatic point of view. The impulse and response form of the AR 1 model shows that when the RTS series is shocked or banged, the effect of the shock is immediate and then tends to persist for a while. The effect decays away at a rate that is given by the AR 1 coefficient. It's not hard to imagine an impulse that would cause an effect like this on demand. Consider a media announcement such as, "Olympic Basketball team proclaims RTS their shoe of choice."

However, it is also not hard to imagine a demand effect in which the AR representation would prove cumbersome. Consider promotions or other impulses that cause demand to build for one or two periods, and then immediately drop back to the mean (for example, a buy-one-get-one-free promotion that runs for three weeks). Because MA models consist of non-correlated components, they can be used to model short-lived or more abrupt correlation patterns in the data.

Consider a simple moving average process–MA 2 represented as weekly sales of paisley tennis shoes (PTS, in equation 8 below).

$$PTS_t = \delta - \theta_2\varepsilon_{t-2} - \theta_1\varepsilon_{t-1} + \varepsilon_t \qquad\qquad 8$$

Although this model is closely related to the AR model, there are important differences.[11] In MA models, the persistence of the effect of any shock on the series is given by the order. Consider the same one-time impulse or shock that is shown in equation 7. For PTS, an MA 2, the shock to the series persists for two periods past the interval in which it occurs. In the time interval after that the series immediately drops back to its mean (holding everything else constant).

Also note that there is a separate coefficient for the two time periods in the correlation interval. This implies that a wide variety of effect shapes (besides jump and decay) can be accommodated. To summarize:

- AR models are primarily useful when the effect on the series is immediate and tends to persist for a while with decay. AR models can accommodate these effects using very few parameters. Equation 7 shows that a long persistence in effect can be accommodated using only one parameter.
- MA models are primarily useful when effects temporarily build in magnitude or when effects come to abrupt stops. MA models can accommodate a wider variety of effects, but more parameters are used. A separate parameter is estimated for each time interval of effect persistence in MA models.

Methods

Model Identification

In this section assume that only MA variation exists. This enables us to focus on the characteristics associated with MA processes in the diagnostic plots. As stated above, the basic framework for applied time series analysis has been set. The MA identification process is very similar to AR identification process.

To review, the following diagnostic plots are presented in the order in which they are generally used and consist of the following:

- the series plot
- the series white noise probability plot
- the autocorrelation function plot
- the partial autocorrelation function plot

The key difference between AR and MA identification is in the use of the correlation function plots for identifying appropriate models; the interpretation of the patterns in the ACF and PACF is exactly the opposite as discussed for AR processes.[12] All of the other interpretations and discussions are basically the same.

Consider the diagnostic plots generated for the PTS series (MA 2) given in equation 8.

The PROC ARIMA code to generate the plots is given below.

Program 8.5 ARIMA procedure code

```
ods graphics on;

proc arima data=work.ma2 plots(unpack)=series(all);
   identify var=PTS nlags=12;
quit;

ods graphics off;
```

You might note that the correlation pattern in the PTS **series plot** appears to be spikier than seen in the RTS series plot, above. There are smaller clusters of observations above or below the mean.

Output 8.20

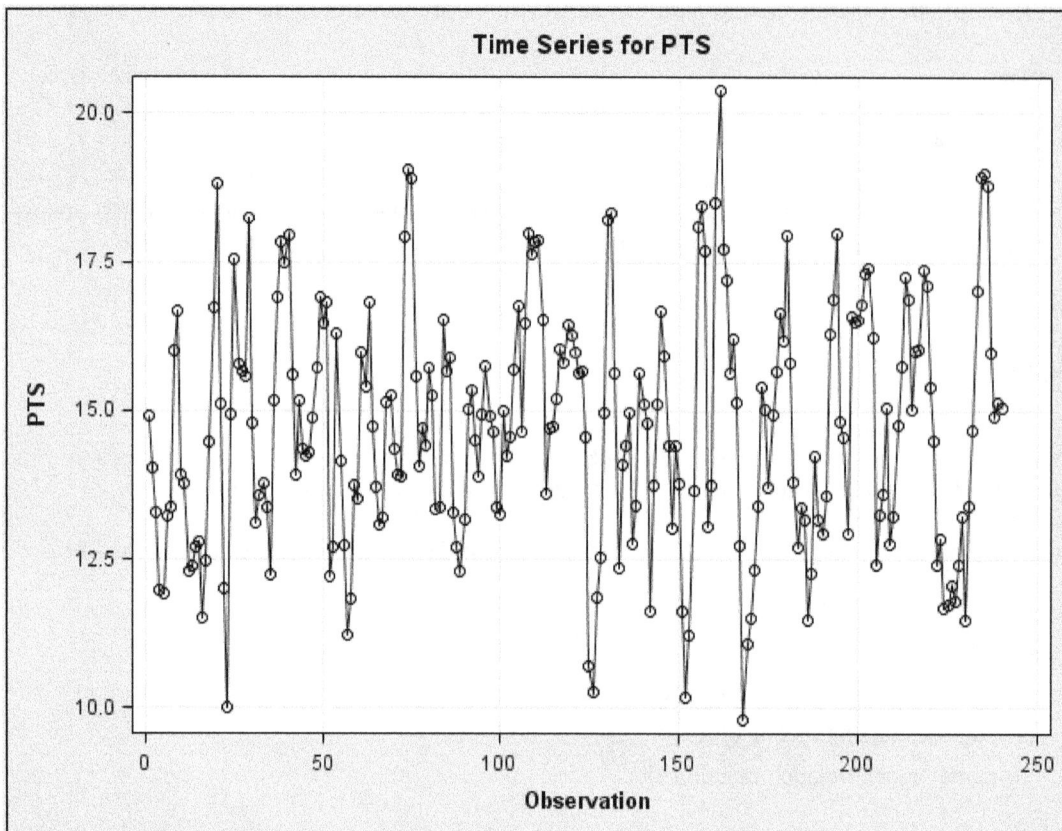

Time Series for PTS

The white noise probability plot (Output 8.20) indicates that the PTS series contains some type of systematic (MA) variation; the null hypothesis of white noise is rejected at all tested lags.

Output 8.21

Autocorrelation Check for White Noise									
To Lag	Chi-Square	DF	Pr > ChiSq	Autocorrelations					
6	129.39	6	<.0001	0.664	0.252	-0.016	-0.056	-0.065	-0.139
12	159.19	12	<.0001	-0.202	-0.220	-0.152	-0.047	-0.027	-0.062

The roles of the ACF and PACF plots for MA identification are the reverse of their use in AR identification; the important plot for MA identification is the ACF plot.

Output 8.22

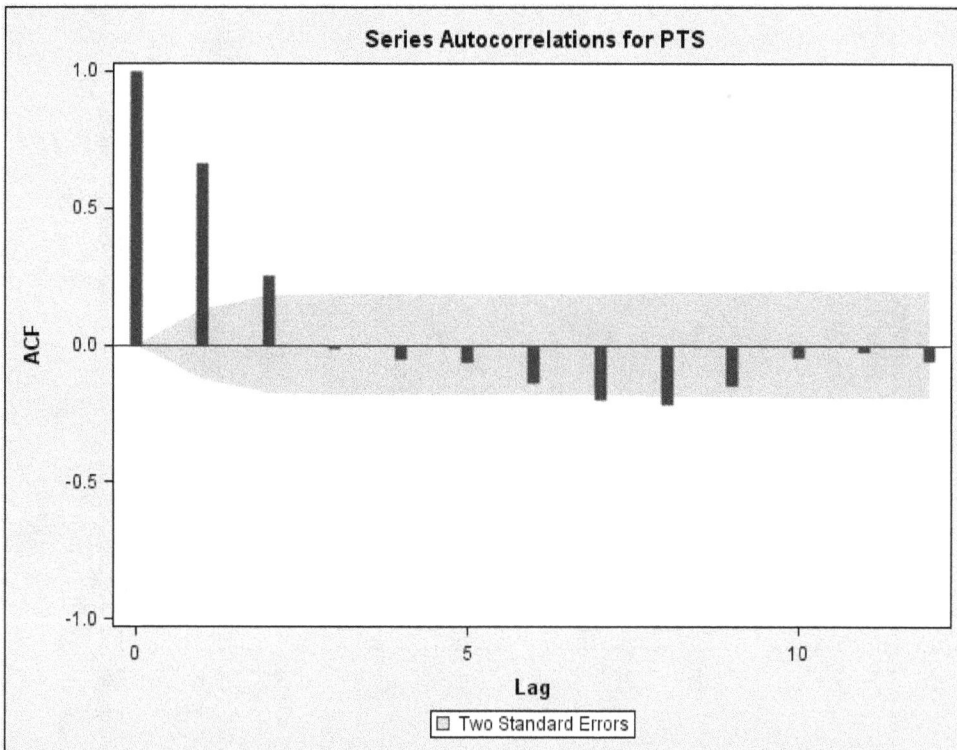

The ACF plot indicates that a reasonable model is MA 2. Note that the above discussion on problems that can occur when conducting lots of hypothesis tests is still relevant here. Some marginally significant higher order spikes exist, but they might be spurious.

For MA processes, the PACF plot suffers from the proximity effect. The classic proximity effect pattern for MA 2 would be significant spikes at lags 1 and 2 in the PACF followed by significant, but diminishing, spikes at higher lags. While this pattern is not evident in the PACF below, spikes do follow a pattern of diminishing significance.

Output 8.23

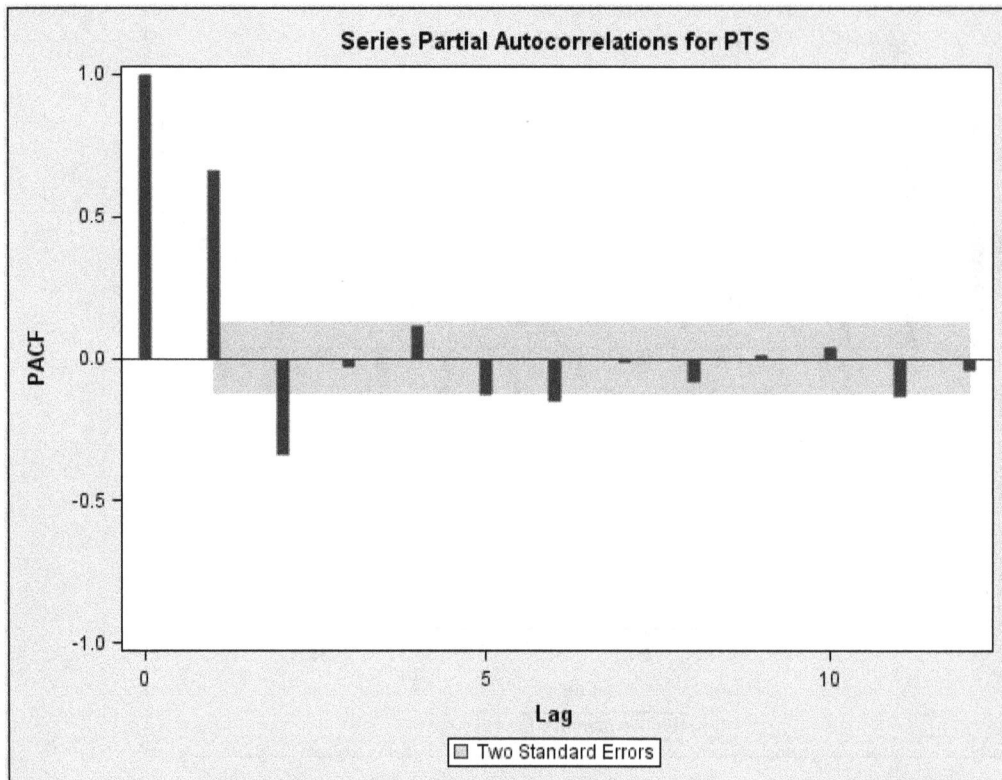

Model Estimation, Selection, and Forecasting

The MA 2 specification is indicated by the diagnostics as the most appropriate for the PTS series. However, given the discussion in the AR identification section above, the prudent and practical way to proceed is to estimate some subset of suggested models. Again, once the models are fit, the best one is selected based on a combination of three criteria:

- individual parameter significance within each model
- fit to the data
- analysis of the models' residuals

An MA 1, MA 2, and an MA 3 model is initially fit to the data. Higher lags are considered if they are warranted, as indicated by the diagnostics. Note that the letter q is used to denote the order of the MA specification in the code below.

Program 8.6

```
ods graphics on;

proc arima data=work.ma2 plots(unpack)=series(all);
   identify var=PTS nlags=12;
   estimate q=1 method=ml plot;
   estimate q=2 method=ml plot;
   estimate q=3 method=ml plot;
quit;

ods graphics off;
```

The MA 1 model

The parameter estimates table indicates that no irrelevant terms are included in the MA 1 model.

Output 8.24

Maximum Likelihood Estimation					
Parameter	Estimate	Standard Error	t Value	Approx Pr > |t|	Lag
MU	14.74231	0.16316	90.35	<.0001	0
MA1,1	-0.62246	0.05094	-12.22	<.0001	1

The reported AIC and SBC provide baseline fit statistics.

Output 8.25

Constant Estimate	14.74231
Variance Estimate	2.440923
Std Error Estimate	1.562345
AIC	897.7426
SBC	904.7039
Number of Residuals	240

The white noise check of residuals indicates that the MA 1 model is inadequately specified; the null hypothesis of white noise is rejected at all tested lags.

Output 8.26

Autocorrelation Check of Residuals									
To Lag	Chi-Square	DF	Pr > ChiSq	Autocorrelations					
6	38.93	5	<.0001	0.239	0.265	-0.149	0.022	-0.052	-0.081
12	54.15	11	<.0001	-0.144	-0.144	-0.130	0.021	-0.045	0.005

The ACF plot on the MA 1 model's residuals indicates that at least an MA 2 term should be added to the model.

Output 8.27

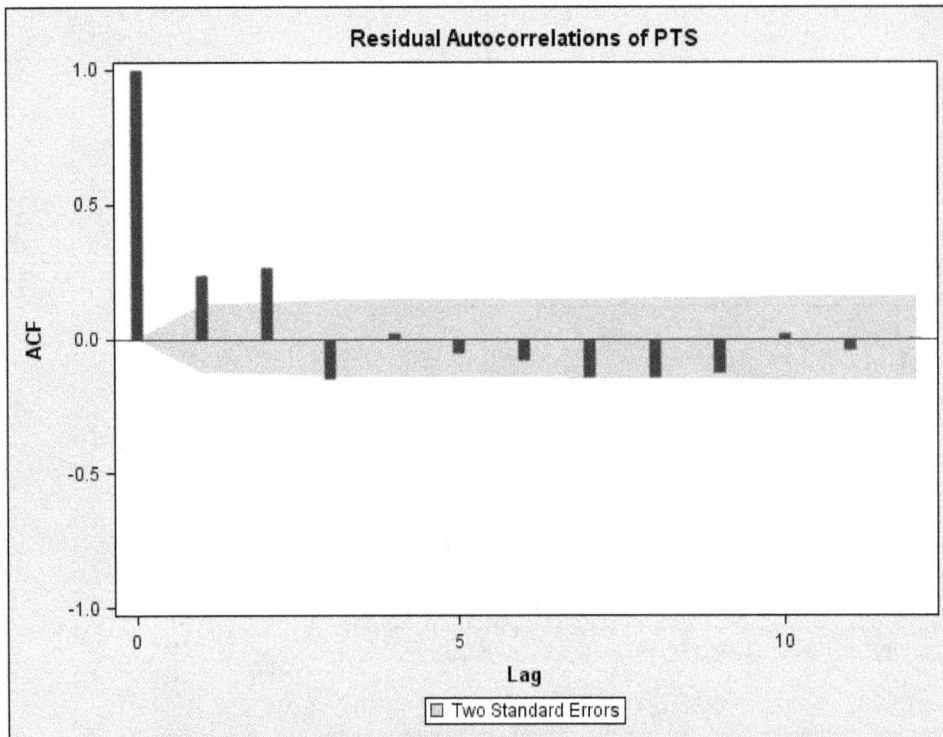

Residual Autocorrelations of PTS

The results for the **MA 2** specification are discussed next. The MA 2 model's parameter estimates table indicates that all included lags are significant.

Output 8.28

Maximum Likelihood Estimation					
Parameter	Estimate	Standard Error	t Value	Approx Pr > |t|	Lag
MU	14.74138	0.21319	69.15	<.0001	0
MA1,1	-0.84803	0.05616	-15.10	<.0001	1
MA1,2	-0.50366	0.05624	-8.95	<.0001	2

The fit statistics for the MA 2 model are better than the MA 1; **AIC** and **SBC** measures are lower.

Output 8.29

Constant Estimate	14.74138
Variance Estimate	1.994157
Std Error Estimate	1.412146
AIC	850.6927
SBC	861.1346
Number of Residuals	240

The **white noise check of residuals** indicates that the MA 2 model is adequately specified; the null hypothesis of white noise is not rejected at all tested lags.

Output 8.30

Autocorrelation Check of Residuals									
To Lag	Chi-Square	DF	Pr > ChiSq	Autocorrelations					
6	4.51	4	0.3416	0.041	-0.006	0.014	-0.080	0.054	-0.084
12	12.18	10	0.2732	-0.076	-0.109	-0.101	0.051	-0.011	-0.003

No further model refinement is tried based on the results above; the MA 2 model is selected as the forecast specification.

8.1.3 Auto Regressive Moving Average (ARMA) Models

Introduction

Since we have covered the foundational concepts, the focus of this section will be on combining the AR and MA processes we have already described, and presenting them in a way that links the ideas with dynamic regression models that are presented in the next section. Most of the necessary tools for building rational polynomial transfer function or dynamic regression models are already in their toolbox. The purpose of the conceptual portion of this subsection is to demonstrate this fact.

You might be uneasy about model identification in a more general, ARMA framework. Because the roles of the ACF and PACF plots switch for AR and MA processes, identification becomes more difficult when both AR and MA types of variation can exist in a series. The usual result of this ambiguity is that there are more candidate models to consider and evaluate, and analyst judgment plays a larger role in selecting the forecast specification.

Ideas

To begin, consider the ARMA 1, 2 process in equation 9 for gray tennis shoes (GTS):

$$GTS_t = \phi_0 + \phi_1 GTS_{t-1} + \varepsilon_t - \theta_1 \varepsilon_{t-1} - \theta_2 \varepsilon_{t-2} \qquad\qquad 9$$

Weekly sales of GTS contain two types of systematic variation. Some factors or impulses cause effects that look like immediate jumps from the mean that persist for a while with decay (AR 1). Others cause abrupt and short-lived (two period or MA 2) effects after the shock. Thus, the ARMA 1, 2 specification. Note that ARMA models are usually denoted with the AR order listed first: ARMA 1, 2 or p=1, q=2. A very useful tool in dealing with time series models is the backshift operator. Briefly, the backshift operator works as follows. (See Pyndick and Rubinfeld for more information.) The backshift operator, B, applied to the current value of a variable yields the lag one value of the variable. Polynomials in the backshift operator can be used to apply the backshift operator multiple times. For example, the backshift operator squared is equivalent to applying the backshift operator twice, and results in shifting the current value back two periods as shown below.

$$By_t = y_{t-1}$$
$$B^2 y_t = BBy_t = y_{t-2}$$
$$B^n y_t = BB...By_t = y_{t-n}$$

Below, the backshift operator is applied enough times to the terms in equation 9 to convert all lagged values of the dependent and error terms to their current values. Autoregressive terms are moved to the left side of the equation. Finally, current values of the dependent variable and error term are factored out as shown in equation 10.

$$GTS_t = \phi_1 GTS_{t-1} + \varepsilon_t - \theta_1 \varepsilon_{t-1} - \theta_2 \varepsilon_{t-2} \qquad \rightarrow$$
$$GTS_t = \phi_1 BGTS_t + \varepsilon_t - \theta_1 B\varepsilon_t - \theta_2 B^2 \varepsilon_t \qquad \rightarrow$$
$$GTS_t(1 - \phi_1 B) = \varepsilon_t(1 - \theta_1 B - \theta_2 B^2) \qquad\qquad 10$$

This manipulation represents the ARMA model as a set of polynomials in the backshift operator. What we are working toward is the rational polynomial transfer function. A little more manipulation gives us the rational part, equation 11.

$$GTS_t = [(1 - \theta_1 B - \theta_2 B^2)/(1 - \phi_1)B]\varepsilon_t \qquad\qquad 11$$

Equation 11 is the rational polynomial transfer function model. There are a couple of handy things to keep in mind to prepare for the next section. This ratio of polynomials converts (or transfers or "filters") impulses from the error variable into effects in the dependent variable. The MA terms are in the numerator part of the transfer function and are known as numerator orders. Numerator orders are responsible for converting impulses in the input into short-lived and abrupt persistence effects in the dependent variable, as described in the MA section above. The AR terms are in the denominator part of the transfer function and are known as denominator orders. Denominator orders are responsible for converting impulses from the input into relatively longer-lived, decaying effects in the dependent variable, as we discussed above.

Note that the words error variable and input are used interchangeably to describe the impulse. This is by design; ARMA models are transfer functions with a white noise input or impulse generator. Dynamic regression models are transfer functions with other (for example, interval or binary valued) types of inputs or impulse generators. You already know everything necessary to productively use and understand ARIMAX models.[13]

Methods

ARMA model identification

The zebra-striped tennis shoe sales series (ZTS) are identified in this section. This section is more ambiguous than previous identification sections because the process being identified is more general. There is the possibility of both AR and MA types of systematic variation in the series.

The series plot for ZTS is shown below. Patterns in the data could include short- and long-lived correlation, but, as processes generating data become more general, "eyeballing" the data becomes relatively less useful as an identification tool. Note that looking at a plot of the data to discern outliers, seasonal patterns, and so on is useful regardless of the correlation structure in the data.

Output 8.31

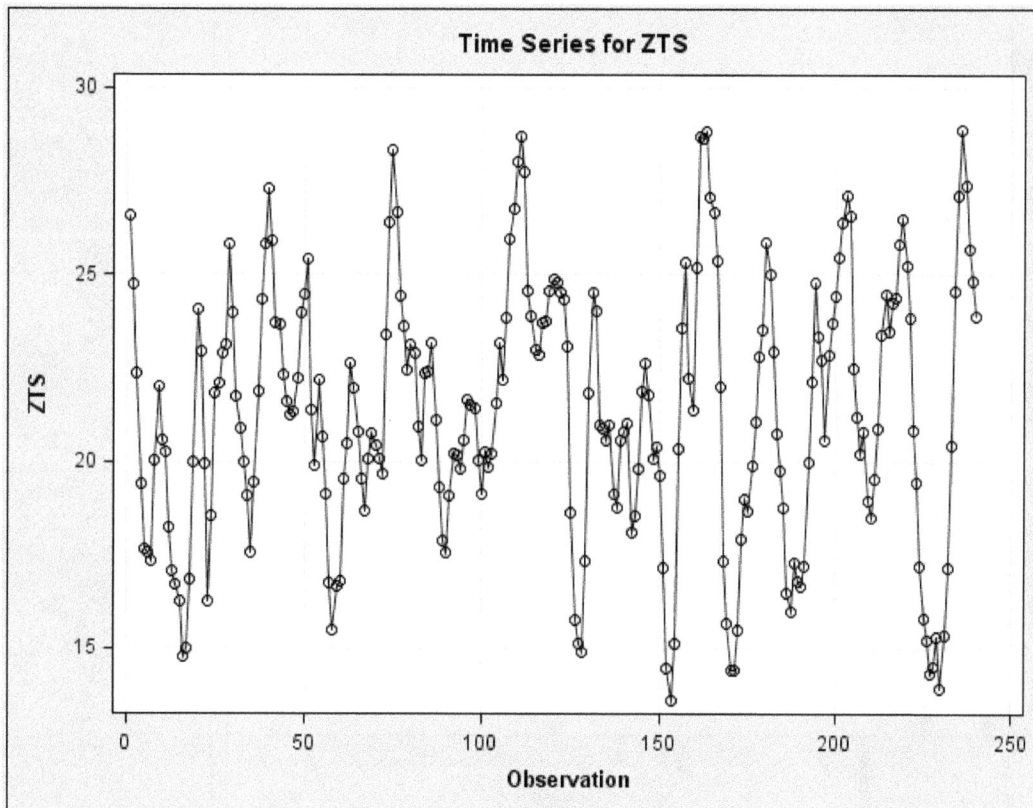

The **white noise probability plot** indicates that the ZTS series is not white noise. The null hypothesis of white noise is rejected at all tested lags (up to lag 12).

Output 8.32

	Autocorrelation Check for White Noise								
To Lag	Chi-Square	DF	Pr > ChiSq	Autocorrelations					
6	316.11	6	<.0001	0.866	0.617	0.357	0.153	-0.001	-0.128
12	389.21	12	<.0001	-0.213	-0.253	-0.249	-0.216	-0.196	-0.185

The ACF and PACF plots are presented next.

Output 8.33

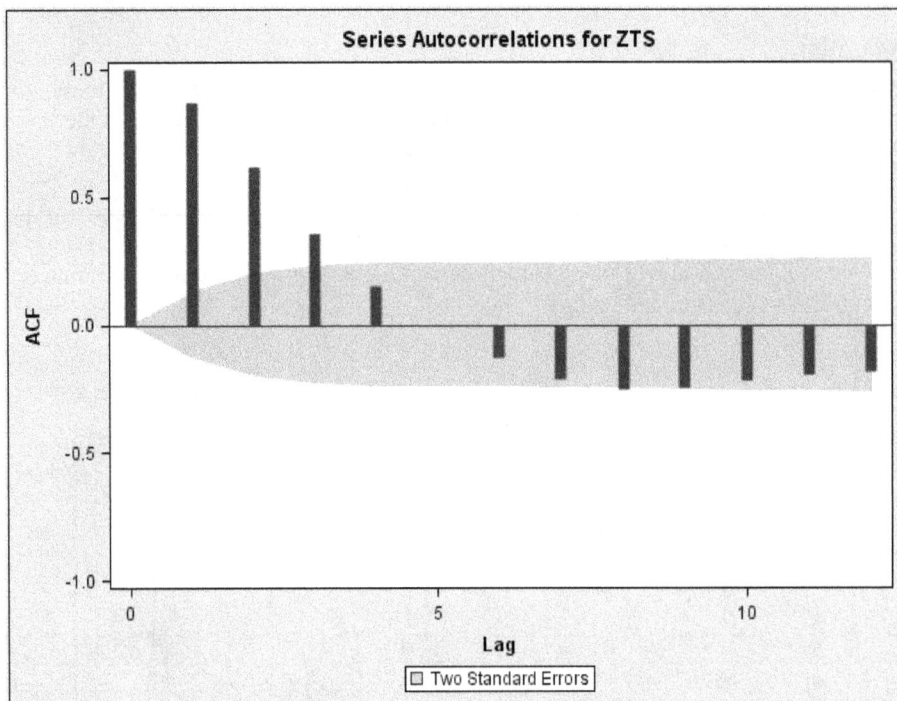

Output 8.34

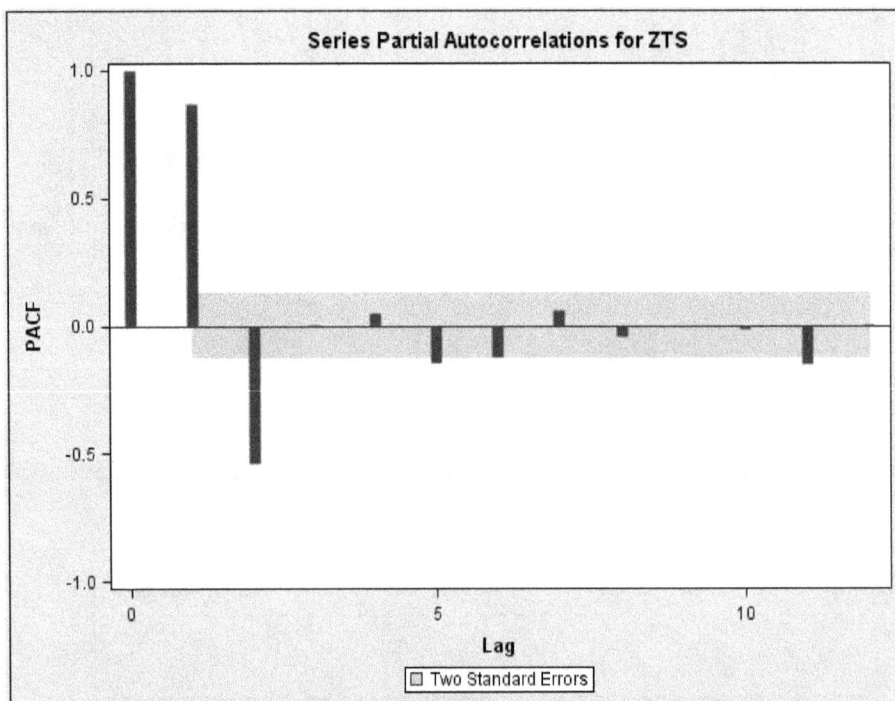

Considering the discussions above about patterns associated with MA and AR processes; there are three distinct subgroups of model specifications that could be suitable for the ZTS series.

1. The ZTS series contains only AR correlation: in this case the significant spikes at lags 1 and 2 in the PACF are probably real, and spikes larger than 2 in the ACF are spurious. The indicated model is AR 2, MA 0, or ARMA 2, 0.

2. The ZTS series contains only MA correlation: the significant spikes at lags 1, 2 and 3 in the ACF are probably real. The indicated model is ARMA 0, 3.

3. The ZTS series contains both AR and MA correlation: there is a mixture of real, significant spikes and spurious, proximity effect spikes in both the ACF and PACF. The indicated range for reasonable AR orders is: 0 to 2. The indicated range for reasonable MA orders is 0 to 3.

Of course it is not known which of the three scenarios is valid. The scenarios were considered to illustrate two points. First, although we have not arrived at the forecast specification, there is now only a reasonably sized subset of models to evaluate. The process of identification starts with a really big number of models to consider, and tries to pick a much smaller subset of reasonable candidates based on sound methodology. Second, because the process generates reasonable candidates, they are also defensible candidates. Unforeseen future events happen and current forecasts go awry. A sound basis for model selection is the best defense, and the creation of the subset of models described above is the first step in the process.

ARMA model estimation, selection, and forecasting

Based on the findings in the identification step, the PROC ARIMA code is augmented to fit the described candidate models.

Program 8.7

```
ods graphics on;

proc arima data=work.arma12 plots(unpack)=series(all);
   identify var=ZTS nlags=12;
     estimate p=0 q=1 method=ml plot;
     estimate p=0 q=2 method=ml plot;
     estimate p=0 q=3 method=ml plot;
     estimate p=1 q=0 method=ml plot;
     estimate p=1 q=1 method=ml plot;
     estimate p=1 q=2 method=ml plot;
     estimate p=1 q=3 method=ml plot;
     estimate p=2 q=0 method=ml plot;
     estimate p=2 q=1 method=ml plot;
     estimate p=2 q=2 method=ml plot;
     estimate p=2 q=3 method=ml plot;

   quit;

ods graphics off;
```

Note that the p=0, q=0 model was omitted based on the results of the white noise test above.

Note: The program is intended to illustrate the process, but it can be tedious to code something like the above each time it becomes necessary to consider candidate models. SAS macros are pieces of reusable SAS code that efficiently handle redundant data processing and analytic chores. "Forecasting Using SAS Software; a Programming Approach" is a SAS Education course containing several forecasting-related custom macros, including the %AUTOARIMA macro, which embeds code like the above. When doing large-scale forecasting, there can be thousands of individual series to identify. In this context, a tool like SAS Forecast Studio provides an automated and robust way to select good ARMA orders for each series, and is the best and most efficient way to create good models in this context.

The estimate results are scanned to find the best models based on AIC. The four best models are listed in the table below.

Output 8.35

Model	AIC	SBC
ARMA 1 2	855.8	869.7
ARMA 1 3	854.5	871.8
ARMA 2 0	854.5	865.2
ARMA 2 2	854.8	872.2

The four best models are approximately equivalent in terms of the reported fit diagnostics. However, further elimination of models might be possible based on other results. The white noise probability tables of each model's residuals were scrutinized. The ARMA 1 2, ARMA 1, 3 and ARMA 2, 2 models' residuals are clean. However, the null hypothesis of white noise is marginally rejected for the ARMA 2, 0 model at lags 1–6, as shown below.

Output 8.36

				Autocorrelation Check of Residuals					
To Lag	Chi-Square	DF	Pr > ChiSq			Autocorrelations			
6	9.30	4	0.0540	0.006	0.007	-0.121	0.023	0.148	-0.026
12	16.34	10	0.0903	-0.031	-0.108	-0.102	0.067	-0.022	-0.001

Based on this result, the ARMA 2, 0 model is eliminated from consideration. This leaves the following subset of candidate models.

Output 8.37

Model	AIC	SBC
ARMA 1 2	855.8	869.7
ARMA 1 3	854.5	871.8
ARMA 2 2	854.8	872.2

The remaining candidate specifications' parameter estimates will be evaluated.

The ARMA 1, 2 model's parameter estimates are shown below. No irrelevant lags seem to be included in the specification.

Output 8.38

	Maximum Likelihood Estimation				
Parameter	Estimate	Standard Error	t Value	Approx Pr > |t|	Lag
MU	21.39252	0.59168	36.16	<.0001	0
MA1,1	-0.67568	0.07083	-9.54	<.0001	1
MA1,2	-0.41449	0.06832	-6.07	<.0001	2
AR1,1	0.68135	0.05844	11.66	<.0001	1

The ARMA 1, 3 model's parameter estimates table indicates that the MA 3 term might be irrelevant.

Output 8.39

	Maximum Likelihood Estimation				
Parameter	Estimate	Standard Error	t Value	Approx Pr > |t|	Lag
MU	21.37591	0.54143	39.48	<.0001	0
MA1,1	-0.82456	0.10633	-7.75	<.0001	1
MA1,2	-0.57988	0.12105	-4.79	<.0001	2
MA1,3	-0.17310	0.09155	-1.89	0.0587	3
AR1,1	0.57176	0.09393	6.09	<.0001	1

The ARMA 2, 2 model's parameter estimates table indicates that the AR lag 2 term might be irrelevant.

Output 8.40

	Maximum Likelihood Estimation				
Parameter	Estimate	Standard Error	t Value	Approx Pr > |t|	Lag
MU	21.37054	0.52468	40.73	<.0001	0
MA1,1	-0.42811	0.16502	-2.59	0.0095	1
MA1,2	-0.30502	0.10017	-3.05	0.0023	2
AR1,1	0.96151	0.16870	5.70	<.0001	1
AR1,2	-0.25914	0.14477	-1.79	0.0734	2

Based on these results, the ARMA 1, 2 model is selected as the forecast model.

To summarize, there is no one diagnostic that can be relied on to select the best model. Model selection depends on three related criteria.[14]

- fit to the data

- adequate specification as indicated by testing the models' residuals

- absence of irrelevant terms or parsimony

Appendix 1: Useful Technical Details

This appendix provides some technical details about the ideas presented in this chapter. It is intended to provide you with further information and is not rigorous. More detail can be found in three excellent references that you should have on your bookshelf Hamilton (1994), Box and Jenkins (1972), and Pindyck and Rubinfeld (1991).

1) How do denominator order polynomials work? You might recall the simple relationship shown in equation 1 below. This relationship maps directly to the way a denominator polynomial order works. Consider the resemblance between equation 1 and the denominator order 1 polynomial shown in equation 2.

Assume that $|\omega| < 1$

$$1 + \omega + \omega^2 + \omega^3 + \omega^4 ...$$

$$\rightarrow$$

$$\sum_{i=0}^{\infty} \omega^i = 1/(1-\omega) \qquad\qquad 1$$

Appy this relaionship to the models.

$$y_t = \gamma/(1 - \delta B) \cdot x_t \qquad\qquad 2$$

$$\rightarrow$$

$$= \gamma \cdot (1 + \delta B + \delta^2 B^2 + \delta^3 B^3 + ...) x_t$$
$$= \gamma \cdot x_t + \gamma\delta \cdot x_{t-1} + \gamma\delta^2 \cdot x_{t-2} + \gamma\delta^3 \cdot x_{t-3} ...$$

Where γ is the scale effect, and B is the backshift operator. Note, above $\Rightarrow |\delta| < 1$.

2) ARMA models as ratios of polynomials in the backshift operator.

Consider the general ARMA (p, q) specification, below.

$$y_t = \phi_1 y_{t-1} + \phi_2 y_{t-2} + \phi_3 y_{t-3} + ... \phi_p y_{t-p}$$
$$+ \xi_t - \theta_1 \xi_{t-1} - \theta_2 \xi_{t-2} - ... - \theta_q \xi_{t-q}$$

Rearrange.

$$y_t - \phi_1 y_{t-1} - \phi_2 y_{t-2} - ... \phi_p y_{t-p} = \xi_t - \theta_1 \xi_{t-1} - \theta_2 \xi_{t-2} - ... \theta_q \xi_{t-q}$$

Apply the backshift operator and rearranging some more.

$$y_t \cdot (1 - \phi_1 B - \phi_2 B^2 - ... \phi_p B^p) = \xi_t \cdot (1 - \theta_1 B - \theta_2 B^2 - ... \theta_q B^q)$$

To save typing, use the following to denote general polynomials in the backshift operator:

$$(\phi B), (\theta B) .$$
$$\rightarrow y_t(\phi B) = (\theta B)\xi_t \quad \rightarrow y_t = (\theta B)/(\phi B)\xi_t$$

The ratio of polynomials creates a "filter" to translate variation in the white noise error term into variation in the dependent variable Y.

3) Why do AR effects persist a long time and MA effects do not?

Note that MA terms are not autocorrelated by definition.

For an MA 1:

$$y_t = \mu - \theta\varepsilon_{t-1} + \varepsilon_t$$

Covariance at lag 0 (variance of y):

$$\gamma_0 = (y_t - \mu)^2 = (\varepsilon_t - \theta\varepsilon_{t-1})^2$$
$$= \varepsilon_t^2 + \theta^2\varepsilon_{t-1}^2 - 2\theta\varepsilon_t\varepsilon_{t-1}$$

Cross terms drop because error terms are independent:

$$E[\varepsilon_t\varepsilon_{t-k}] = 0 \qquad \forall k \neq 0$$
$$\gamma_0 = \sigma^2 + \theta^2\sigma^2$$
$$= \sigma^2(1 + \theta^2)$$

Covariance at lag 1:

$$\gamma_1 = (y_t - \mu)(y_{t-1} - \mu) = (\varepsilon_t - \theta\varepsilon_{t-1})(\varepsilon_{t-1} - \theta\varepsilon_{t-2})$$
$$= \varepsilon_t\varepsilon_{t-1} - \theta\varepsilon_{t-1}^2 - \varepsilon_t\theta\varepsilon_{t-2} + \theta^2\varepsilon_{t-1}\varepsilon_{t-2}$$

Cross terms drop:

$$\gamma_1 = -\theta\sigma^2$$

Covariance at lag 2 (and all higher lags):

$$\gamma_2 = (y_t - \mu)(y_{t-2} - \mu) = (\varepsilon_t - \theta\varepsilon_{t-1})(\varepsilon_{t-2} - \theta\varepsilon_{t-3})$$
$$= \varepsilon_t\varepsilon_{t-2} - \theta\varepsilon_{t-1}\varepsilon_{t-2} - \varepsilon_t\theta\varepsilon_{t-3} + \theta^2\varepsilon_{t-1}\varepsilon_{t-3}$$

Cross terms drop:

$$\gamma_2 = 0$$

We see that for MA 1 innovations persist for one interval. This correspondence between order and persistence holds for all MA specifications; the length of persistence of innovations is given by the MA order.

Now, contrast this with an autoregressive model of order 1.

The covariance at lag 0 of the AR 1 process is shown below. Note that in the following it is assumed the mean is zero. This minimizes clutter, and, in practice, represents a re-scaling of the series.

$$\gamma_0 = E(y_t)^2 = E(\phi y_{t-1} + \varepsilon_t)^2$$
$$\rightarrow E(\phi^2 y_{t-1}^2 + \varepsilon_t^2 + 2\phi y_{t-1}\varepsilon_t) = \phi^2 \gamma_0 + \sigma^2$$
$$\gamma_0 = \sigma^2 / (1 - \phi^2)$$

The covariance at lags 1 and 2 of the AR 1 process are derived next.

$$\gamma_1 = E(y_t y_{t-1}) = E(y_{t-1}(\phi y_{t-1} + \varepsilon_t))$$
$$\rightarrow E(\phi y_{t-1}^2 + y_{t-1}\varepsilon_t) = \phi \gamma_0$$
$$\gamma_1 = \phi \sigma^2 / (1 - \phi^2)$$

$$\gamma_2 = E(y_t y_{t-2}) = E((\phi y_{t-1} + \varepsilon_t)y_{t-2})$$
$$\rightarrow E(\phi y_{t-1} y_{t-2} + \varepsilon_t y_{t-2}) = \phi \gamma_1$$
$$\gamma_2 = \phi^2 \sigma^2 / (1 - \phi^2)$$

$$\Rightarrow \gamma_n = \phi^n \sigma^2 / (1 - \phi^2)$$

The series is stationary, so the AR 1 parameter is less than 1 in absolute value. We could continue to derive the auto-covariance function for longer lags, but the above shows the general idea. Innovations in AR 1 series persist, but diminish over time. The rate of decay is regulated by the AR 1 coefficient.

Appendix 2: The "I" in ARIMA

This section broadens the modeling discussion to accommodate nonstationary series in the rational polynomial, transfer function modeling framework. The discussion is limited to the classic, or ARIMA, time series model. Recall that the classic time series model does not include structural variables.

Ideas

Nonstationary series

In section 8.1, when we presented most of the necessary information about nonstationarity, we noted that the most obvious characteristic of a stationary series is a well-defined mean. It is also true that the most obvious characteristic of a nonstationary series is that its mean is not well defined, or that it changes over time.

Recall these equations:

$$(RTS_t - \mu) = \phi_1(RTS_{t-1} - \mu) + \varepsilon_t \qquad\qquad 1$$
$$\rightarrow RTS_t = \phi_0 + \phi_1 RTS_{t-1} + \varepsilon_t \qquad\qquad 2$$
$$\phi_0 = \mu(1 - \phi_1) \rightarrow \mu = \phi_0 / (1 - \phi_1) \qquad 3$$

The mean of the red tennis shoes series (defined on the right side of equation 3) is a function of the intercept and the AR 1 parameter. This mean is undefined when the AR 1 parameter is equal to 1. This condition is also known as a "unit root." The stationarity diagnostic discussed below tests whether the AR 1 parameter is equal to 1 (null hypothesis → nonstationary) or less than one (alternative hypothesis → stationary).

Note that a preliminary and straightforward diagnostic for stationarity is the magnitude of the spike at lag 1 in the PACF. Recall that this spike corresponds in magnitude to the AR 1 parameter. If this spike is greater than .5, then the steps outlined below should be followed to determine the stationarity of the series before proceeding with modeling.

A good question at this point is: why does an AR 1 parameter of unity imply that the mean of the series changes over time? To answer this, rewrite equation 2 under the assumption of nonstationarity. For now, assume that the intercept is equal to 0.

$$RTS_t = RTS_{t-1} + \varepsilon_t \qquad\qquad 4$$

Equation 4 represents a nonstationary series. It is also known as a *random walk*. The error term still follows the classic assumptions: independence, mean 0, and so on. Consider what happens as this process generates data: the next observation is equal to the current observation plus some random shock. What happens when a big negative or positive shock comes along? The series is pushed to a new level and stays there until another shock comes along. Because the parameter on the AR 1 term is unity, there is no "reversion to the mean" in this series. Another way of thinking about this is that all changes or shocks to the series are permanent.

Equation 4 is deceptively simple, and it has a wide variety of modeling applications. Another representative of a random walk process is shown in equation 5.

$$RTS_t = \phi_0 + RTS_{t-1} + \varepsilon_t \qquad\qquad 5$$

In this process, the next observation is equal to the current observation, a random shock plus a constant term. This process is the *random walk with drift*, and the intercept or constant is the "drift" term. If the constant is positive, the series tends to drift up over time at a rate regulated by the drift term. A negative constant causes downward drift.

Differencing

The processes depicted in equations 4 and 5 have particularly nasty statistical properties. Their means, as we have seen, are undefined, and their variances are infinite (Hamilton 1994). Modeling nonstationary processes poses a problem. This is particularly true if modelers are interested in deriving diagnostics that depend on properties of the series mean and variance.

To model nonstationary series, we first convert them into stationary series. This is usually done by differencing (BoxJenkins 1972). The differenced, now stationary, series can be modeled using the techniques described in previous sections. To see how this works, consider equation 6. Applying differencing yields equation 7.

$$RTS_t = RTS_{t-1} + \varepsilon_t \qquad\qquad 6$$
$$(RTS_t - RTS_{t-1}) = w_t = \varepsilon_t \qquad\qquad 7$$

Most of the textbooks referenced in this chapter provide a good coverage of differencing. What follows are some useful things to keep in mind.

- When differencing, it is useful to think of creating a new series (for example, w) from the original series.

- What is shown above is a first difference. More generally $w_t = (y_t - y_{t-1})$.

- Second differences are also used: $w'_t = (y_t - y_{t-1}) - (y_{t-1} - y_{t-2})$.

- Note that a second difference is not the same as a two difference or two span difference:
  $w^*_t = (y_t - y_{t-2})$.

- Seasonal differences are common. For data in monthly interval a seasonal difference probably means a 12 span difference: $w^s_t = (y_t - y_{t-12})$.

- You can difference as much as you want. For example, 20th differences, 42 span differences, and so on can be created in the software shown in this book, but you should difference cautiously.

o   This last point deserves more discussion. Differencing should be done cautiously because differencing implies a loss of information. To see why, consider a series with *n* observations. Differencing it once implies the loss of one observation. This is because only *n*–1 pairs of adjacent observations exist. (See the second bullet above.) Following the same logic, the seasonal difference shown above entails the loss of 12 observations.

o   If series are of adequate length, you might be tempted to think that this loss doesn't matter—they are lost from the first part of the series, and these observations cannot reveal much about what is going to happen tomorrow. However, differencing can "remove" or impact stationary components in the series as well as nonstationary components. This last point is tricky and is discussed later. The rule of thumb is to difference only when you have to. That is, difference only when the series being modeled is nonstationary.

- Finally, the theory is that a series is differenced until what remains is stationary. In application, the first differences are common, second differences are rare and differences of higher order are unheard of. Note that this does not apply to "span" differences; seasonal span differences are also common.

Returning to equation 7, we see that it is stationary. In this case the differenced series is equal to the error series whose properties are well behaved, the mean is defined (0), and the variance is some constant that is less than infinity. Although the differencing has converted a nonstationary series, RTS, to a stationary one, w, you might be dubious about the usefulness of the exercise. We are left with a series that is white noise by definition.

The I in ARIMA

When differencing we hope that what is created, the stationary series, has some systematic component that can be captured by AR or MA terms or both. We want to use exactly the same approach that was outlined in the beginning of the chapter. In general, a nonstationary series can be thought of as having (it is hoped) two components: stationary and nonstationary. Differencing can be thought of as getting the nonstationary component out of the way so that the stationary component can be identified and estimated.

However, consider a nonstationary series as represented by equation 5. If the series has, for example, positive drift, we would want this component in our final forecast, too. That is, the final forecast should be on the scale of the original series, and not the differenced series. The I in ARIMA stands for integrated. This is another way of saying adding the series back up or undoing the difference prior to forecasting.

Diagnosing Nonstationary Series

The TS_NA (tennis shoe sales in North America) series that is used in this section is shown below. Over the time range considered, the series mean certainly looks like it changes over time. To assess the stationarity of the series, a diagnostic called the Augmented Dickey, Fuller test (ADF) is used (Dickey et al. 1979).

Output 8.41

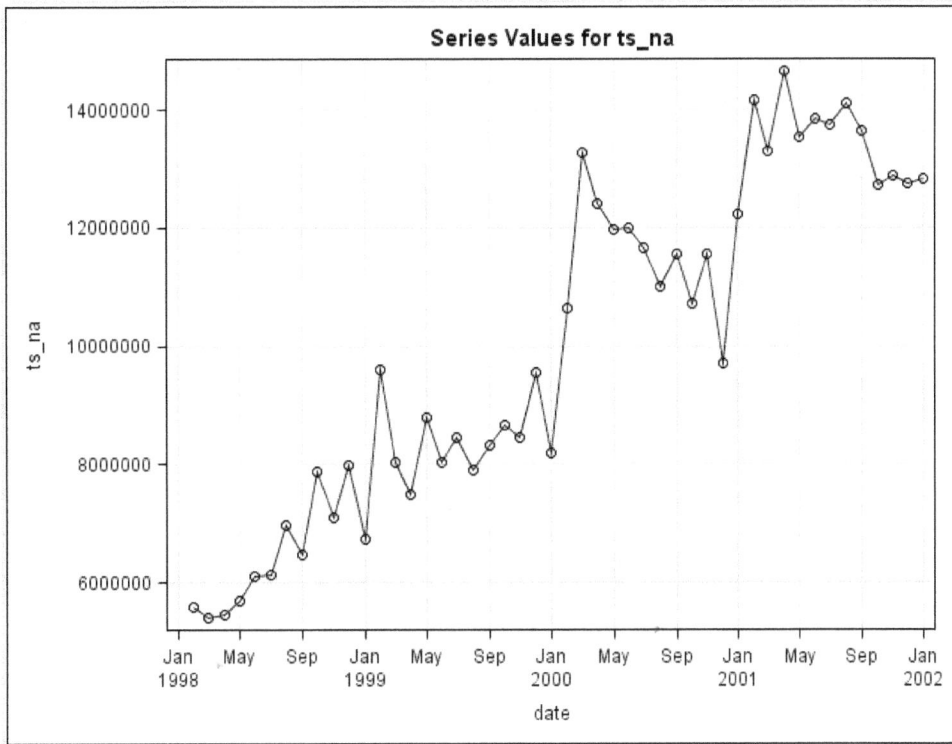

The ARIMA procedure generates the ADF test results below.

Program 8.8

```
proc arima data=nts_1;
i var=ts_na stationarity=(adf=(0 1 2 3 4 5 6 7));
run;
```

New syntax consists of the stationarity option that invokes the ADF test. The numbers listed after ADF correspond to the augmenting lags tested, and they are listed under the lags column in the output below.

Output 8.42

		Augmented Dickey-Fuller Unit Root Tests					
Type	Lags	Rho	Pr < Rho	Tau	Pr < Tau	F	Pr > F
Zero Mean	0	0.3541	0.7630	0.48	0.8139		
	1	0.5041	0.8009	1.01	0.9147		
	2	0.5153	0.8035	1.08	0.9252		
	3	0.4951	0.7982	0.99	0.9123		
	4	0.4896	0.7967	1.01	0.9151		
	5	0.4686	0.7912	0.90	0.8981		
	6	0.4954	0.7977	1.12	0.9296		
	7	0.5034	0.7995	1.12	0.9287		
Single Mean	0	-4.2918	0.4908	-1.62	0.4658	1.78	0.6275
	1	-2.5979	0.6962	-1.39	0.5789	2.02	0.5673
	2	-2.5342	0.7041	-1.43	0.5614	2.22	0.5204
	3	-2.7007	0.6826	-1.44	0.5524	2.10	0.5495
	4	-2.4743	0.7112	-1.36	0.5950	1.96	0.5818
	5	-2.7306	0.6783	-1.39	0.5782	1.87	0.6054
	6	-1.8916	0.7836	-1.14	0.6925	1.78	0.6270
	7	-2.0348	0.7658	-1.22	0.6571	1.93	0.5906
Trend	0	-25.7005	0.0082	-3.93	0.0184	7.80	0.0251
	1	-17.5442	0.0751	-2.62	0.2749	3.65	0.4691
	2	-21.9835	0.0229	-2.53	0.3121	3.53	0.4926
	3	-55.7459	<.0001	-2.87	0.1821	4.46	0.3166
	4	-1839.84	0.0001	-2.94	0.1600	4.66	0.2801
	5	41.1241	0.9999	-3.45	0.0585	6.32	0.0682
	6	28.2353	0.9999	-3.20	0.0987	5.37	0.1454
	7	17.3174	0.9999	-3.64	0.0392	6.99	0.0439

The wealth of information presented can be overwhelming. What follows are some things to keep in mind when reading the table.

The *null hypothesis* of all tests presented in the table is that the series tested is nonstationary.

- p values (columns headed with Pr) greater than 0.05 indicate nonstationarity.
- p values less than 0.05 indicate stationarity.

Notes:

- The non-ODS graphics version of the ADF output is presented because it is most compact.
- The 0.05 threshold is a widely used rule of thumb, but it is not appropriate under all sample sizes or conditions.
- The null hypothesis tests for a unit root at lag 1.
- p values greater than 0.05 indicate that a first difference is needed, p values less than 0.05 indicate that no differencing is needed.
- If the series has already been differenced, it is customary to re-run the test on the differenced series to confirm stationarity. If p values fail to reject on the differenced series, a second difference is indicated.

There are three versions or types of the test because the test is relatively low powered (an artifact of the statistical properties of nonstationary series). One way to think about the versions is that the more we can tell the test about the process generating the data, the better the chance of rejecting (Dickey, 2008). The three versions are outlined next.

- The *zero mean* version is appropriate when the series has no apparent trend, and it is centered around zero. The null hypothesis for this test is a random walk. (See equation 4.) The alternative hypothesis is a stationary AR process (for example, AR 1).
- The *single mean* version is appropriate when the series has no apparent trend, and it is centered around a number that is not zero. The null hypothesis for this test is a random walk. The alternative hypothesis is a stationary AR process.

- The *trend test* is appropriate when the series has an apparent trend. The null hypothesis is a random walk with drift. (See equation 5.) The alternative hypothesis is stationary variation around a deterministic linear trend. An example of modeling under the alternative hypothesis is given in a following section.

There are *three test statistics* to consider: F, Rho and Tau. The Tau test seems to have the most power to reject under the most general circumstances. This implies that our attention is focused on the column headed "Pr < Tau" when deciding whether to take a difference.

The data (see Output 8.41) seems to have a distinct upward trend over the range of the sample. Combining this information with the guidelines given above restricts our attention to the shaded portion of the table shown below.

Output 8.43

		Augmented Dickey-Fuller Unit Root Tests					
Type	Lags	Rho	Pr < Rho	Tau	Pr < Tau	F	Pr > F
Zero Mean	0	0.3541	0.7630	0.48	0.8139		
	1	0.5041	0.8009	1.01	0.9147		
	2	0.5153	0.8035	1.08	0.9252		
	3	0.4951	0.7982	0.99	0.9123		
	4	0.4896	0.7967	1.01	0.9151		
	5	0.4686	0.7912	0.90	0.8981		
	6	0.4954	0.7977	1.12	0.9296		
	7	0.5034	0.7995	1.12	0.9287		
Single Mean	0	-4.2918	0.4908	-1.62	0.4658	1.78	0.6275
	1	-2.5979	0.6962	-1.39	0.5789	2.02	0.5673
	2	-2.5342	0.7041	-1.43	0.5614	2.22	0.5204
	3	-2.7007	0.6826	-1.44	0.5524	2.10	0.5495
	4	-2.4743	0.7112	-1.36	0.5950	1.96	0.5818
	5	-2.7306	0.6783	-1.39	0.5782	1.87	0.6054
	6	-1.8916	0.7836	-1.14	0.6925	1.78	0.6270
	7	-2.0348	0.7658	-1.22	0.6571	1.93	0.5906
Trend	0	-25.7005	0.0082	-3.93	0.0184	7.80	0.0251
	1	-17.5442	0.0751	-2.62	0.2749	3.65	0.4691
	2	-21.9835	0.0229	-2.53	0.3121	3.53	0.4926
	3	-55.7459	<.0001	-2.87	0.1821	4.46	0.3166
	4	-1839.84	0.0001	-2.94	0.1600	4.66	0.2801
	5	41.1241	0.9999	-3.45	0.0585	6.32	0.0682
	6	28.2353	0.9999	-3.20	0.0987	5.37	0.1454
	7	17.3174	0.9999	-3.64	0.0392	6.99	0.0439

Recall that the null hypothesis tested is a unit root at lag one (random walk with drift version), or an AR 1 parameter equal to one. Rejecting the null hypothesis indicates a difference is not necessary. Failure to reject indicates a difference is needed.

The best case scenario is rejection at all tested lags in the shaded region of the table, in this case 0–7. This would be a strong indication of stationarity. An alternative scenario is failure to reject at all tested lags. This would be a strong indication of nonstationarity. However, what is shown in the shaded region is rejection at lag 0, a failure to reject at lags 1–4, and marginal rejection at lags 5–7. This leads to the last guideline for using the ADF diagnostic.

The lags, more properly called *augmenting lags*, are included in the test for the same reason that there are three versions or types of the test. The more information the user can provide about the process generating the data, the more power the test has to reject. The augmenting lags correspond to the stationary part of the series.

To find out more about the stationary part of the series, users can look at the PACF (Dickey 2008). The PACF generated by the PROC ARIMA code above is shown next.

Output 8.44

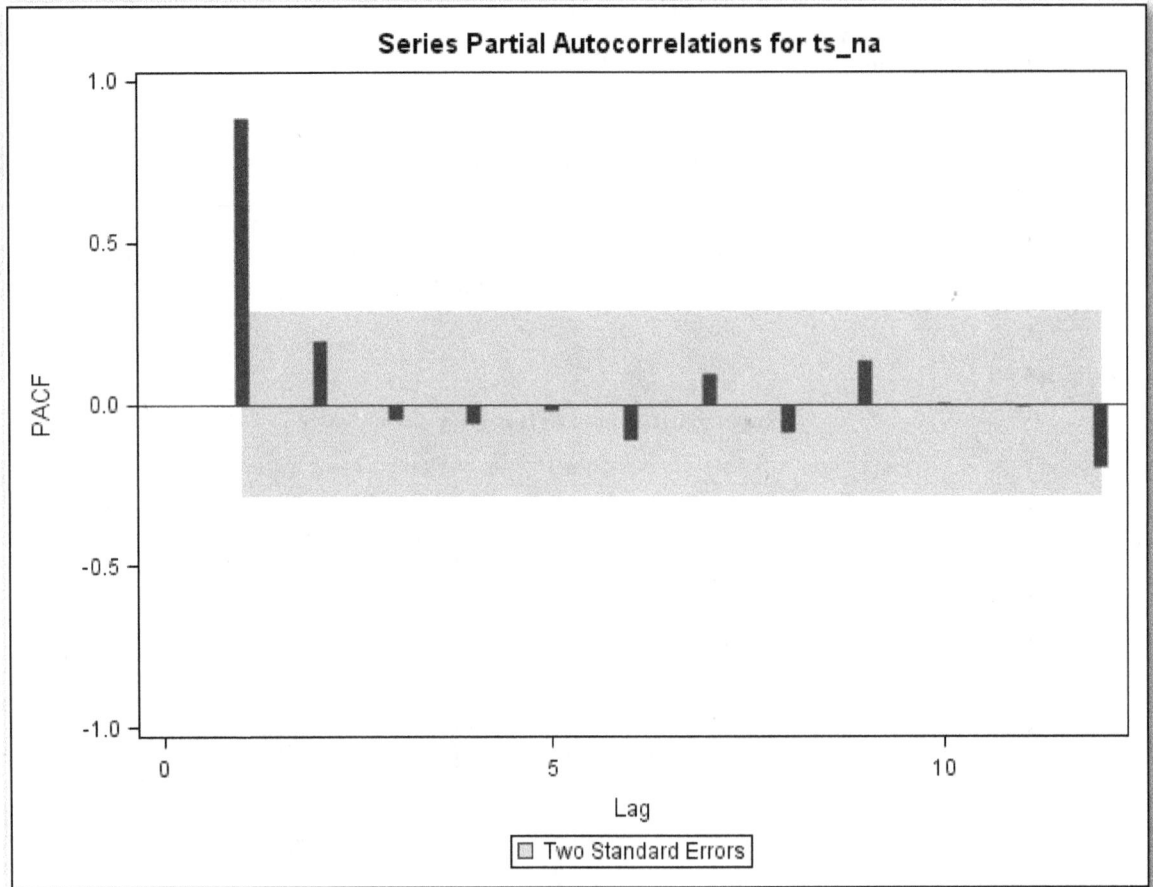

Note that the augmenting lags begin counting at zero. This means that a significant spike at lag 1 in the PACF corresponds to augmenting lag 0. If the last significant spike was at lag 3, then the p value at augmenting lag 2 would determine stationarity. Because the ADF test rejects at lag 0, stationarity is indicated. More specifically, no differencing is indicated, the series should be modeled using a linear trend, and an ARMA model should be used to fit the residuals if appropriate.

Additional points to consider with nonstationarity

There are a couple more points to consider when discussing nonstationarity diagnostics..

All of the stationarity tests considered in the table above focused on a unit root at lag 1, and the process generating the data was assumed to be AR1. If the process is more general, say AR4, the testing process outlined above stays the same, but it is a characteristic of a combination of the AR coefficients that determines stationarity. See Hamilton (1994) for further details. An alternative manifestation of nonstationarity is a unit root at a seasonal lag. For example, in monthly intervaled data, this would usually be an AR 12 parameter equal to one. However, in terms of testing and modeling, the lag length is really the only difference. That is, everything discussed above is still relevant in nonstationary seasonality. The only differences are that a seasonal lag is tested, and if the test fails to reject a seasonal difference is indicated.

The ADF test for a seasonal lag can be generated by adding an option to the PROC ARIMA syntax. Note that there is no trend test in the seasonal version of the ADF test. It is assumed that if a unit root at lag 1 is an issue, appropriate differences have been taken before running the seasonal test. The seasonal test should be considered when the estimate of the AR parameter at the seasonal lag exceeds 0.5.

The multiple comparisons problem, discussed earlier in the chapter, impacts the ADF test as well. In situations where the ADF test results are mixed, as above, a widely used best practice is to fit models under both the null and alternative hypotheses. Either AIC or an out-of-sample fit statistic can be used to pick the champion model.

Because an example of modeling under the alternative hypothesis for the trend test is shown later, this section concludes with an example of modeling the differenced series. In PROC ARIMA, the series can be differenced using a parenthetic list in the IDENTIFY statement.

Program 8.9

```
proc arima data=nts_1 plots(unpack)=series(all);
i var=ts_na(1) stationarity=(adf=(0 1 2 3 4 5 6 7));
run;
```

The plot below shows what the TS_NA series looks like after being differenced. Note that differencing is a useful tool for detrending and deseasonalizing series.

Output 8.45

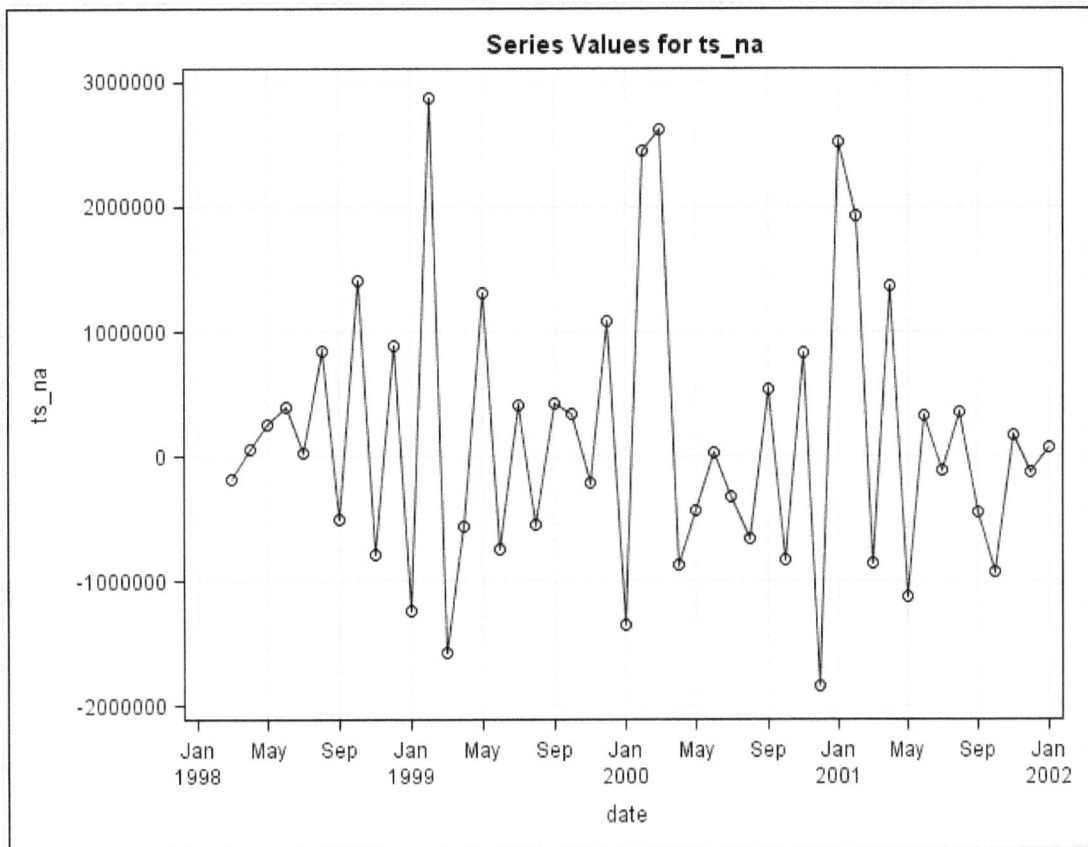

For the differenced series, the appropriate test appears to be the zero mean type.

Output 8.46

Augmented Dickey-Fuller Unit Root Tests							
Type	Lags	Rho	Pr < Rho	Tau	Pr < Tau	F	Pr > F
Zero Mean	0	-62.9415	<.0001	-9.88	<.0001		
	1	-67.3457	<.0001	-5.67	<.0001		
	2	-54.9767	<.0001	-3.97	0.0002		
	3	-60.0477	<.0001	-3.38	0.0012		
	4	-40.9406	<.0001	-2.70	0.0080		
	5	-126.069	0.0001	-2.78	0.0066		
	6	-81.2826	<.0001	-2.35	0.0196		
	7	85.6936	0.9999	-2.56	0.0117		
Single Mean	0	-64.2125	0.0004	-10.10	0.0001	50.96	0.0010
	1	-76.0602	0.0004	-5.96	0.0001	17.76	0.0010
	2	-73.9978	0.0003	-4.27	0.0014	9.12	0.0010
	3	-120.748	0.0001	-3.71	0.0071	6.89	0.0043
	4	-118.316	0.0001	-3.03	0.0400	4.60	0.0639
	5	92.3556	0.9999	-3.19	0.0277	5.12	0.0431
	6	56.0829	0.9999	-2.84	0.0619	4.05	0.0940
	7	23.4256	0.9999	-3.18	0.0287	5.13	0.0429
Trend	0	-64.4270	<.0001	-10.05	<.0001	50.47	0.0010
	1	-77.8247	<.0001	-5.98	<.0001	17.87	0.0010
	2	-79.2419	<.0001	-4.31	0.0070	9.32	0.0010
	3	-140.578	0.0001	-3.76	0.0289	7.07	0.0421
	4	-146.254	0.0001	-3.09	0.1213	4.82	0.2489
	5	85.8055	0.9999	-3.22	0.0953	5.19	0.1800
	6	52.9609	0.9999	-2.89	0.1763	4.24	0.3583
	7	23.2544	0.9999	-3.18	0.1024	5.08	0.2007

The ADF test rejects at all tested lags (shaded area). This indicates that no further differencing is necessary. We can now proceed with modeling using the techniques presented in earlier sections. The ACF and PACF plots of the differenced series are presented next.

Output 8.47

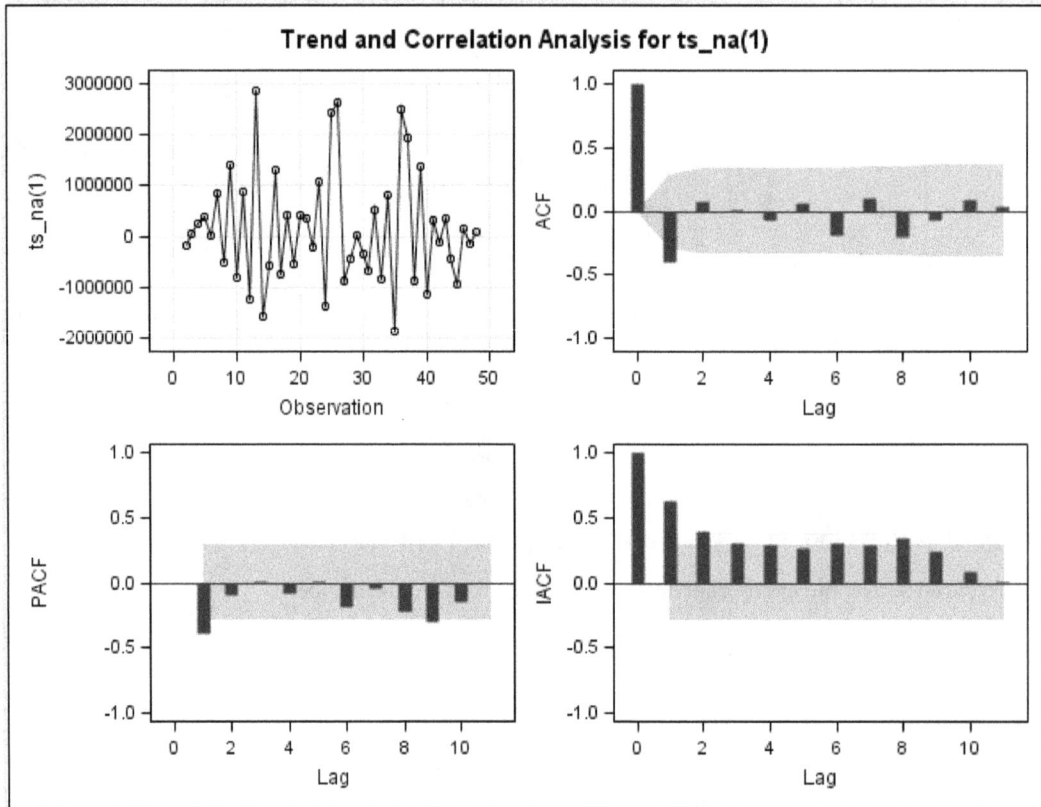

Trend and Correlation Analysis for ts_na(1)

An AR 1 model seems appropriate for the differenced TS_NA series. The PROC ARIMA syntax is modified to fit this model below. Forecasts based on this model are also generated.

Program 8.10

```
proc arima data=nts_1 plots;
i var=ts_na(1) stationarity=(adf=(0 1 2 3 4 5 6 7));
estimate p=1 plot method=ml;
forecast id=date interval=month lead=12;
run;
```

Below, the AR 1 parameter estimate is significant, and the white noise test on the residuals of the model indicates that the model is adequately specified.

Output 8.48

Maximum Likelihood Estimation					
Parameter	Estimate	Standard Error	t Value	Approx Pr > |t|	Lag
MU	156799.5	108939.3	1.44	0.1501	0
AR1,1	-0.38854	0.13662	-2.84	0.0045	1

Output 8.49

Autocorrelation Check of Residuals										
To Lag	Chi-Square	DF	Pr > ChiSq	Autocorrelations					
6	2.43	5	0.7869	-0.044	-0.074	0.007	-0.072	-0.032	-0.175
12	9.78	11	0.5501	-0.033	-0.255	-0.143	0.100	0.106	0.115
18	14.84	17	0.6073	0.170	0.037	-0.121	0.157	0.022	-0.030
24	23.16	23	0.4517	-0.114	-0.055	0.070	-0.250	0.087	-0.024

There a couple of interesting points to note about the forecast. First, it is on the original, or not differenced, scale. This is the "I" in ARIMA. The second is that the lead forecast seems to preserve the trend. This is an artifact of the nonstationarity assumption. There is no mean reversion in ARIMA forecasts based on differenced series. That is, when the order of integration is nonzero.

Output 8.50

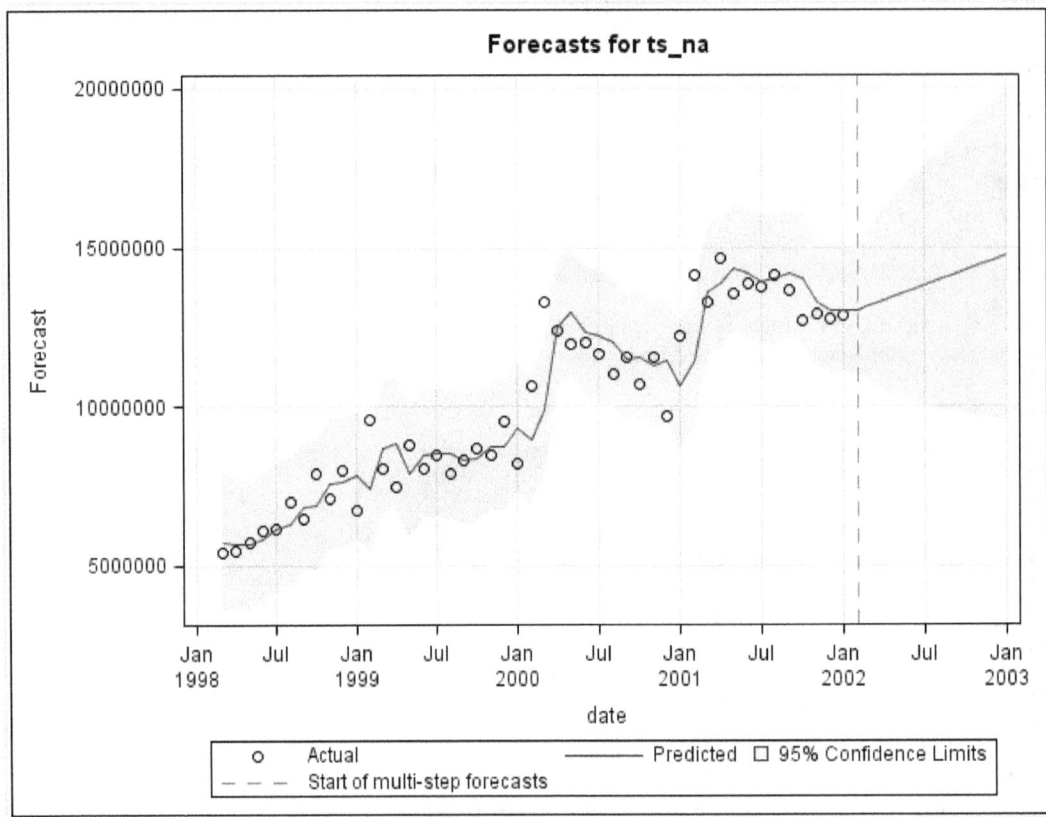

Forecasts for ts_na

---

1 If the series to be modeled is nonstationary, the same ideas and techniques discussed here can be applied to the series after it has been made stationary by appropriate differencing.

2 Unobserved Component Models (UCM), discussed in Chapter 11, relax this restriction.

3 What is described is actually positive autocorrelation. Negative autocorrelation, observations above the mean that tend to be followed by observations below it, is also possible but not as common.

4 There are several other tools in SAS and in other software packages that accommodate rational polynomial transfer function models. PROC ARIMA was chosen to illustrate the modeling process in the majority of this chapter because, in the opinion of the authors, it provides the most straightforward implementation of the models discussed here and in textbooks on this topic. That said, once analysts learn to implement the modeling process in PROC ARIMA, implementing these models using other tools is reasonably simple. Note that summary details on the usage of the procedure are presented. For full details about the PROC ARIMA syntax, see the SAS/ETS documentation.

5 PROC ARIMA has three automatic order determining algorithms; ESACF, MINIC and SCAN. These algorithms are useful for augmenting the information found in the autocorrelation function plots. Also, in a large-scale forecasting scenario, analysts might not have time to peruse the autocorrelation function plots for each series. Order-determining algorithms might be the only choice for model identification in situations where there are a lot of series to consider and time is short. For more information, see the ARIMA procedure in the SAS/ETS documentation. In a large-scale forecasting scenario, the number of series to be identified can run into the thousands. The best alternative in this context might be to have an algorithm determine the order for you on the majority of the series. SAS Forecast Server is a useful and robust large-scale forecasting tool.

6 A third plot, the Inverse Autocorrelation Function (IACF) plot, contains essentially the same information as the PACF, but it is not as easily interpreted. Having both the PACF and the IACF can be an advantage when the stationarity of the series is in question. See the ARIMA procedure's detail section in the SAS/ETS documentation for more information about the IACF.]

7 Statisticians call this the Multiple Comparisons Problem. See Miller, 1981 for further details.

8 The important topic of using holdout samples or model selection based on extrapolation performance is covered in Chapter 10.

9 Both diagnostics have built-in penalty terms. For AIC or SBC to improve, or get smaller, a variable entering the model has to improve the overall fit enough to overcome the penalty. Because SBC has a bigger penalty, specifications chosen based on SBC tend to have fewer terms than specifications chosen based on AIC. SBC does a better job at eliminating irrelevant terms,

but using it implies a relatively greater risk of omitting an important term. The authors think that omitting an important variable is more problematic than including an irrelevant one and prefer AIC.

[10] To eliminate clutter the intercept has been dropped from the RTS equation shown earlier in the chapter (2a, 2b). In practice this would represent a rescaling of the RTS series to have a zero mean.

[11] Another difference (shown in equation 8) is that, by convention, MA coefficients enter the model with negative signs.

[12] See Chapter 3 of Brocklebank and Dickey (2003) for more information.

[13] A more general representation of the transfer function and background information on how denominator order polynomials work are found in Appendix 1.

[14] SAS Forecast Server contains patented algorithms that automate the model selection process described

Chapter 9: Model Building: ARIMAX or Dynamic Regression Modes

Introduction

This chapter presents the rational polynomial transfer function framework. It begins with the ordinary regression model in a time series setting. Extensions to the regression framework illustrate how dynamic relationships between inputs and the target can be accommodated using the rational polynomial transfer function. Dynamic relationships include lags, shifts, and persistence effects. Determining the relationships between a model's inputs and the target is called the model identification process. Another topic is accommodating trend and seasonal variation in the transfer function framework. The appendix provides details and a discussion of issues associated with stochastic input variables.

The focus in this chapter is on using structural or explanatory variables in the rational polynomial transfer function framework. From a conceptual point of view, this involves replacing the white noise input variable with an explanatory input variable. However, using explanatory variables puts more structure on the problem, and some practical concerns need to be addressed.

In Chapter 8, relationships between structural variables and the dependent variable are not explicitly specified. The factors causing systematic patterns in the data for the dependent variable are unknown. In this chapter, some knowledge of the structural variables and their relationships to the dependent variable are assumed.

The analyst can now specify more precise and more parsimonious representations of the relationships between the impulse and dependent variables in the model building process. For some influential variables, knowledge of these relationships might come from business expertise, familiarity with the data, and so on. For other variables, not much might be known except for information such as "variable X seems to have a positive impact on the forecast variable."

The cross-correlation function (CCF) plot provides information about the relationships between the dependent and structural variables. It helps determine whether there is a significant relationship between the candidate independent and forecast variable. It answers the questions, "Is the relationship best modeled as a numerator (similar to MA) order, a denominator (similar to AR) order, or a mixture (ARMA) process? Is there contemporaneous correlation between the input and dependent variable, or should a shift relationship be added to the transfer function?"

There are other practical issues with using input variables in the rational polynomial transfer function framework. Input variables come in a variety of measurement scales, and input variables might be stochastic. Input variable types include binary, categorical, and interval. Simple inputs like binary promotion flags or deterministic trend variables can be used in a relatively straightforward way in the rational polynomial framework. Other interval-valued variables include regional or macro-economic indicators, prices of substitute or complimentary goods, and so on. These variables might be stochastic.

Stochastic input variables cause problems in two areas of applied forecasting. First, because these variables might be autocorrelated, model identification becomes more difficult. Second, in a forecasting scenario, lead values for model inputs must be provided. Because future values for stochastic inputs are not known, future values for these variables must be forecasted before creating the forecast for the dependent variable. This introduces another source of noise or error into the forecast for the dependent variable.

The distinction between stochastic and deterministic inputs provides a framework for the applied parts of this chapter. After an introduction to dynamic regression models and the main diagnostic used in building them, time series models with deterministic input variables are considered. This provides an uncluttered basis for considering additional diagnostics and most of the ideas related to building transfer function models with structural variables.

9.1 ARIMAX Concepts

The starting point for this section is the ordinary or static regression model. In a time series setting, a static model looks like the one in equation 1 for weekly sales of blue tennis shoes (BTS):

$$BTS_t = \alpha + \omega_0 P_t + \varepsilon_t \qquad\qquad 1$$

BTS sales are specified as a function of an interval valued input—the current unit (pair) price of BTS. This model is not too different from the models previously introduced. In fact, it is a very simple version of a rational polynomial transfer function model—a numerator order 0 (zero).

Order 0 indicates that there is no lag relationship between the input and dependent variable. Only contemporaneous correlation exists between these two variables. Variation in price during the current week only impacts sales in the current week.

As a more dynamic example, consider the model for the weekly sales of Argyle tennis shoes (ATS) in equation 2. B is used to denote the backshift operator.

$$ATS_t = \alpha + \omega_0 P_t + \omega_1 P_{t-1} + \omega_2 P_{t-2} + \varepsilon_t \qquad \rightarrow$$
$$ATS_t = \alpha + (\omega_0 + \omega_1 B + \omega_2 B^2) P_t + \varepsilon_t \qquad\qquad 2$$

Price is correlated with sales at lags 0, 1, and 2, and the model is a numerator order 2. This indicates that variation in price in the current period is correlated with demand in the current period and in the following two periods.

Even when the input is interval valued and might be varying continuously, it is still useful to think in terms of the impulse response relationship. A one-time price jump in the current interval affects sales of ATS in the current period and in the following two periods.

If the relationship between price and ATS sales was longer and followed a pattern of decay, it could be rewritten as a denominator order 1, as in equation 3. (For additional details, see Appendix 1 in Chapter 8.)

$$ATS_t = \alpha + (1/1 - \delta B) P_t + \varepsilon_t \qquad\qquad 3$$

Equations 1 through 3 demonstrate that the choice of impulse or input does not affect the rational polynomial transfer function framework. However, more structure means more specific types of relationships to accommodate.

Shift is one of these relationships. Consider the effect of a coupon mailing promotion on the sales of red tennis shoes (RTS). The AR 1 representation of RTS can be modified as shown in equation 4:

$$RTS_t = \varphi_0 + \delta_1 CM_{t-1} + \delta_2 CM_{t-2} + \delta_3 CM_{t-3} + \varphi_1 RTS_{t-1} + \varepsilon_t \quad \rightarrow$$

$$RTS_t = \varphi_0 + (\delta_1 B + \delta_2 B^2 + \delta_3 B^3)CM_t + \varphi_1 RTS_{t-1} + \varepsilon_t \qquad \qquad 4$$

The coupon promotion input in CM_t would usually be included in the data as a binary variable. In the model, the input has no contemporaneous effect on the demand for RTS. This results from a natural time lag between the interval that the mailing is initiated and the interval that RTS demand is impacted.

Customers do not begin redeeming their coupons until the week following a mailing. The coupons continue to have an effect for the following two weeks. The lag length might be set based on the coupon's expiration date. The effect might follow a quadratic shape. For example, a few early birds $\rightarrow$ the majority of the redeemers $\rightarrow$ a few procrastinators. However, if the good is non-perishable, like RTS, the coupons could cause buyers to shift demand or stock up. In this case, one or both δ_2, δ_3 might be negative.

Usually, the term "lag" implies correlation that persists for some time after the contemporaneous (lag 0) interval. For the ATS model in equation 2, the numerator 2 or lag=2 model implies correlation at lags 0, 1, and 2. The term "shift" implies the number of periods after the lag 0 interval that the correlation starts. The RTS model in equation 4 can be described as a shift=1 lag=2 model. The correlation starts at lag 1 and then persists for two more intervals.

A relevant question at this point is, "What lag or shift structure should be used to represent the correlation between a given input and the dependent variable?" That is, how is the relationship between candidate inputs and the dependent variable identified? The remainder of this section discusses the primary diagnostic used to answer these questions—the cross-correlation function (CCF) plot. General ideas and best practices are presented. The CCF is put into practice in the modeling process in the following section.

The CCF plot depicts the dynamic relationship between a dependent variable Y and an independent or exogenous variable X. This plot is similar to an autocorrelation function plot. The bars or spikes represent the correlation at each lag, and the shaded band represents the 95% rejection region. However, in the CCF plot, the bars represent the correlation between Y and X at each lag.

In this section, all identified input variables are non-autocorrelated or white noise. When input variables are not white noise, model identification becomes less clear. See the appendix for how to identify autocorrelated input variables.

Patterns in the significant lags of the CCF suggest reasonable candidate models. Some general rules associated with the CCF plot are summarized. (For additional details, see Box and Jenkins 1972.)

- A significant spike at lag 0 suggests significant contemporaneous correlation between Y and X. This implies that shift=0.
- Spikes at positive lags represent lag relationships between Y and X. Variation in current values of X are correlated with variation in future values of Y.[1]
- Builds in spikes at positive lags suggest numerator orders to try.
- Positive spikes that gradually decay to non-significance suggest that a denominator order 1 or 2 term should be tried.[2]

Output 9.1

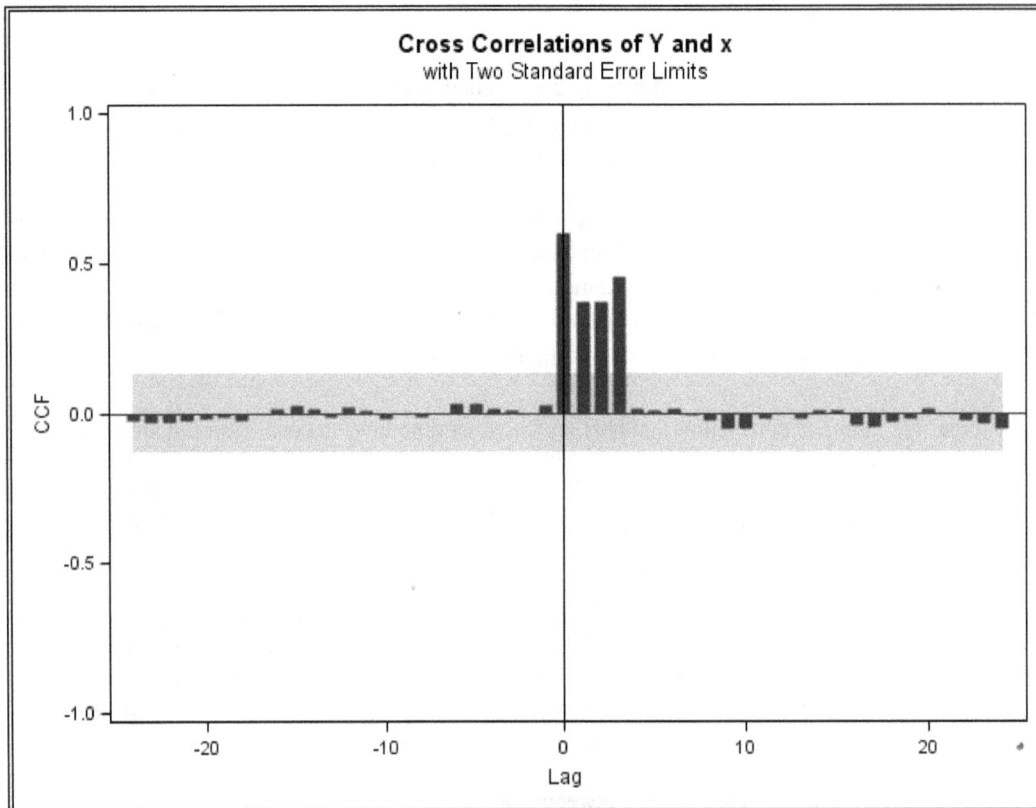

Output 9.1 shows the relationship between Y and a white noise (non-autocorrelated) input variable X. Considering the pattern in the significant spikes of the CCF plot, the rules suggest the following:

- Significant correlation at lag 0 → shift=0.

- A build in significant spikes at lags 1 through 3 → numerator=3.

- No decay or immediate drop back to non-significance after lag 3 → denominator=0.

PROC ARIMA code can accommodate this model.

Program 9.1

```
ods graphics on;

proc arima data=ccfplot1;
identify var=y crosscorr=(x);
estimate input=((1 2 3)x) method=ML; run;

ods graphics off;
```

New PROC ARIMA syntax provides additional options. The CROSSCORR option lists candidate input variables for a cross-correlation analysis with the dependent variable listed in the VAR option. In addition, it generates the CCF plot. Candidate independent variables must be listed in the CROSSCORR option to be used in subsequent ESTIMATE statements. The INPUT option specifies the correlation structure between Y and X. Above, X enters the model as a numerator order 3 term as specified by the numbers preceding X in the parentheses. An implication that is not obvious in the syntax is that the lag 0 term is always implied in the model as shown in the parameter estimates table. Subsequent models illustrate how to specify shift terms.

Output 9.2

	Maximum Likelihood Estimation						
Parameter	Estimate	Standard Error	t Value	Approx Pr > |t|	Lag	Variable	Shift
MU	-132.57204	8.69381	-15.25	<.0001	0	Y	0
NUM1	0.82828	0.04525	18.30	<.0001	0	x	0
NUM1,1	-0.41873	0.04545	-9.21	<.0001	1	x	0
NUM1,2	-0.44737	0.04551	-9.83	<.0001	2	x	0
NUM1,3	-0.61956	0.04532	-13.67	<.0001	3	x	0

The information in the parameter estimates table is a preliminary indication that the model suggested in the CCF plot is appropriate. Variation in X in the current period is significantly correlated with variation with Y in the current period and for the following three periods. An analysis of residuals and refinement might be necessary before declaring that the model is an adequate representation of the systematic variation in Y.

Output 9.3

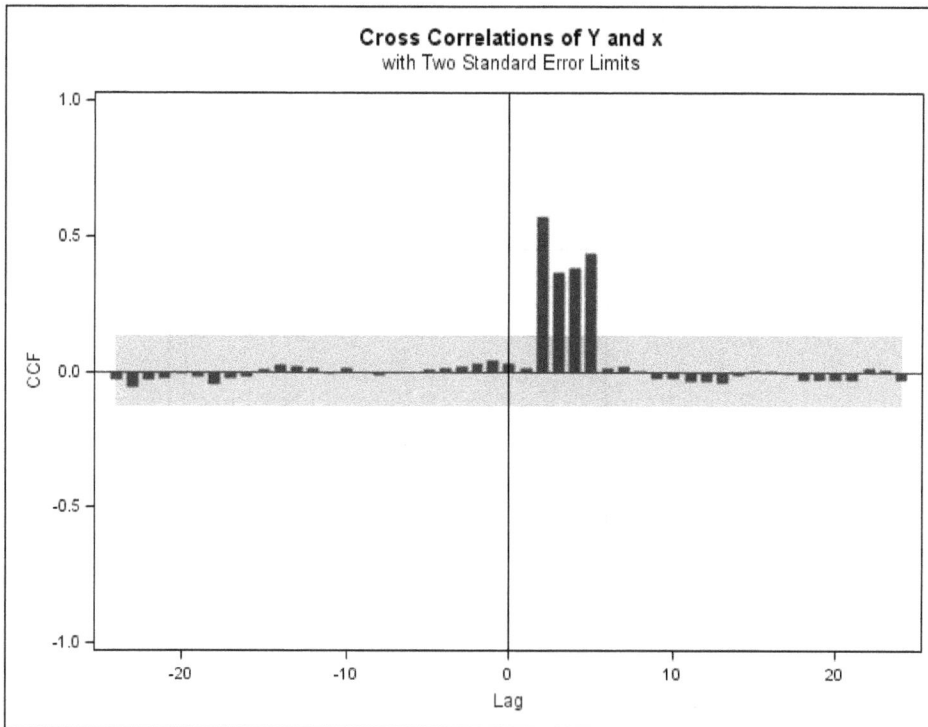

The pattern shown in CCF plot 2 indicates that there is no contemporaneous correlation between the Y and X series. Rules indicate the following model might be appropriate:

- Two intervals (0 and 1) before the first significant spike → shift=2.
- A three-spike build following the first significant spike[3] → numerator=3.
- No decay. An immediate drop to non-significance → denominator=0.

PROC ARIMA syntax can accommodate this model.

Program 9.2

```
ods graphics on;

proc arima data=ccfplot2;
identify var=y crosscorr=(x); run;
estimate input=(2$(1 2 3)x) method=ML; run;

ods graphics off;
```

The shift is specified in the ESTIMATE statement. Numbers preceding the parenthetical list of numerator orders indicate the shift. Above, a $ is used to further differentiate the shift order from the numerator orders. The $ is optional. Parameter estimates agree with the model suggested by the CCF plot.

Output 9.4

	Maximum Likelihood Estimation						
Parameter	Estimate	Standard Error	t Value	Approx Pr > |t|	Lag	Variable	Shift
MU	-132.01095	9.51143	-13.88	<.0001	0	Y	0
NUM1	0.80270	0.04943	16.24	<.0001	0	x	2
NUM1,1	-0.42131	0.04964	-8.49	<.0001	1	x	2
NUM1,2	-0.47614	0.04964	-9.59	<.0001	2	x	2
NUM1,3	-0.60821	0.04943	-12.30	<.0001	3	x	2

Output 9.5

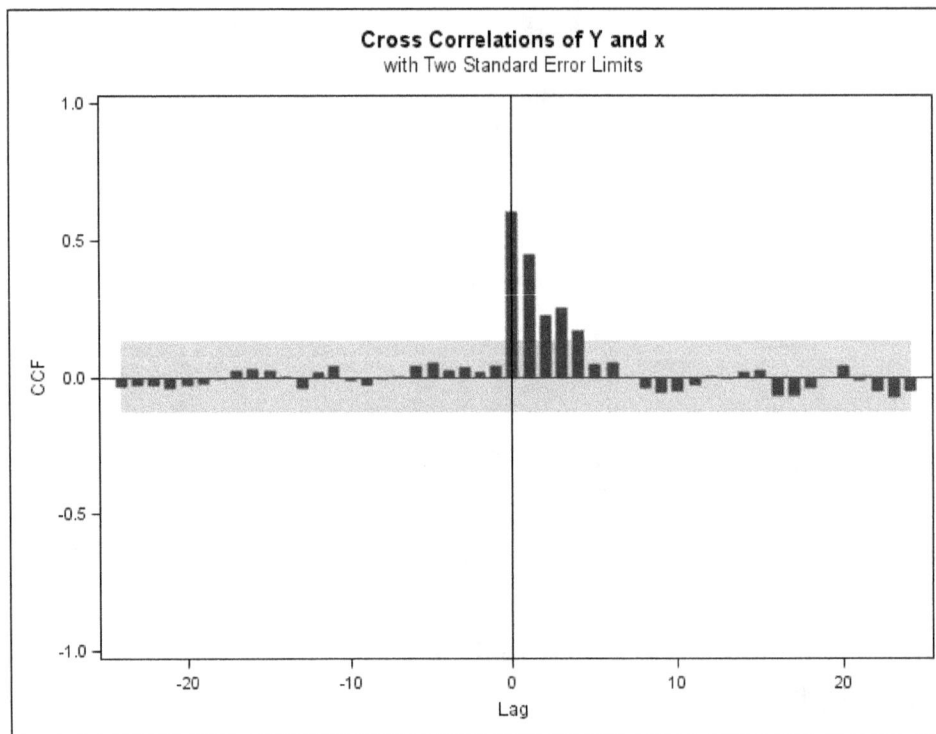

CCF plot 3 shows evidence of decaying significant spikes after lag 0. The following model might be appropriate.

- Significant contemporaneous correlation → shift=0.

- No substantial builds after lag 0 → numerator=0.

- Decaying spikes[4] past lag 1 → denominator=1.

PROC ARIMA syntax can accommodate the suggested model.

Program 9.3

```
ods graphics on;

proc arima data=ccfplot3;
identify var=y cross=(x);
estimate input=(/(1)x) method=ML; run;  quit;

ods graphics off;
```

The denominator order one term is specified using the forward slash in the ESTIMATE statement. The parameter estimates table indicates that the model suggested by the CCF plot is probably appropriate. The DEN1,1 parameter represents the trajectory of decay in the contemporaneous effect of the input. Because interpreting estimated parameters is problematic when the model's residuals might not be white noise, further consideration of the estimated denominator parameter is deferred to the next section.

Output 9.6

Maximum Likelihood Estimation							
Parameter	Estimate	Standard Error	t Value	Approx Pr > |t|	Lag	Variable	Shift
MU	-148.17588	20.09465	-7.37	<.0001	0	Y	0
NUM1	0.98338	0.07216	13.63	<.0001	0	x	0
DEN1,1	0.60236	0.03796	15.87	<.0001	1	x	0

9.2 ARIMAX Applications

This section considers additional sources of systematic variation to be accommodated in the model. A step-by-step, manual approach to building a time series regression model is presented. In other chapters of this book, forecasting tools are discussed that handle this process automatically. Why should you spend valuable time and effort learning the steps and building intuition when algorithms exist that do the job for you? The first reason is that algorithms do not always work. Default settings that are inappropriate for the data, convergence issues, and more can cause algorithms to fail. Using the tools presented here, you can conduct a manual forecasting analysis or repair failing parts of a system. Second, analysts have information about the data that algorithms do not have. Knowledge of all parts of the process and how they fit together enables analysts to leverage their information about the data into the forecasting system and to refine models when appropriate. Finally, good automatic forecasting systems are very expensive relative to forecasting tools presented in this and the preceding chapter. A manual analysis might be the only alternative available to you.

The NTS series contains a positive trend and a mid-series shock or event. The series might also contain AR and MA variation. Systematic variation attributed to ARMA terms is usually called *irregular variation* in this setting. When sources of variation include both input variables and irregular terms, the approach to identifying models becomes more ambiguous. The approach used here is to fit the regression portion of the model first, and then refine the regression models' residuals with ARMA terms if necessary. This approach is widely used because most analysts start with a good idea of which variables are important in the model. Fitting the regression model *first* provides the best chance of obtaining a model that is consistent with theory or business knowledge, generating marginal effects of interest, and so on.

However, an equally valid approach is to fit the irregular model first and then to fit the regression terms to its residuals. It is not known ahead of time which approach produces the best model. SAS Forecast Studio is used in the other chapters of this book. SAS Forecast Studio fits models using both approaches and it generates forecasts automatically using the best candidate from an automatic model selection process.

The series to be identified is the monthly revenue generated by aggregate monthly tennis shoe sales in the Northern region (or NTS). The plot of the NTS series is shown below. The plot shows a strong, approximately linear trend in sales over the sample. There is at least one event, shown in the circled area, beginning approximately in September 2003.

Figure 9.1

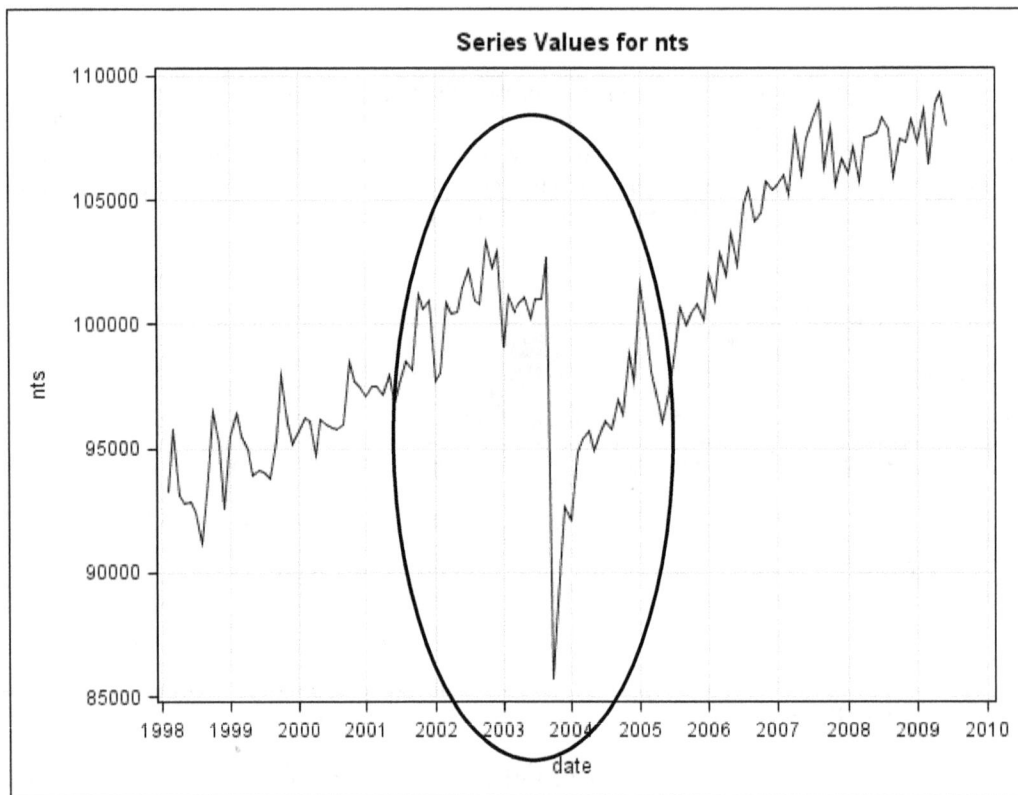

The best way to model the September 2003 event is ambiguous. It could be modeled using dummy variables either as a permanent shift in the intercept or as a temporary change with some persistence.

A preliminary model that could accommodate either type of event variation described is given in equation 17.

$$NTS_t = \delta_0 + \eta_1 D1_t + \delta_1 t \qquad\qquad 17$$

In the preliminary model, the NTS series is modeled as a function of an intercept, a binary or dummy variable $D1_t$ that captures the variation associated with the 2003 event, and a deterministic trend variable t. The table shows a portion of the NTS data set. The data set has been augmented with two candidate event variables and trend variable.

Figure 9.2

	date	nts	t	step_sep03	pulse_sep03
64	05/01/03	101062.25	64	0	0
65	06/01/03	100243.5	65	0	0
66	07/01/03	101034.18182	66	0	0
67	08/01/03	101037.27273	67	0	0
68	09/01/03	102658.36364	68	1	1
69	10/01/03	85741.545455	69	1	0
70	11/01/03	89729.909091	70	1	0
71	12/01/03	92661.272727	71	1	0
72	01/01/04	92136.636364	72	1	0
73	02/01/04	94831.5	73	1	0
74	03/01/04	95368	74	1	0
75	04/01/04	95680.083333	75	1	0

The intention is to present you with a reasonably complete example of a widely used time series analysis technique involving deterministic inputs interrupted time series.[5] Keep in mind the following facts about augmented data.

- Both candidate binary variables are flags that shift the intercept. In equation 17, they enter the specification in the $D1_t$ term.

- The STEP_SEP03 variable characterizes the event as an immediate and permanent shift of the intercept. It is equal to 0 before September 2003 and 1 after.

- The PULSE_SEP03 variable characterizes the event as an immediate and temporary change in the intercept. It is equal to 1 on September 2003 and 0 elsewhere.

- The linear trend variable, T, is a counter that begins at the first observation with a value of 1. Other deterministic trends are possible and are constructed in a similar way.

- The final observation of NTS is on June 2006. Lead values of the date variable and the input variables have been added to the data set.

The IDENTIFY statement syntax is similar to the syntax described in the previous section. The candidate input variables are listed in the CROSSCORR option. This defines input variables for subsequent analysis and provides the CCF diagnostics.

Program 9.4

```
ods graphics on;

proc arima data=nts plots(unpack)=series(all);
identify var=nts crosscorr=(t pulse_sep03 step_sep03); run;
quit;

ods graphics off;
```

The CCF plot depicting the correlation between NTS sales and the pulse event variable is shown below. The indicated correlation could be described as diminishing with no contemporaneous correlation. Shift=1, Num=0, Den=1. The insignificance of the correlation at the lag 0 interval could represent a shift relationship, or, in monthly data, could be caused by an event that occurs late in the month.

Output 9.7

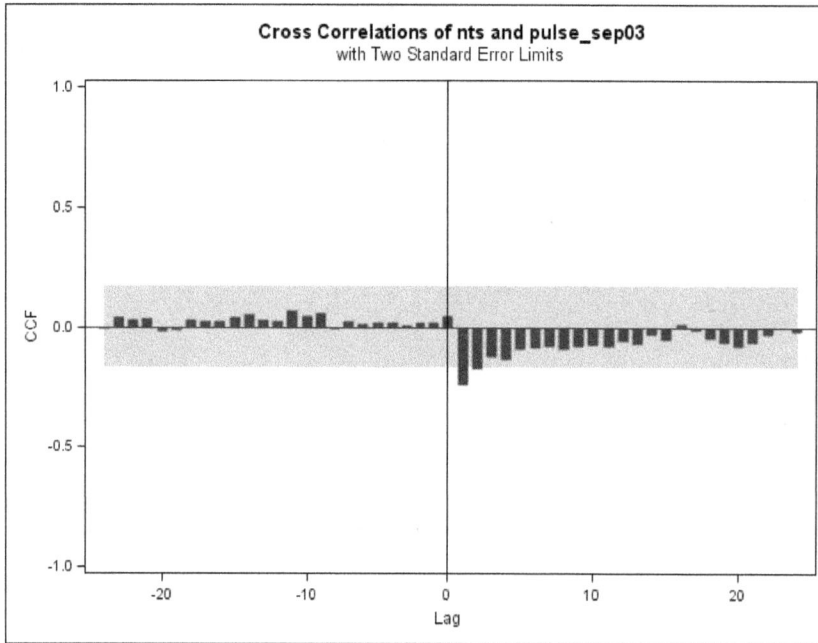

In the CCF plot above, there is a weak correlation pattern given the magnitude of the event in the series plot. This could be the result of the confounding of variation associated with the trend and the event. In general forecasting analyses, a widely used rule is to detrend and de-seasonalize the interval-valued series before model identification. In this example, the trend term could be fit first, and then the event variable analysis could be performed on the residuals of this model. Another alternative to detrending is taking the first difference of the series. This technique is demonstrated in Appendix 2 in Chapter 8 (note that the model specified above represents the alternative to the Random Walk with Drift hypothesis model mentioned in the Appendix: the I in ARIMA). The CCF plot depicting the correlation between NTS and the step event (Output 9.8, below) shows further evidence of problems with correlation between the trend and the event in the data.

Output 9.8

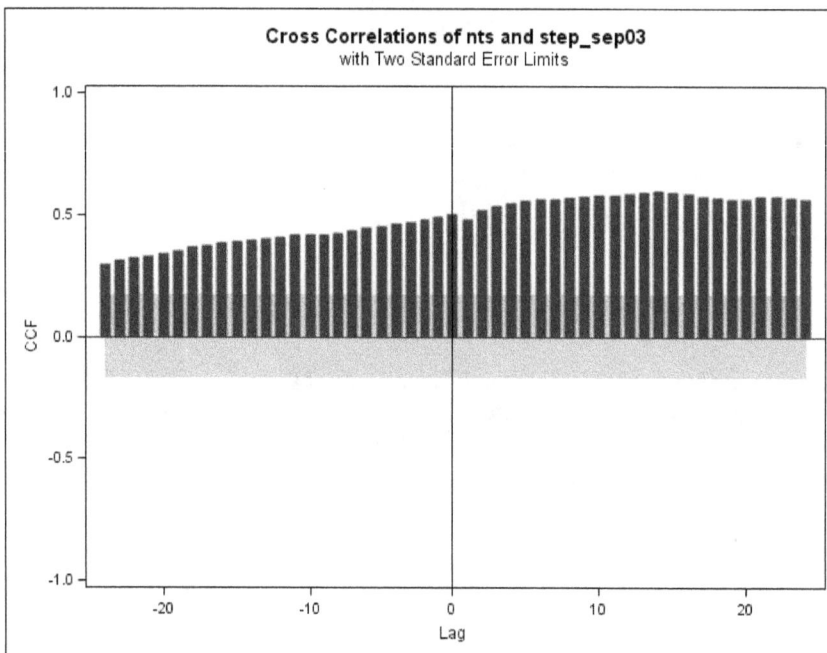

Despite the unsatisfactory event model identification, the two preliminary models are specified in the ESTIMATE statements based on visual scrutiny of the plot of the data (Figure 9.1). These are designated the Temporary Change model and the Permanent Change model.

Program 9.5

```
ods graphics on;

proc arima data=nts plots(unpack)=series(all);
identify var=nts crosscorr=(t pulse_sep03 step_sep03);
estimate input=(t 1$/(1)pulse_sep03) method=ML plot;
estimate input=(t step_sep03) method=ML plot;
run; quit;

ods graphics off;
```

Refining the Residuals of the Temporary Change Model

The parameter estimates table for the Temporary Change model indicates that no insignificant terms have been included.

Output 9.9

	Maximum Likelihood Estimation						
Parameter	Estimate	Standard Error	t Value	Approx Pr > |t|	Lag	Variable	Shift
MU	93181.6	258.46021	360.53	<.0001	0	nts	0
NUM1	121.76866	4.26760	28.53	<.0001	0	t	0
NUM2	-11511.6	623.59877	-18.46	<.0001	0	pulse_sep03	1
DEN1,1	0.94890	0.0059101	160.56	<.0001	1	pulse_sep03	1

The autocorrelation check of residuals indicates that the residuals are not white noise.

Output 9.10

	Autocorrelation Check of Residuals								
To Lag	Chi-Square	DF	Pr > ChiSq	Autocorrelations					
6	36.32	6	<.0001	0.394	0.261	0.131	0.137	0.040	0.029
12	41.64	12	<.0001	0.103	0.049	0.107	-0.022	0.018	0.104
18	74.32	18	<.0001	-0.122	-0.168	-0.336	-0.156	-0.147	-0.098
24	80.13	24	<.0001	-0.092	-0.065	0.036	-0.134	-0.048	-0.040

Using rules presented in Chapter 8 , the ACF and PACF plots of the residuals indicate that reasonable candidate models are either p=(1 12) or q=2. An autoregressive lag of 12 makes sense for monthly data—this is the seasonal lag. Significant spikes at lags higher than 12 should be suspected of being spurious unless knowledge of the data indicates that they are meaningful. Further significant lags will be considered if necessary after the residuals of these models are assessed.

Output 9.11

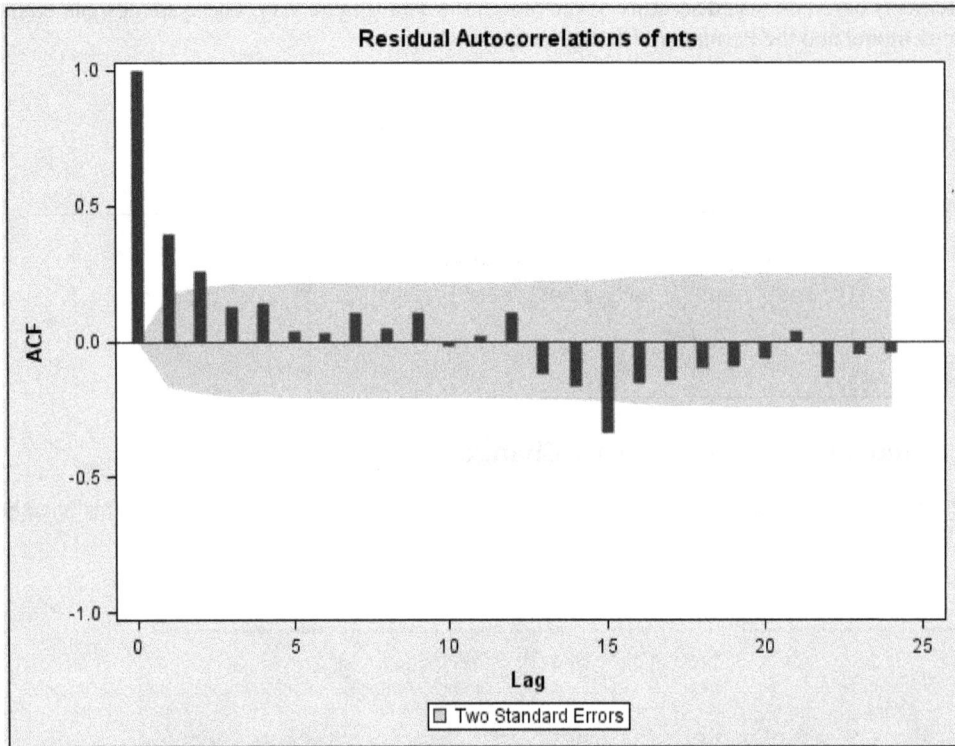

Output 9.12

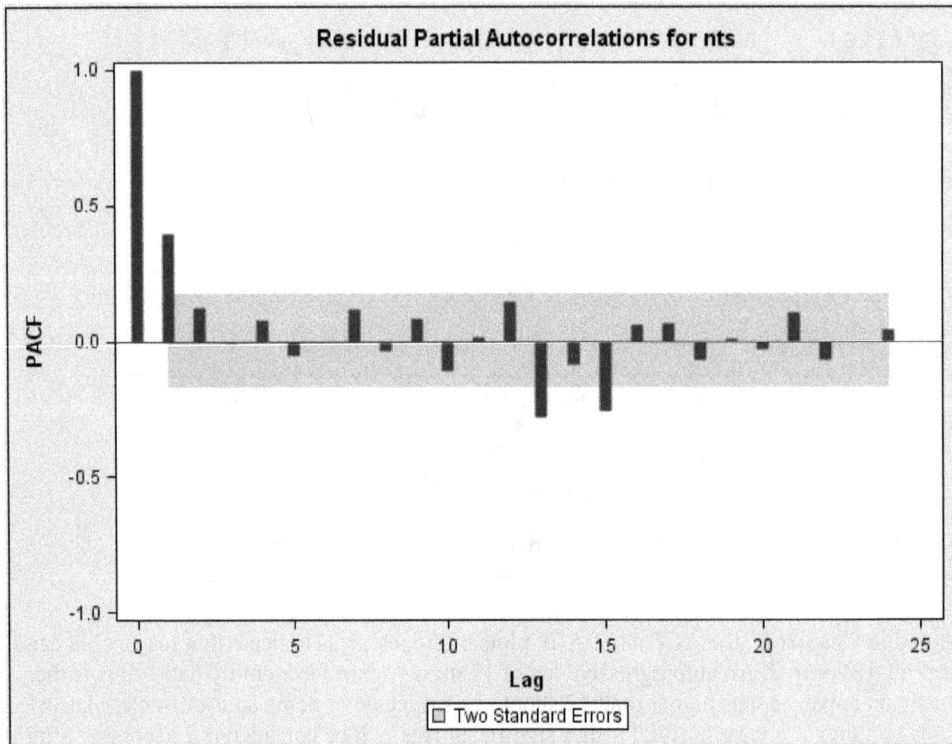

PROC ARIMA syntax is modified to augment the Temporary Change model. The Permanent Change model has been temporarily commented out.

Program 9.6

```
ods graphics on;

proc arima data=nts plots(unpack)=series(all);
identify var=nts crosscorr=(t pulse_sep03 step_sep03);
estimate input=(t 1$/(1)pulse_sep03) q=(1 2) method=ML plot;
estimate input=(t 1$/(1)pulse_sep03) p=(1 12) method=ML plot;
*estimate input=(t step_sep03) method=ML plot;
run; quit;

ods graphics off;
```

The autocorrelation check of residuals of the Temporary Change model augmented with MA terms is shown below.

Output 9.13

Autocorrelation Check of Residuals									
To Lag	Chi-Square	DF	Pr > ChiSq	Autocorrelations					
6	5.06	4	0.2816	0.013	0.057	0.103	0.147	-0.014	0.000
12	16.97	10	0.0750	0.105	-0.007	0.138	-0.088	-0.004	0.206
18	30.42	16	0.0159	-0.103	-0.040	-0.260	0.005	-0.074	-0.042
24	38.84	22	0.0147	-0.068	-0.051	0.130	-0.165	-0.016	-0.011

The null hypothesis of white noise is rejected when tested at lags 1 through 18. Note the relatively large estimated autocorrelations at lags 12 and 15.

The autocorrelation check of residuals of the Temporary Change model augmented with AR terms is shown below.

Output 9.14

Autocorrelation Check of Residuals									
To Lag	Chi-Square	DF	Pr > ChiSq	Autocorrelations					
6	7.92	4	0.0944	-0.042	0.171	0.027	0.157	-0.007	-0.004
12	15.64	10	0.1103	0.128	-0.029	0.121	-0.077	-0.001	0.120
18	35.99	16	0.0029	-0.173	-0.018	-0.299	0.018	-0.108	-0.032
24	44.18	22	0.0034	-0.063	-0.073	0.115	-0.159	0.012	-0.048

The null hypothesis of white noise is rejected when tested at lags 1 through 18. Note the relatively large estimated autocorrelation at lag 15.

Both models are augmented with the significant lags. The modified PROC ARIMA code is shown below.

Program 9.7

```
ods graphics on;

proc arima data=nts plots(unpack)=series(all);
identify var=nts crosscorr=(t pulse_sep03 step_sep03);
estimate input=(t 1$/(1)pulse_sep03) q=(1 2 12 15) method=ML plot;
estimate input=(t 1$/(1)pulse_sep03) p=(1 12 15) method=ML plot;
*estimate input=(t step_sep03) method=ML plot;
run; quit;

ods graphics off;
```

The null hypothesis of white noise continues to be rejected at lags 1 through 6 for the Temporary Change model augmented with MA terms as shown below.

Output 9.15

Autocorrelation Check of Residuals									
To Lag	Chi-Square	DF	Pr > ChiSq	Autocorrelations					
6	10.21	2	0.0061	0.054	0.081	0.163	0.165	-0.067	0.068
12	15.41	8	0.0516	0.135	-0.015	0.102	-0.069	0.043	0.004
18	19.49	14	0.1470	-0.056	-0.032	-0.086	-0.115	-0.038	-0.015
24	25.61	20	0.1790	-0.096	-0.081	0.089	-0.103	-0.050	0.032

The residuals of the Temporary Change model augmented with AR terms are clean. The model is adequate and will be used as the Temporary Change representation of the event for the remainder of the analysis.

Output 9.16

Autocorrelation Check of Residuals									
To Lag	Chi-Square	DF	Pr > ChiSq	Autocorrelations					
6	6.11	3	0.1062	0.000	0.126	0.049	0.143	-0.055	0.041
12	12.14	9	0.2056	0.103	0.013	0.133	-0.076	0.005	0.081
18	18.43	15	0.2409	-0.160	-0.048	-0.068	0.089	-0.022	0.020
24	22.58	21	0.3669	-0.067	-0.068	0.078	-0.101	0.008	-0.013

Assessing the Candidate Models

Similar steps were taken to refine the residuals of the Permanent Change model. The final model specification is shown below.

Program 9.8

```
ods graphics on;

proc arima data=nts plots(unpack)=series(all);
identify var=nts crosscorr=(t pulse_sep03 step_sep03);
estimate input=(t 1$/(1)pulse_sep03) p=(1 12 15) method=ML plot;
estimate input=(t step_sep03) p=(1 2 12) method=ML plot;
run; quit;

ods graphics off;
```

The fit of both models and the parameter estimates of the winning representation of the event can now be assessed.

Fit diagnostics of the Temporary Change representation of the event are shown below.

Output 9.17

Constant Estimate	69773.51
Variance Estimate	1439159
Std Error Estimate	1199.65
AIC	2306.443
SBC	2326.78
Number of Residuals	135

Fit diagnostics of the Permanent Change representation of the event are shown below.

Output 9.18

Constant Estimate	30692.37
Variance Estimate	3392481
Std Error Estimate	1841.869
AIC	2455.214
SBC	2472.734
Number of Residuals	137

The Temporary Change representation of the event has a better overall fit to the data based on both AIC and SBC. It is chosen as the forecast model for the data.

The parameter estimates of the Temporary Change representation of the event are shown below.

Output 9.19

	Maximum Likelihood Estimation						
Parameter	Estimate	Standard Error	t Value	Approx Pr > |t|	Lag	Variable	Shift
MU	93316.6	314.36804	296.84	<.0001	0	nts	0
AR1,1	0.37655	0.07646	4.92	<.0001	1	nts	0
AR1,2	0.17297	0.08011	2.16	0.0308	12	nts	0
AR1,3	-0.29723	0.08138	-3.65	0.0003	15	nts	0
NUM1	118.94415	5.16275	23.04	<.0001	0	t	0
NUM2	-12572.0	886.09392	-14.19	<.0001	0	pulse_sep03	1
DEN1,1	0.94054	0.0083565	112.55	<.0001	1	pulse_sep03	1

All estimated coefficients are significant (95% confidence level). The NUM1 coefficient is the estimated slope of the linear trend. The NUM2 coefficient is the estimated effect of the event in November 2003 (shift=1). The DEN1,1 coefficient is the estimated coefficient that regulates the trajectory back to the status quo intercept and trend line after the event. This coefficient can be interpreted as follows:

- The effect of the event in November 2003 is given by the NUM2 coefficient=-12,572.
- The effect of the event in December 2003 is the product of NUM2 and DEN1,1 coefficients: -12,572*0.94=-11,818.
- The effect of the event in January 2004 is the product of NUM2 and DEN1,1 squared: -12,572*0.94*0.94=-11,109.
- The effect of the event in February 2004 is the product of NUM2 and DEN1,1 cubed: -2,572*0.94^3=-10,442, and so on.

The estimated coefficient for DEN1,1 must be less than one in absolute value. Otherwise, the effect of the event would build over time. The series would go off to negative infinity. The closer this estimated coefficient is to one, the longer the persistence of the effect of the event, or the longer it takes the series to transition back to the steady state slope and intercept following the event. The persistence of the effect of the event is long.

PROC ARIMA code is modified to generate forecasts using the Temporary Change event model.

Program 9.9

```
ods graphics on;

proc arima data=nts plots(unpack)=series(all) plots=forecast(all);
identify var=nts crosscorr=(t pulse_sep03 step_sep03);
estimate input=(t 1$/(1)pulse_sep03) p=(1 12 15) method=ML plot;
forecast interval=month id=date lead=18 out=tmp_cngfor printall;
run; quit;

ods graphics off;
```

The FORECAST statement has been added to the code. Lead forecasts are generated 18 months into the future. An output data set, tmp cng for, containing the within sample and lead forecasts, standard errors, and confidence limits, is generated.

Output 9.20

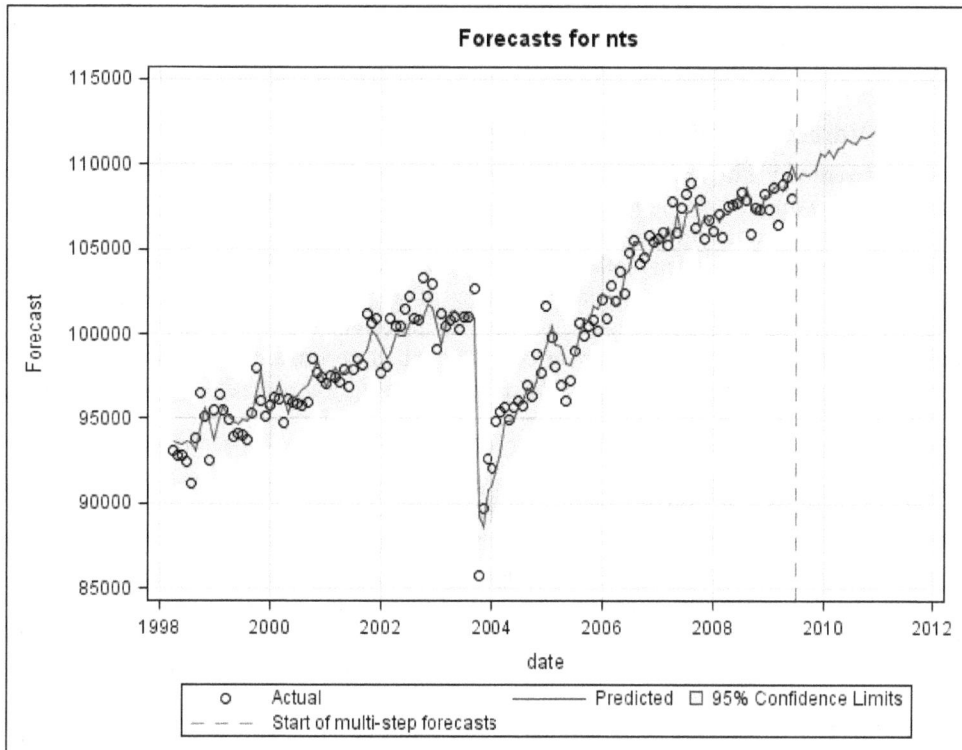

The intention was to present an example of building time series regression models with deterministic inputs. This provided a fairly succinct framework in which to tie together the conceptual and applied elements of time series analysis. However, you might be wondering, "My data has thousands of series to be forecast and hundreds of candidate inputs to consider. Is it necessary to spend weeks looking at CCF plots and refining residuals on perhaps thousands of candidate models to produce good forecasts?" These questions define the large-scale forecasting problem. Algorithms can be extremely useful in this situation.

The most widely used approach when using modeling algorithms in a large-scale forecasting scenario is forecasting by exception (Leonard 2008). This approach implies that algorithms do a good job of model building and forecasting *in the majority of series* in a large-scale forecasting scenario if analysts perform the proper steps in preparing the data and in eliminating irrelevant and redundant input variables. Exceptions to the majority are flagged for intervention and refinement. Flagged series are the result of either a deficiency in accuracy associated with system-generated models, or because the associated value of a series is sufficiently high enough to warrant additional analyst time and expertise. It is on the exceptional series that forecast analysts use the tools presented above and that are now in their toolbox.

Appendix: Prewhitening and Other Topics Associated with Interval-Valued Input Variables

Introduction

Interval or continuously valued input variables add another layer of complexity to the modeling process. This is because these types of inputs can have their own seasonal, trend, and autoregressive components that need to be considered in the identification portion of the modeling process.

When inputs are autocorrelated, over-parameterized models are usually indicated by the diagnostics that are used to identify the appropriate model specification. The technique of prewhitening inputs or stripping out their autoregressive component before model identification is a widely used correction for model identification diagnostics.

Global Systematic Variation and Further Topics

It is relatively straightforward to identify how white noise input variables like dummy variables enter model specification. In contrast, interval-valued candidate input series might have embedded systematic variation, which makes the task of model identification more complex. Consider two interval-valued series that happen to have similar cycles. A widely used example is solar activity and US GNP. Both of these series contain a cycle that repeats about every seven years. A cross-correlation analysis of the two series would likely show significant correlation at some lag. However, the structural causes underlying the cycles and other variation in each series are different. After removing the seven-year cycles from both series, any remaining correlation would likely be spurious. Not many macro-economists have sunspot activity as a structural component in their GNP forecast models.

As a further example, consider building a forecast model for the national sales revenue of a firm that produces peanut butter. Consumer income is a candidate independent variable. Assume that the data are taken from a time when inflation is increasing. An examination of the data will likely reveal that both revenue and income are trending upward as a result of the inflation effect on wages and the prices that the firm charges. The two series appear to be highly correlated. It is desirable to remove the effect of inflation on revenues and income before model identification. Variation in income might actually have very little effect on peanut butter sales, but it will be hard to discern the true relationship until the trend in both series has been removed.

Global variation components in series are usually defined as large or obvious series characteristics. Trend, cycle, and seasonal variation are the most common forms of global variation. When identifying how input variables should enter a forecast specification, global variation in the input and target series is typically considered to be nuisance variation. This leads to a straightforward rule for forecast model identification with interval-valued candidate input variables. Target and candidate independent variables should be detrended, de-seasonalized, and de-cycled before model identification. A common method for doing this is to take appropriate differences of the series. (See Appendix 2 in Chapter 8 for a discussion on differencing.)

Two related ideas deserve discussion. First, differencing has been criticized as an overly harsh way to remove global systematic components from series. Second, when many analysts think of detrending, it is in the context of a deterministic trend. Stochastic trend or non-stationarity in the target and input variables is also a concern. (See Appendix 2 in Chapter 8 for a discussion of stationarity.) Regressing one nonstationary series on another can lead to spurious results. The exception to this is when the two nonstationary series are cointegrated. (See Granger 1969.) The basic idea of cointegration is that two nonstationary series, y and x, are cointegrated if a linear combination of them is stationary. In practice, the residuals of a regression of y on x are the linear combination that is tested. If the residual series meets the criteria of white noise, then valid results can be obtained from the regression. Cointegration might provide a preferred alternative to differencing when the target and input are nonstationary.

Prewhitening

A more subtle problem in model identification arises when the candidate input variable is autocorrelated. Autocorrelated input variables (that are also correlated with the target) usually generate lots of significant spikes in cross-correlation plots. Most of these spikes are spurious from a model identification point of view. To see why this happens, consider the plots shown below.

In the first pair of plots, a white noise candidate input variable, x1, pulses at time=11. This is followed by a spike in the dependent variable, y, at time=16. The relationship between the candidate input variable and the target is easy to see—a shift 5 relationship.

Figure 9.3

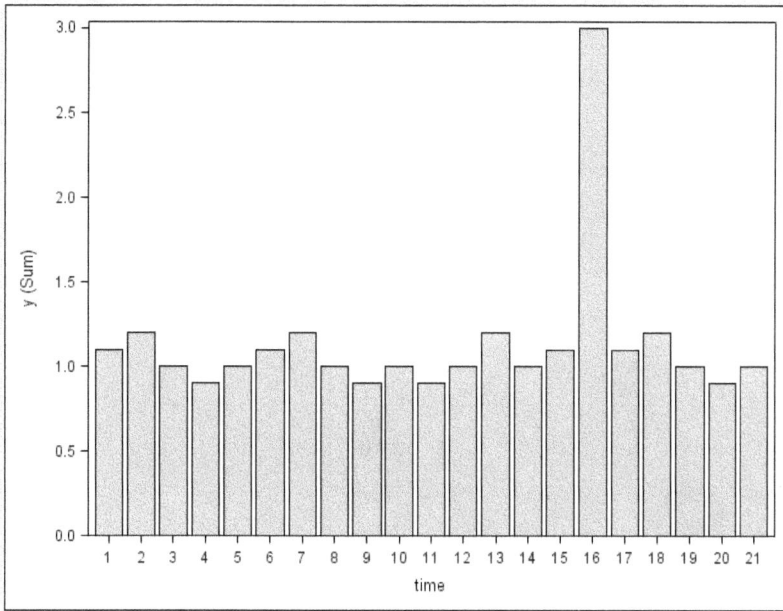

Figure 9.4

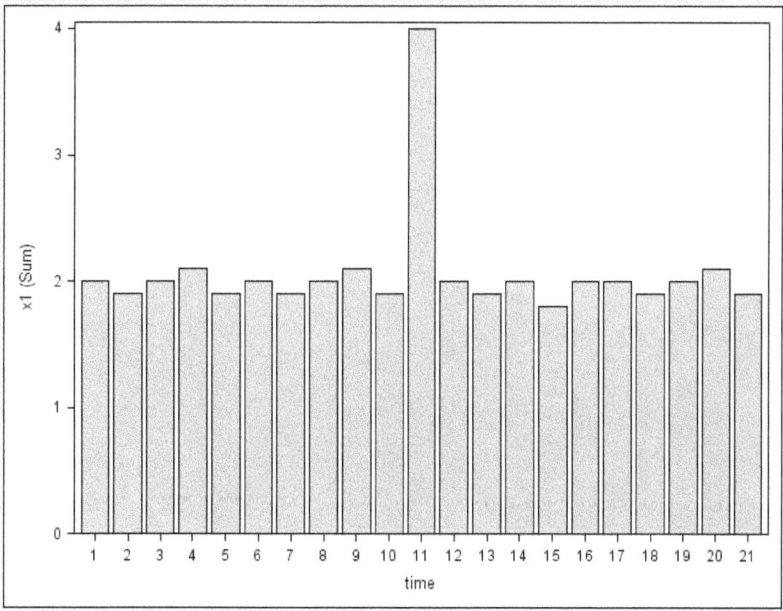

In the second pair of plots, the same target series is shown, but the candidate input is x2. The input x2 fits the description of a stationary series with an ARMA component. It jumps above its mean at time=11 and stays above its mean for about 4 periods before reverting. The relationship between y and x2 is difficult to discern. Is the x2 impulse that is correlated with the spike in y at lag 11, lag 13, or elsewhere?

Figure 9.5

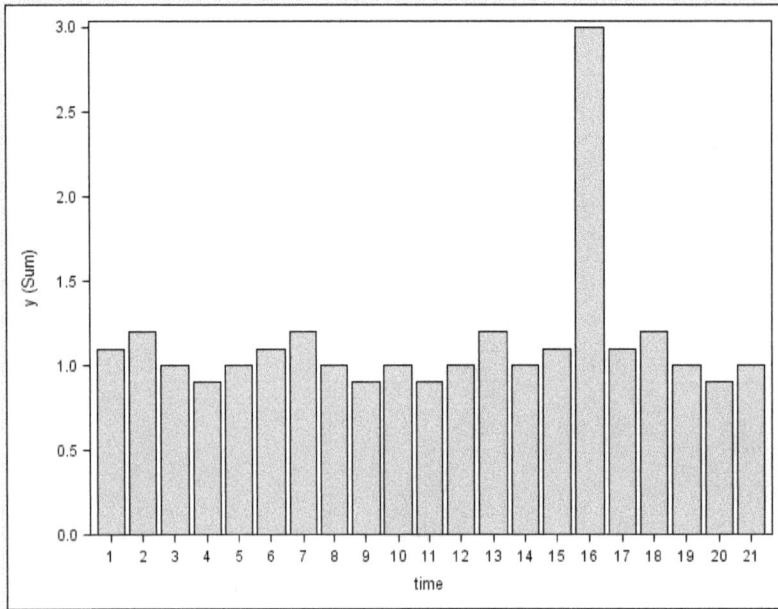

Figure 9.6

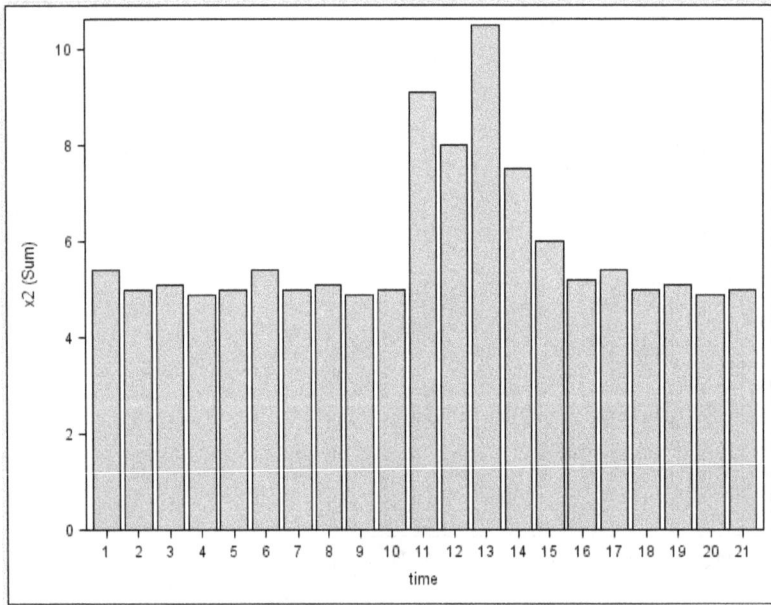

Estimating the cross-correlation at various lags and then plotting them does not help the problem. The cross-correlation between y and x2 is done pairwise with respect to the lags. The cross-correlation function plot will likely have significant spikes corresponding to all of the substantial jumps in the x2 plot (lag 5, lag 4, lag 3, and lag 2).

A widely used approach for identifying autocorrelated candidate input variables is called *prewhitening*. Prewhitening helps solve the problem by stripping out the ARMA component of the candidate input series. It reduces the input series to white noise, and it should make the task of model identification easier. Here are the steps:

1. If the target and input series contain trend, seasonal, or cycle variation, perform appropriate steps to remove or accommodate these components.

2. Fit and estimate the ARMA model on the candidate input series and determine whether the residuals of the model are white noise. This ARMA specification is called the *prewhitening filter*.

3. Apply this filter to the dependent or target series, y.

4. Cross-correlate the residuals of the y model with the filtered candidate input series generated in step 2.

Methods

SAS Forecast Server does the prewhitening on interval-valued candidate inputs before the automated process of model identification, estimation, selection, and forecasting. However, many analysts want to manually identify models for series that are considered high value or when information about the marginal effects of input variables contained in a specification is of interest.

The ARIMA procedure has built-in functionality that relieves analysts of the more tedious steps in the prewhitening process. All that the analyst needs to do is specify the prewhitening filter for each candidate interval-valued input variable before identifying the target variable.

In the syntax below, the target variable, y2, is identified. A cross-correlation analysis with an interval-valued input variable, y1, without prewhitening, is requested. (See the ARMA identification portion of this chapter and SAS/ETS for explanations of the syntax used.[6])

Program 9.10

```
ods graphics on;
ods html;

proc arima data=varmax_sim plots=all;
    identify var=y2 crosscorr=(y1);
run;
quit;
```

The generated cross-correlation function plot indicates that the two series are correlated. However, many parameters are needed to accommodate all of the significant positive lags.

Output 9.21

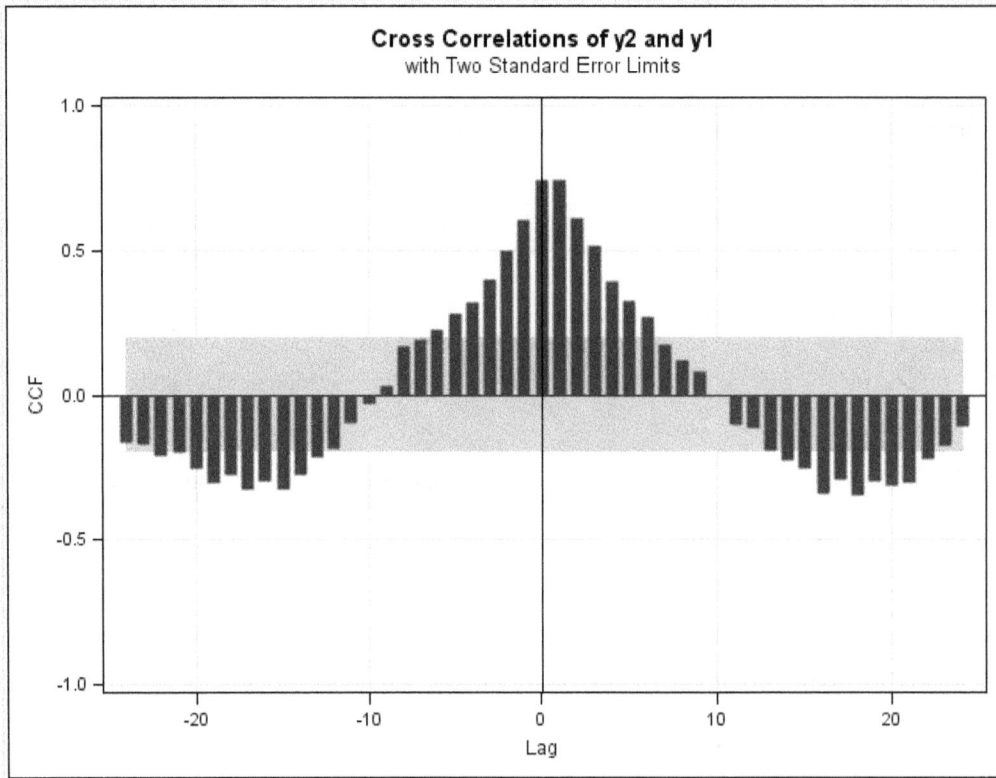

Cross Correlations of y2 and y1
with Two Standard Error Limits

Because the y1 series is interval valued, it is suspected that some of the significant spikes in the cross-correlation function plot are a result of autocorrelation in y1 and are spurious. The next block of syntax investigates this by identifying the y1 variable. (ODS GRAPHICS statements are not shown to eliminate clutter, and the y2 variable identification step has been commented out.)

Program 9.11

```
proc arima data=varmax_sim plots=all;
    identify var=y1;

    *identify var=y2 crosscorr=(y1);
run;
quit;
```

The generated diagnostic plots indicate that the y1 series is likely an AR1.

Output 9.22

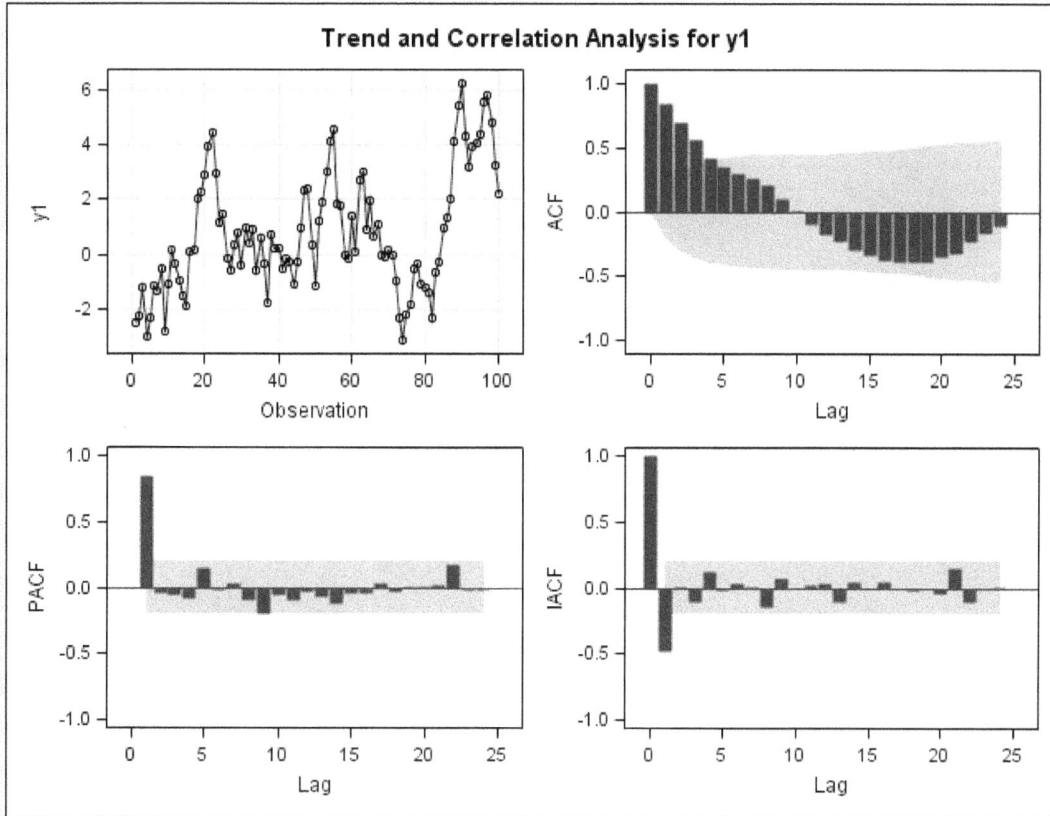

Trend and Correlation Analysis for y1

An appropriate model is fitted to the y1 series in the following syntax.

Program 9.12

```
proc arima data=varmax_sim plots=all;
    identify var=y1;
    estimate p=1 plot;

    *identify var=y2 crosscorr=(y1);
run;
quit;
```

Diagnostics on the y1 model residuals confirm that they are white noise. The AR1 model is the prewhitening filter for the y1 variable.

Output 9.23

To Lag	Chi-Square	DF	Pr > ChiSq	Autocorrelations					
6	3.88	5	0.5662	0.001	0.021	0.008	-0.183	0.035	-0.037
12	9.41	11	0.5837	0.087	0.183	0.005	0.051	-0.051	-0.056
18	10.71	17	0.8714	0.044	-0.070	0.026	-0.053	-0.011	-0.022
24	14.90	23	0.8982	-0.075	-0.013	-0.145	0.033	0.038	0.055

Autocorrelation Check of Residuals

The final step in identifying the y2 variable is shown below.

Program 9.13

```
proc arima data=varmax_sim plots=all;

/* step 1; generate the prewhitening filter for the candiate
           input variable, y1                                 */

    identify var=y1;
    estimate p=1 plot;

/* step 2; identify the target variable, y2, and generate the
           CCF plot for y1 & y2                                */

    identify var=y2 crosscorr=(y1);
run;
quit;
```

The generated cross-correlation function plot shows the prewhitened relationship between the two variables. Prewhitening y1 before cross-correlating with y2 has altered the estimated relationship substantially. A reasonable specification for y2 could include y1 at lag 0 and lag 1.

Output 9.24

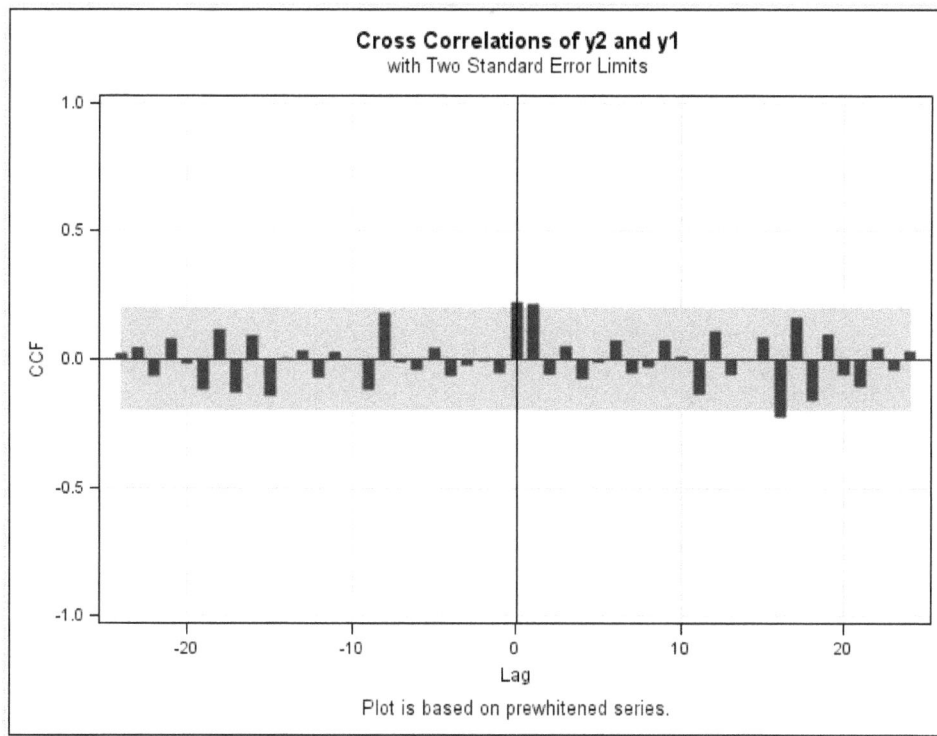

You should note the text at the bottom of the CCF plot. It indicates that once the prewhitening filter on the independent variable is specified in PROC ARIMA, it is automatically used to filter the target variable before series identification. This functionality makes the analyst's life easier, particularly when several candidate independent and target variables are being considered. All that the analyst needs to do is specify the filter for each candidate independent variable.[7] Specified filters are then automatically applied to the target variables before identification when any of these candidate independent variables appear in the CROSSCORR option.

---

[1] Spikes at negative lags are ignored. These represent lead relationships between Y and X. Variation in current values of Y are correlated with variation in future values of X. Lead relationships between Y and X are also known as feedback effects. These cannot be accommodated in the modeling framework. The Vector Auto-Regression (VAR) framework does accommodate feedback effects.

[2] A denominator order 1 implies a smooth transition back to the status quo. A denominator order 2 implies an oscillating transition back to the status quo. The remainder of this section presents three CCF plots and discusses models that would be reasonable to specify based on the rules. All caveats about multiple comparison problems with cross-correlation function plots are still relevant. The models suggested are reasonable and defensible, but there is no guarantee that the proposed models are correct.

[3] Start counting at the first significant spike when specifying numerator orders.

[4] A smooth decay can be interpreted as evidence of a denominator order 1. An up-and-down decay could be interpreted as evidence of a denominator order 2.

[5] Further understanding of modeling interrupted time series and using deterministic trend variables can be gained by taking the SAS Education course, "Forecasting Using SAS Software: A Programming Approach."

[6] The input and target series are simulated. They are both stationary. No detrending is necessary, but they are both strongly autocorrelated at lag 1. The CCF plots in the Series Analysis View of SAS Forecast Studio are not based on prewhitened relationships.

[7] It should be emphasized that for analysts in a large-scale forecasting environment, the above steps are not feasible for the majority of their target and candidate input variables. There are too many series to manually process and scrutinize. In this environment, the steps outlined should be reserved for high-value series or when the precision of marginal effects of inputs are valuable. SAS Forecast Server provides prewhitening functionality for model identification in a large-scale forecasting environment.

Chapter 10: Model Building: Further Modeling Topics

Introduction

This chapter presents additional modeling topics. The focus shifts to a collection of issues that arise frequently in applied forecasting. The topics in the first part explain and provide solutions to issues associated with a large-scale forecasting scenario. The example data sets are small and manageable, but the concepts and techniques can be used with data sets of a more realistic size for large-scale forecasting. The topics in the second part describe solutions to other topics that arise in applied forecasting. The methods and strategies are modeling tools related to the following topics:

- creating time series data and data hierarchies using accumulation and aggregation methods
- reconciling statistical forecasts
- intermittent data
- high-frequency data and mixed-frequency forecasting
- holdout samples and forecast model selection in time series
- planning vs. forecasting and manual overrides
- scenario-based forecasting
- new product forecasting

10.1 Creating Time Series Data and Data Hierarchies Using Accumulation and Aggregation Methods

Introduction

This section presents tools and ideas associated with building time series data and data hierarchies from transactional series data. A time series is a set of equally spaced observations with respect to time. Data hierarchies are nested representations of several time series, and their structure is determined by whatever is logical for the entity generating the data. For example, we saw in a previous chapter that tennis shoe sales seem to be naturally arranged by the locations in which the sales occurred. So, a data hierarchy for tennis shoe sales might consist of individual shoe type sales by store location at the bottom and total sales across all regions and shoe types at the top. Several intermediate or middle levels are possible—sales by city, by state, by region, and so on.

There are many ways to roll up transactional data into time series and time series into a hierarchy. However, the primary idea that an analyst needs to consider is that the process of creating the data and hierarchy has an impact on the raw transactional data that flows into the process. *The methods used to create the data and hierarchy should maximize the strength of the signal that the raw data has.* Traditional product-based business structures might not be aligned properly to the market. Market attribute-based hierarchies provide higher-quality forecasts.

There are tradeoffs implicit in data-handling choices. For example, rolling up the raw data into a monthly interval might give the best look at any annual seasonal cycle that the data has, but it obliterates any day-of-the-week effect that might be present.

Next we will introduce tools for creating time series data and data hierarchies from transactional time-stamped data.[1] We will emphasize the idea of rolling up data to reveal patterns of interest.

Many analysts might be thinking, "These ideas are fine in theory, but my CFO insists on forecasts that are arranged in such a way that the net present value can be easily extracted." This thought is addressed in post-forecasting, post-processing steps. The primary goal of an analyst should be to create a data structure in which components of variation in the data that drive value can form a basis for subsequent modeling steps in a straight forward way.

Creating Time Series Data Using Accumulation Methods

The example data so far has been time series data. The definition of a time series is a set of *equally spaced,* with respect to an index or time interval, observations. Recall how AR models work. Systematic correlation between observations (for example, three intervals apart) is accommodated and quantified. If the data is not a time series, the parameters of the model are meaningless.

Raw data for forecasting is usually transactional or time-stamped. The data is not equally spaced with respect to a time interval. The choice of an interval is made by the analyst, and an interval must be selected and applied to the raw data. We will discuss the methods for rolling up or accumulating raw transactional data in an equally spaced time interval. The choices of an interval and accumulation method impact the systematic components that the data contains. You should carefully choose an accumulation method and time interval that can improve the accuracy and usefulness of generated forecasts.

The following data is a portion of raw transactional data to be used in a forecasting analysis. The data is the sales for a line of tennis shoes mysteriously branded with the single letter "Y." A preliminary look at the data reveals that some week intervals have more than one sales value, and some week intervals have no sales values or missing values for sales.

Figure 10.1

.	15JAN10:00
208252.09004	18JAN10:00
275633	22JAN10:00
224585.56241	23JAN10:00
.	28JAN10:00
302603	29JAN10:00
233416.4816	02FEB10:00
310442	05FEB10:00
234513.28122	07FEB10:00
248115.93616	12FEB10:00
293082.76072	17FEB10:00
281421	19FEB10:00
300031.72167	22FEB10:00
300451	26FEB10:00
.	27FEB10:00
269455.50391	04MAR10:00
275633	05MAR10:00
224552.75321	09MAR10:00

A scatterplot shows that the range of the data is from about September 2009 to September 2010. Sales appear to be tracking downward, but no other patterns are obvious.

Output 10.1

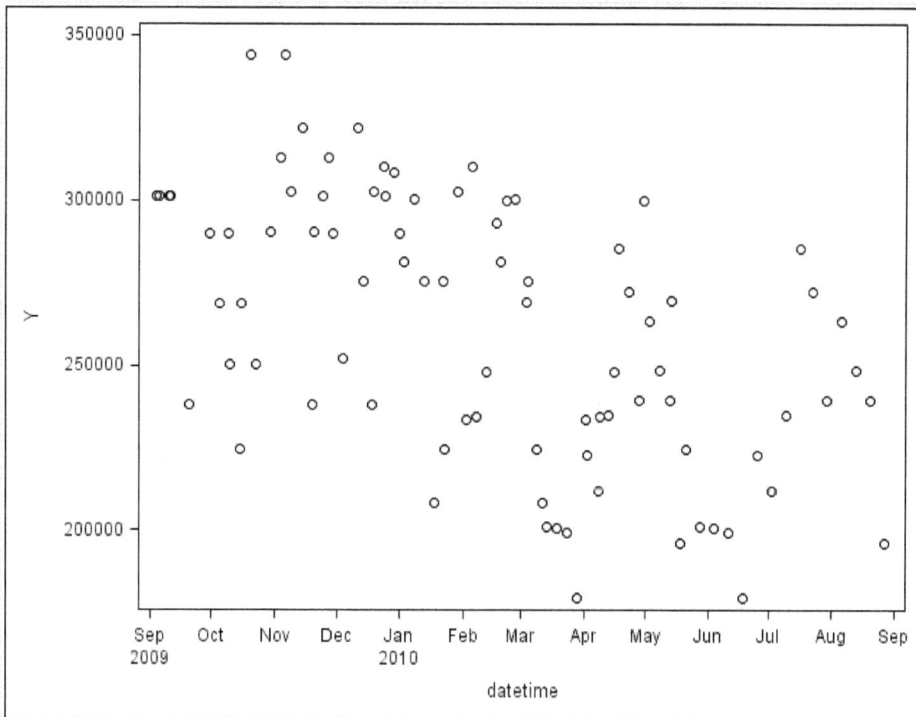

Methods

A useful tool for time series data handling is PROC TIMESERIES. PROC TIMESERIES is used to create time series data from raw transactional data. The initial interval selected is week.

Program 10.1

```
ods html;
ods graphics on;
options helpbrowser=sas;

proc timeseries data=y_sales out=y_week plots=(series);
id datetime interval=dtweek accumulate=total setmissing=0;
var y;
run;
```

A summary of the syntax follows. Additional details are found in PROC TIMESERIES documentation for SAS/ETS.

- The ODS statements turn on the plotting functionality.
- The DATA= option references the input data set containing the transactional data.
- The OUT= option names the output data set containing the accumulated weekly interval data.
- The PLOTS= option indicates the types of plots to be generated.
- The ID statement lists the ID variable in the input data set and the appropriate interval for accumulation.[2]
- An accumulation method TOTAL is used to create equally spaced observations. This means that the sales observation for each week is the sum of all sales that occurred in that week.
- SETMISSING=0 indicates the imputation method used for missing values of Y sales. A zero is plugged in for any week interval with a missing value for Y sales. The assumption is that weeks with no sales observations are weeks in which no sales of Y occurred.

Here is the generated plot of the accumulated series:

Output 10.2

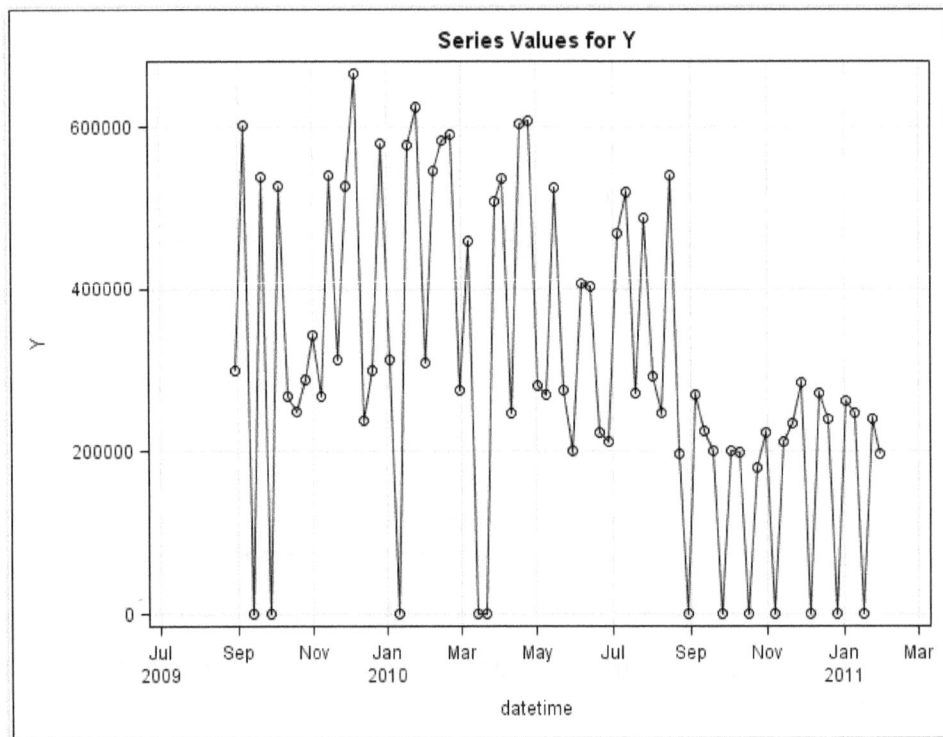

There are at least two concerns with the methods that roll up the data. First, imputing zeros for missing values does not seem appropriate. Second, there is a shift in the data starting around September 2010 that was not obvious in the plot of the transactional series. This is a pattern that emerged in the accumulation method. One explanation is that many weeks before September 2010 had more than one sales observation, and most weeks after had one or zero sales observations. Sales have approximately the same magnitude, but become more spread out in recent data.

Although this pattern might be valuable for total sales, care should be used in choosing the total accumulation method. For example, accumulation using total is generally not appropriate for a price series or series that is based on a measure besides count (for example, percent).

Using another method of accumulation (and imputation) emphasizes different types of patterns in the data. (ODS statements are not shown to eliminate clutter.)

Program 10.2

```
proc timeseries data=y_sales out=y_week plots=(series);
id datetime interval=dtweek accumulate=average setmissing=average;
var y;
run;
```

Output 10.3

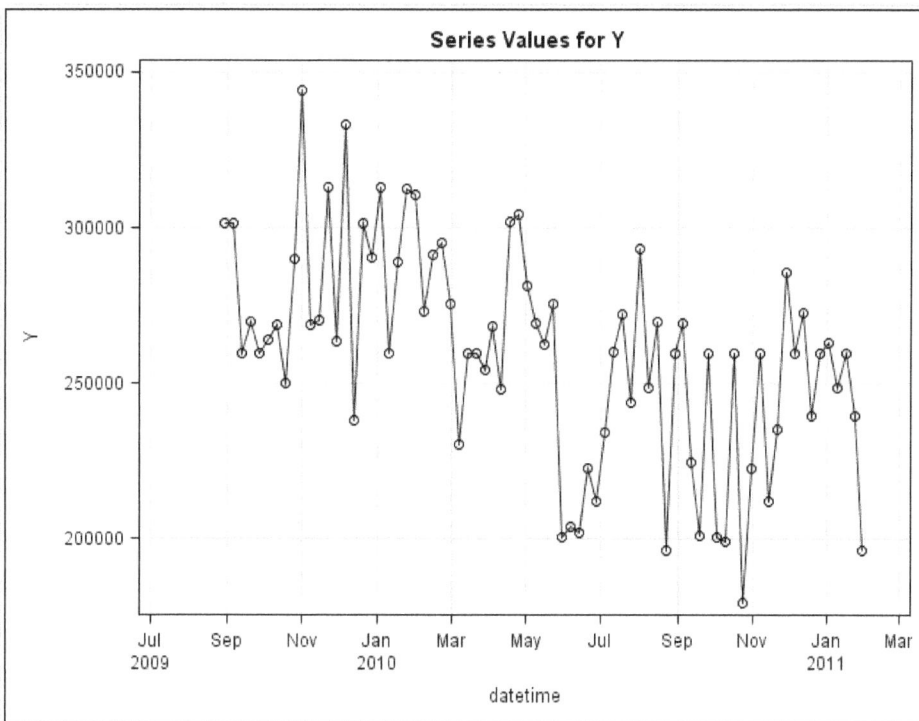

Although the imputation method, AVERAGE, might be problematic, it emphasizes the overall cycle and the trend patterns in Y sales over the data.

If cycle and trend patterns in the data are of primary interest, one method of examining them is to roll up the data into a lower frequency (for example, monthly interval). The imputation method NEXT is used to handle missing values of Y. This means that any missing values are changed to the next nonmissing value of Y in the series.

Program 10.3

```
proc timeseries data=y_sales out=y_week plots=(series);
id datetime interval=dtmonth accumulate=average setmissing=next;
var y;
run;
```

Here is the generated plot:

Output 10.4

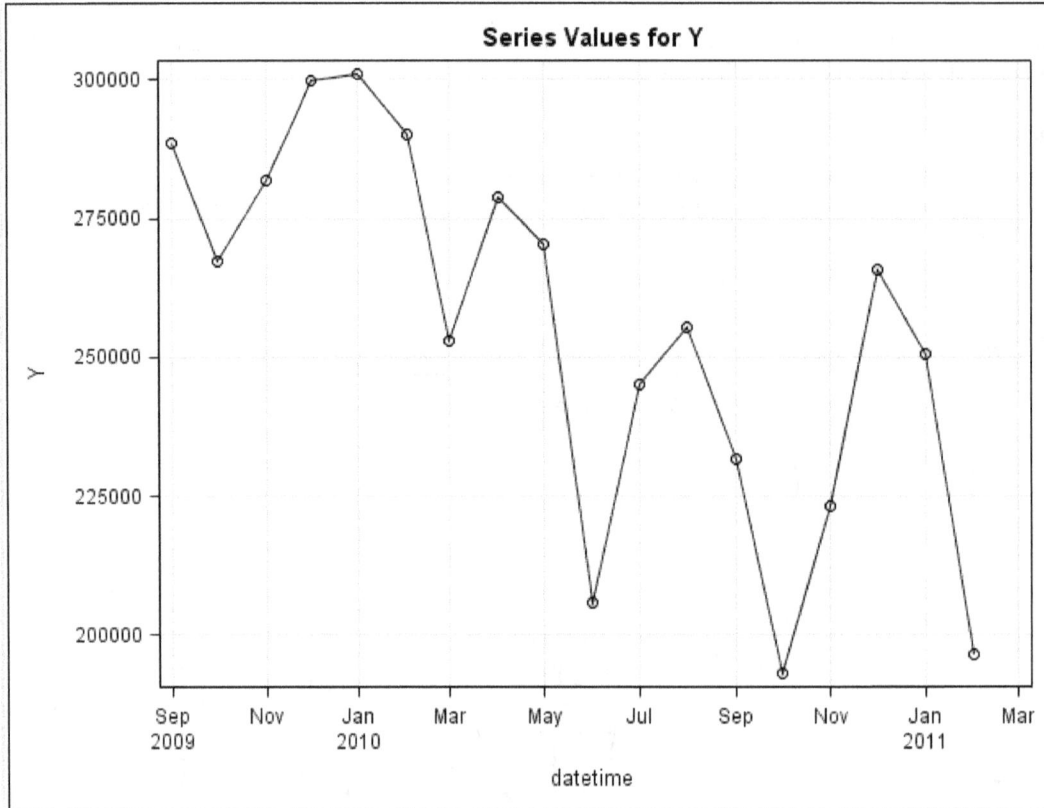

It appears that average sales have been dropping steadily over the range of the data. There is evidence of an annual seasonal pattern that peaks in December or January.

Commonly used accumulation methods available in PROC TIMESERIES are NONE (default), TOTAL, AVERAGE, MAXIMUM, MINIMUM, MEDIAN, NEXT, and PREVIOUS. (See SAS/ETS documentation for additional details and less widely used accumulation options.)

Creating Data Hierarchies Using Aggregation Methods

Aggregation means creating data hierarchies from time series data. Data hierarchies are nested representations of time series data. Because the representations are nested, only one forecast or dependent variable is allowed. This fact is not as restrictive as it sounds. All of the tennis shoe series can be accommodated in a data hierarchy by using a single dependent variable (sales or units). You can differentiate between brands with an additional variable that codes shoe types. This additional subsetting variable is called a BY variable in SAS.

The levels of the hierarchy are constructed or aggregated using a BY variable. In the previous example, an additional BY column that flags sales by store generates a second level in the hierarchy. Using BY variables, a three-level hierarchy can be created: shoe sales by store and type on the bottom, sales for each store in the middle, and sales across all stores in each time interval on the top.

Accumulation and aggregation share identical methods—TOTAL, AVERAGE, and so on. They comprise the two steps of building a data hierarchy. A representative raw data set in large-scale forecasting usually consists of a time stamp or ID column, a dependent variable column (for example, sales), several candidate input variable columns, and several BY variable columns.

The first step in creating the data hierarchy is accumulating the transactional series to time series data at the lowest level of the hierarchy. In the previous example, this involves accumulating the data by the store and then by brand. The bottom level is listed last and consists of sales of brand and store pairs. Subsequent aggregation steps consist of rolling up the bottom-level data or aggregating to create the middle and top levels in the hierarchy. Following the previous example, the middle-level data in the hierarchy is aggregated by store. This data consists of store sales for all brands.

The only place the observations (actual data on sales) directly flow into the hierarchy is in the (accumulated) bottom level. All middle- and upper-level data is constructed from this data.

Because of reconciliation and other considerations, forecast consumers might advocate building a super-hierarchy that contains all items, distribution centers, regions, and so on, in the data. This is not a good idea. A forecasting best practice is to build smaller hierarchies that contain like items, distribution centers, regions, and so on. To see why, consider creating a super-hierarchy in the current example. A business might have a mixture of product types: new products that are trending up quickly, mature products that vary around some level, and end-of-life-cycle products that are trending down. If these series are aggregated, the middle- and upper-level data will be composed of a consensus of the different signals in the product types. Potentially interesting trends and other signals in lower-level data will be obliterated using the super-hierarchy approach. If C-level executives want forecast results that reconcile to total company sales, this can be accomplished as a post-processing step after forecasts are created from appropriately aggregated data using tools presented in this and other chapters of this book.

Methods

The data shown below is a portion of the raw transactional data that will be used to illustrate accumulation and aggregation methods. It is the basis for the data hierarchy created. The time ID variable is DATE, the dependent variable is VALUE, and the two BY variables index value by TYPE (region) and SKU (brand). There are two SKU types; BTS (black tennis shoes) and RTS (red tennis shoes). There are four sales regions: REG1 through REG4.

Figure 10.2

date	value	type	sku
02JUN2007	1334	REG1	BTS
03JUL2007	1022	REG1	BTS
31JUL2007	1041	REG1	BTS
31AUG2007	975	REG1	BTS
30SEP2007	987	REG1	BTS
31OCT2007	873	REG1	BTS
30NOV2007	977	REG1	BTS
31DEC2007	875	REG1	BTS
01MAR2008	906	REG1	BTS
01APR2008	978	REG1	BTS
01MAY2008	1186	REG1	BTS
01JUN2008	1190	REG1	BTS
02JUL2008	1008	REG1	BTS
30JUL2008	.	REG1	BTS
29SEP2008	980	REG1	BTS

The data must first be accumulated to an equally spaced time interval. A monthly interval and total accumulation are feasible preliminary choices based on the denomination of the dependent variable and the spacing of the observations. The code accomplishes the accumulation and plots the series. The first step in building a hierarchy is sorting the data.

Program 10.4

```
proc sort data=ag;
by type sku date;

proc timeseries data=ag out=ag_month_base plots=(series);
    id date interval=month accumulate=total setmissing=next notsorted;
    var value;
    by type sku;
run;
```

New syntax includes the BY statement. The BY statement lists the subsetting variables in the input data set. Above, the raw input data is accumulated by TYPE (region), and then by SKU. The ID statement contains the NOTSORTED option. This indicates that the data set is in a BY variable format. There are several sequences of the time ID stacked on top of each other.

A portion of the accumulated data, ag_month_base, comprises the bottom level of the hierarchy.

Output 10.5

type	sku	date	value
REG1	BTS	JUN2007	1334
REG1	BTS	JUL2007	2063
REG1	BTS	AUG2007	975
REG1	BTS	SEP2007	987
REG1	BTS	OCT2007	873
REG1	BTS	NOV2007	977
REG1	BTS	DEC2007	875
REG1	BTS	JAN2008	906
REG1	BTS	FEB2008	906
REG1	BTS	MAR2008	906
REG1	BTS	APR2008	978
REG1	BTS	MAY2008	1186
REG1	BTS	JUN2008	1190
REG1	BTS	JUL2008	1008
REG1	BTS	AUG2008	980
REG1	BTS	SEP2008	980

The plot depicts the eight series at the base level of the hierarchy.[3]

Output 10.6

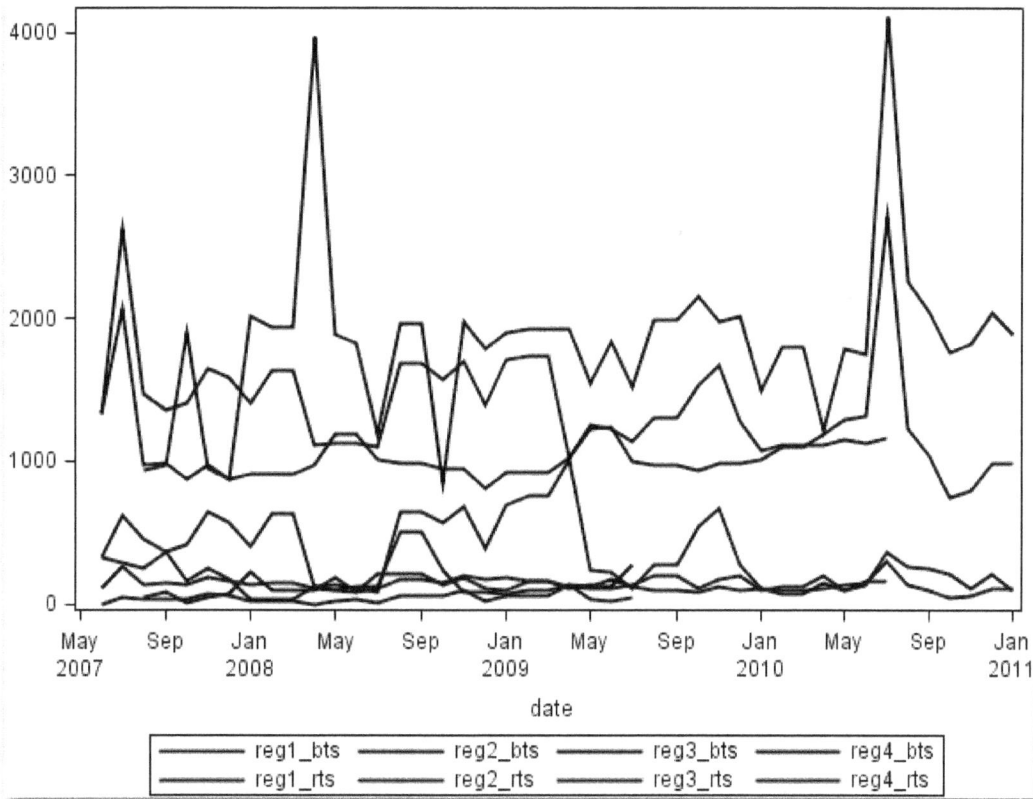

Program 10.5

```
proc timeseries data=ag out=ag_month_middle plots=(series);
   id date interval=month accumulate=total setmissing=next notsorted;
   var value;
   by type;
run;
```

A portion of the aggregated data in the middle level of the hierarchy, ag_month_middle, is shown. VALUE observations are the total of RTS and BTS sales (values) in each region.

Output 10.7

type	date	value
REG1	JUN2007	1663
REG1	JUL2007	2346
REG1	AUG2007	1229
REG1	SEP2007	1355
REG1	OCT2007	1026
REG1	NOV2007	1231
REG1	DEC2007	1052
REG1	JAN2008	938
REG1	FEB2008	938
REG1	MAR2008	938
REG1	APR2008	1098
REG1	MAY2008	1286
REG1	JUN2008	1317

The plot shows VALUE observations for the four regions in the data.

Output 10.8

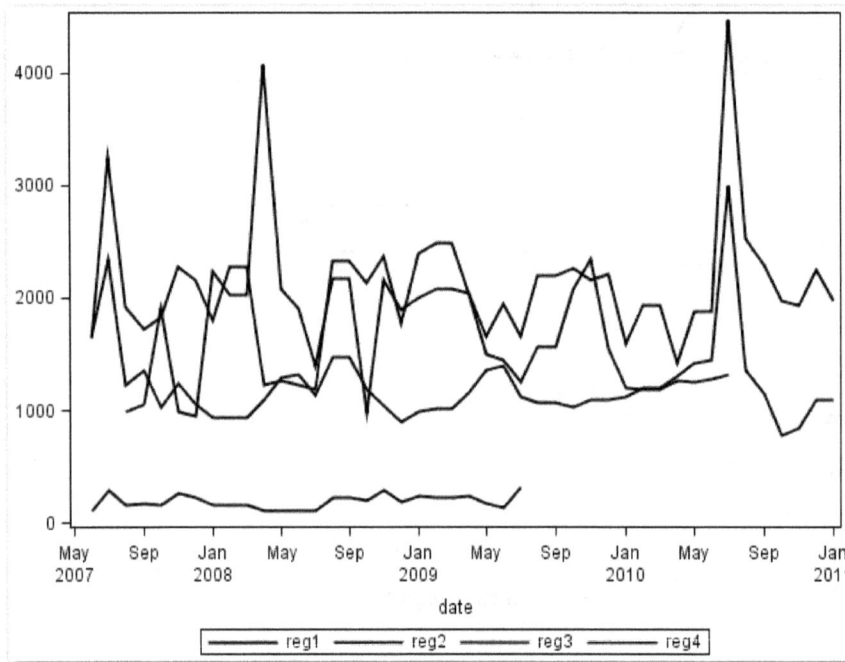

Finally, the syntax creates the data at the top level of the hierarchy. Note the absence of the BY statement.

Program 10.6

```
proc timeseries data=ag out=ag_month_top plots=(series);
    id date interval=month accumulate=total setmissing=next notsorted;
    var value;
run;
```

Top-level observations represent total VALUE across all regions and SKUs for each time interval.

Output 10.9

date	value
JUN2007	3416
JUL2007	5884
AUG2007	4295
SEP2007	4306
OCT2007	4916
NOV2007	4762
DEC2007	4375
JAN2008	4192
FEB2008	5398
MAR2008	5398
APR2008	6498
MAY2008	4739

The top-level series, ag_month_top, is plotted.

Output 10.10

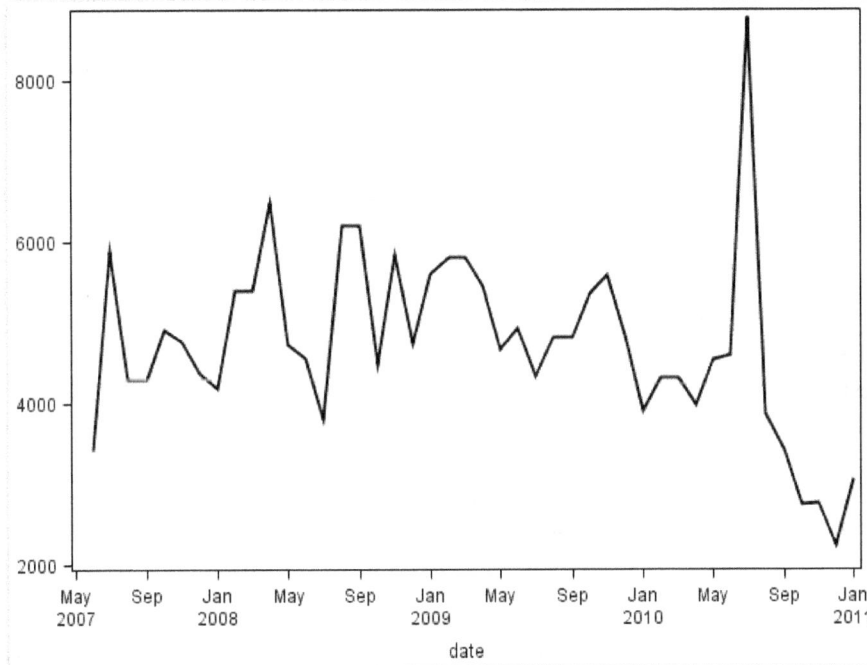

The data hierarchy consists of three data sets. In general, there is one data set for each level of the hierarchy created. Because the modeling techniques used in this book are one-series-at-a-time techniques, there is no connection between the data sets after they are created. In the forecasts created later in this book with SAS Forecast Server, each series in each data set is treated independently. There is no connection between series and levels of a hierarchy unless the analyst chooses to reconcile the statistical forecasts at the end of the process.

Data must be equally spaced to analyze time series with the models. Because most data that analysts encounter are transactional, some accumulation method must be chosen. This section emphasizes that the accumulation and aggregation methods chosen have impacts on the series to be modeled. Choosing data-handling methods should not be haphazard or left to default settings. Analysts should start with an understanding of the data that they have to work with and the types of patterns that they are interested in exploring.

10.2 Statistical Forecast Reconciliation

Ideas

The last subsection introduced ideas for handling transactional series flowing into a forecasting system. This section considers the other end of the process: reconciling statistical forecasts that flow out of the forecasting system. While the focus and mechanics shift, the approach is similar to the one presented above. Because the process of reconciliation changes the generated forecasts, it can degrade or improve the generated forecast's signal and accuracy. Analysts should carefully consider the choice of reconciliation method and its effects before and during implementation.

Reconciliation is related to the nested data hierarchy. Once models are built and forecasts are generated for each series in the hierarchy, the forecasts can be reconciled or made to add up for each time interval. Consider a three-series hierarchy with store 1 sales at the top and sales of RTS and WTS at the bottom. Assume that the data has a monthly interval and that forecasted sales for next January are as follows: store1=110, RTS=80, and WTS=20.[4] The sum of individual tennis shoe sales that flow through store 1 is forecast to be 100, which leaves a 10-unit discrepancy. One way to accomplish reconciliation for this time interval is to change the store 1 January forecast to 100. Another way to reconcile is to increase the bottom-level forecasts by a total of 10 as follows: RTS=88, WTS=22.

In adjusting the bottom-level forecasts, 8 units were added to RTS and 2 to WTS to accomplish reconciliation. This implementation is called top-down reconciliation using forecast proportions. The bottom-level forecasts are adjusted to reconcile the top level, and the proportions of the bottom-level reconciled forecasts are kept as close as possible to the proportions of the original statistical forecasts. Forecast proportions is the default method for dividing the positive or negative excess among lower-level forecasts in SAS Forecast Server. An equal split of the difference between lower-level forecasts is also feasible.

The method of reconciliation usually refers to the level of the hierarchy that all other levels must reconcile to. There are three methods.

- Top Down indicates that the top-level statistical forecast will not change. Statistical forecasts in the middle level and base level of the hierarchy are adjusted to accomplish reconciliation for each time interval in the hierarchy.

- Middle Out indicates that the selected middle-level statistical forecasts will not change. Statistical forecasts in levels above and below the selected middle level are adjusted to accomplish reconciliation for each time interval in the hierarchy.

- Bottom Up indicates that the bottom-level statistical forecasts will not change. Statistical forecasts in the middle level and top level are adjusted to accomplish reconciliation for each time interval in the hierarchy.

A statistical forecast is an objective best guess about what will happen. Reconciliation changes these best guesses and, if it is done in a haphazard way, can substantially degrade the forecast accuracy and usefulness of reconciled series. Methods for mitigating reconciliation damage to forecasts are discussed next.

One way to mitigate the costs of diminished accuracy associated with reconciliation is to reconcile to the most important level of the hierarchy in business value terms. For example, if supply chain decisions are made at the distribution center level, then that level of the hierarchy should be chosen as the reconciliation method. By doing this, statistical forecasts at the business decision level do not change. All changes are apportioned to less important levels.

Another idea to consider is that the process of aggregation or building the hierarchy has a smoothing effect on the data. Data at the bottom level are usually sparse and especially prone to random shocks. As the data is rolled up, it becomes less sparse and shocks tend to cancel out. However, as the data continues to be rolled and creates the upper middle and top levels, there is a danger of over-smoothing or washing out valuable signal. To mitigate this, one widely used method is to reconcile to a selected middle level in the hierarchy because it gives the models the best look at any signal the data contains. Statistical forecasts associated with the middle-level data tend to provide the best combination of accuracy and accommodation of valuable systematic patterns in the data.

Extending on the idea that small focused hierarchies are a best practice, reconciliation can actually improve the usefulness of generated forecasts. For example, consider a hierarchy containing only the sales of new products that are thought to be trending up. Sales for new products can be sporadic, and the best model for some base-level series might be a random walk or a mean. However, as the sales data is rolled up, trend and other patterns emerge in the middle-level data. In this example, the sum of forecasts at the bottom level will, on average, be lower than the forecasts for stores, regions, and so on, that they flow through during months that contain seasonal peaks. Also, the sum of lower-level sales will on average be higher than middle- and upper-level sales during months that contain seasonal troughs. By reconciling middle out (top down from the middle to the bottom), trends and other patterns found in the middle-level forecasts can be pushed down to augment the lower-level forecasts.

Methods

The last section showed the data sets that form a three-level hierarchy. Models were fit and estimated, and forecasts were produced for all series in the hierarchy using SAS Forecast Studio. This process produces three output forecast data sets; forecast_level2, forecast_level1, and forecast_level0.

Output 10.11 Top-Level Forecast Data Set forecast_level2

Variable Name	date	Predicted Values
value	JUL2009	4193.5449449
value	AUG2009	6729.3042904
value	SEP2009	4819.5878455
value	OCT2009	3107.1023237
value	NOV2009	6720.1074676
value	DEC2009	4496.0058397
value	JAN2010	5716.0679066
value	FEB2010	4104.1592265
value	MAR2010	4312.4708171
value	APR2010	3954.5774391
value	MAY2010	3193.4318959

Output 10.12 Middle-Level Forecast Data Set forecast_level1

type	Variable Name	date	Predicted Values
REG4	value	JUN2009	208.43378731
REG4	value	JUL2009	182.46940346
REG4	value	AUG2009	232.45549944
REG4	value	SEP2009	232.45549944
REG4	value	OCT2009	232.45549944
REG4	value	NOV2009	232.45549944
REG4	value	DEC2009	232.45549944
REG4	value	JAN2010	232.45549944
REG4	value	FEB2010	232.45549944
REG4	value	MAR2010	232.45549944
REG4	value	APR2010	232.45549944
REG4	value	MAY2010	232.45549944

Output 10.13 Bottom-Level Forecast Data Set forecast_level0 (two portions of the same data set are shown)

type	sku	Variable Name	date	Predicted Values	type	sku	Variable Name	date	Predicted Values
REG4	RTS	value	MAY2009	103.38541949	REG4	BTS	value	MAY2009	77.6512354
REG4	RTS	value	JUN2009	135.46646737	REG4	BTS	value	JUN2009	56.228666745
REG4	RTS	value	JUL2009	114.92725463	REG4	BTS	value	JUL2009	50.812922974
REG4	RTS	value	AUG2009	273.35679227	REG4	BTS	value	AUG2009	-1.141585662
REG4	RTS	value	SEP2009	273.35679227	REG4	BTS	value	SEP2009	78.021588287
REG4	RTS	value	OCT2009	273.35679227	REG4	BTS	value	OCT2009	65.668103062
REG4	RTS	value	NOV2009	273.35679227	REG4	BTS	value	NOV2009	33.459929153
REG4	RTS	value	DEC2009	273.35679227	REG4	BTS	value	DEC2009	79.170352984
REG4	RTS	value	JAN2010	273.35679227	REG4	BTS	value	JAN2010	18.097336673
REG4	RTS	value	FEB2010	273.35679227	REG4	BTS	value	FEB2010	40.428919251
REG4	RTS	value	MAR2010	273.35679227	REG4	BTS	value	MAR2010	66.738674529
REG4	RTS	value	APR2010	273.35679227	REG4	BTS	value	APR2010	22.62645461
REG4	RTS	value	MAY2010	273.35679227	REG4	BTS	value	MAY2010	72.489426866
REG4	RTS	value	JUN2010	273.35679227	REG4	BTS	value	JUN2010	48.16447123
REG4	RTS	value	JUL2010	273.35679227	REG4	BTS	value	JUL2010	

The statistical forecasts do not reconcile for all time intervals. For example, SKU sales for RTS and BTS that flow through region 4 are forecast to be 103.4 and 77.6 for May 2009. The sum of the SKU-level forecasts for this date is 181. The forecast for sales in region 4 for May 2009 is 208.4. There is a discrepancy of 27.4 units.

The tool for reconciliation is a SAS Forecast Server procedure, PROC HPFRECONCILE.[5] Complete information about this procedure can be found in the documentation for SAS High-Performance Forecasting.

- Bottom up (BU) and top down (TD) are the available reconciliation directions. Middle out reconciliation is accomplished using a mixture of the two directions in subsequent runs.

- Reconciliation is accomplished pairwise between levels (data sets) of the hierarchy.

- The DISAGGDATA option always indicates the data set in the pair that is lower in the hierarchy, regardless of the direction of aggregation.

- The AGGDATA option always indicates the data set in the pair that is higher in the hierarchy, regardless of the direction of aggregation.

- The variables listed in the BY statement always point to the disaggregated data level of the hierarchy. It is constructed exactly as it would be in PROC TIMESERIES if the disaggregated data level of the hierarchy were being generated.

The reconciliation method chosen is middle out. This means that statistical forecasts in the middle level of the hierarchy will remain unchanged. Statistical forecasts in the top and bottom levels are adjusted to accomplish reconciliation.

The first step in this reconciliation process is reconciling the top-level forecasts to the middle-level forecasts. Bottom up (BU) reconciliation is the direction used. The disaggregated data in this pair of data sets is the middle-level forecast data, and the BY statement points to this data set.

Program 10.7

```
proc hpfreconcile aggdata=forecast_level2 disaggdata=forecast_level1
        direction=bu
        outfor=level2_recfor
        recdiff;
    id date interval=month;
    by type;
run;
```

The output or outfor data set, level2_recfor, contains the reconciled top-level forecasts. The RECDIFF option in the syntax puts the Reconciliation Difference column in the outfor data set. This column represents the impact of reconciliation on underlying statistical forecasts. Many forecasters track this column and manually scrutinize and re-adjust reconciled forecasts in intervals that the reconciliation difference exceeds some threshold.

Output 10.14

Variable Name	date	Reconciled Predicted Values	Reconciliation Difference
value	JUL2009	4246.1137281	52.568783215
value	AUG2009	5121.11582	-1608.18847
value	SEP2009	5034.6614796	215.07363412
value	OCT2009	3549.0405047	441.93818107
value	NOV2009	5333.9419441	-1386.165523
value	DEC2009	5423.0542546	927.04841485
value	JAN2010	4957.4810105	-758.5868962
value	FEB2010	4664.2884832	560.12925662
value	MAR2010	4691.9354501	379.46463298
value	APR2010	4791.0453977	836.46795863
value	MAY2010	4648.9565831	1455.5246872
value	JUN2010	4896.6347336	80.025025458

The second step is reconciling the bottom-level forecasts to the middle-level forecasts. The aggregated data is the middle-level forecasts, the direction is top down, and the BY statement points to the bottom-level forecasts.

Program 10.8

```
proc hpfreconcile aggdata=forecast_level1 disaggdata=forecast_level0
        direction=td
        outfor=level0_recfor
        recdiff;
    id date interval=month;
    by type sku;
run;
```

The outfor data set, level0_recfor, contains the reconciled forecasts for the bottom level of the hierarchy. The portion shown contains reconciled forecasts for RTS series that flow through region 4.

Output 10.15

type	sku	Variable Name	date	Reconciled Predicted Values	Reconciliation Difference
REG4	RTS	value	MAY2009	128.05110381	24.665684314
REG4	RTS	value	JUN2009	143.83579397	8.3693265979
REG4	RTS	value	JUL2009	123.29186756	8.3646129293
REG4	RTS	value	AUG2009	253.47693869	-19.87985359
REG4	RTS	value	SEP2009	213.89535171	-59.46144056
REG4	RTS	value	OCT2009	220.07209433	-53.28469795
REG4	RTS	value	NOV2009	236.17618128	-37.18061099
REG4	RTS	value	DEC2009	213.32096937	-60.03582291
REG4	RTS	value	JAN2010	243.85747752	-29.49931475
REG4	RTS	value	FEB2010	232.69168623	-40.66510604
REG4	RTS	value	MAR2010	219.53680859	-53.81998368
REG4	RTS	value	APR2010	241.59291855	-31.76387372
REG4	RTS	value	MAY2010	216.66143242	-56.69535985
REG4	RTS	value	JUN2010	228.82391024	-44.53288203

The pairwise data set reconciliation method extends to larger hierarchies. Nothing really changes, there are just more pairs to process and more associated runs of PROC HPFRECONCILE. PROC HPFRECONCILE does not distinguish what procedure or process generated the aggregated and disaggregated data sets. It assumes that the forecasts are in the nested hierarchical framework. So, if the CFO wants forecasts that reconcile to some total level for reporting purposes, forecasts can be pulled and aggregated across projects or hierarchies, and then made to reconcile to the requested level as part of forecast post-processing steps.

10.3 Intermittent Demand

In time series applications, intermittent demand means that there are substantial periods of the demand history of a series in which no demand occurs. Modeling approaches have been developed that do the best job possible in forecasting these problematic series. (See Croston 1972 and Willemain et al. 1994.) In a large-scale forecasting scenario, a substantial portion of the series to be processed, modeled, forecast, and reconciled can be intermittent. A good solution for intermittent series is to roll them up to a lower-frequency interval. A series that is intermittent on a daily interval is usually not intermittent on a quarterly interval. However, this solution is not always feasible. This section outlines a traditional strategy for directly modeling intermittent series.

Ideas

Intermittent demand usually manifests itself as lots of zeros in the series being modeled. This condition is likely to occur for luxury items, new products, and especially highly seasonal items. For example, in the southeastern United States, most consumers buy Ice Melt salt only when the local weatherman forecasts that an ice storm is imminent. The few times that an ice storm is imminent, Ice Melt is sold in large quantities. At other times, none is sold.

Models described previously predict the future using weighted averages of the historical values of a series. A weighted average of mostly zeros tends to be a number very close to zero. This can be a problem for the hardware store owner that sells Ice Melt. She wants to have some stock on hand in case an ice storm is forecast, but inventory is expensive, and there are opportunity costs involved. Stocking lots of Ice Melt means not stocking other items.

There is no magic algorithm for creating precise forecasts from data in which the nonzero demand is regulated by highly stochastic events. The models that are designed to be used with intermittent data generate forecasts that are predictions of average nonzero demand across the series (the average stock of Ice Melt the hardware store owner should keep in the back room). The intermittent demand model illustrated in this section is one of the most widely used—Croston's method (Croston 1972).

Croston's method and other traditionally used intermittent demand models are extensions of the exponential smoothing models. However, they depend on pre-processing the intermittent series.

The following series fits the description of intermittent demand.

Figure 10.3

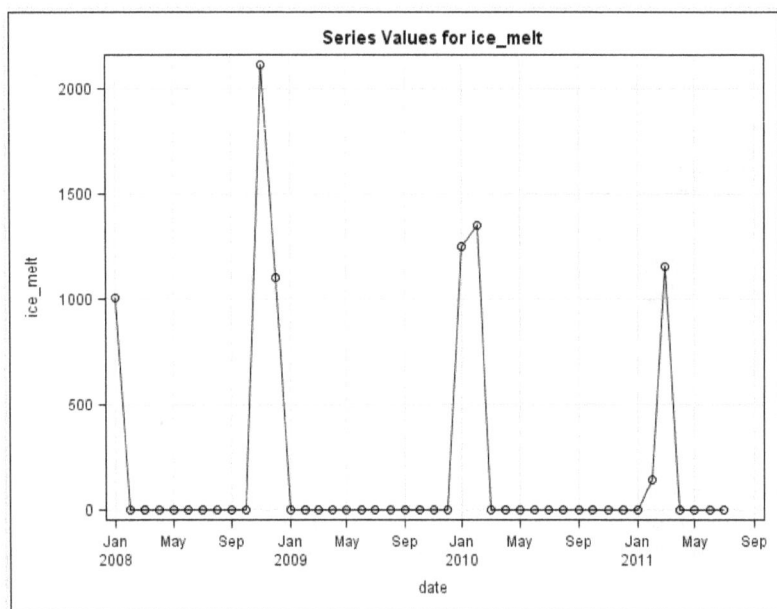

The first step in preparing the series to be modeled is to split or decompose the series into two parts: the size of nonzero demands and the length of interval between nonzero demands.

Figure 10.4

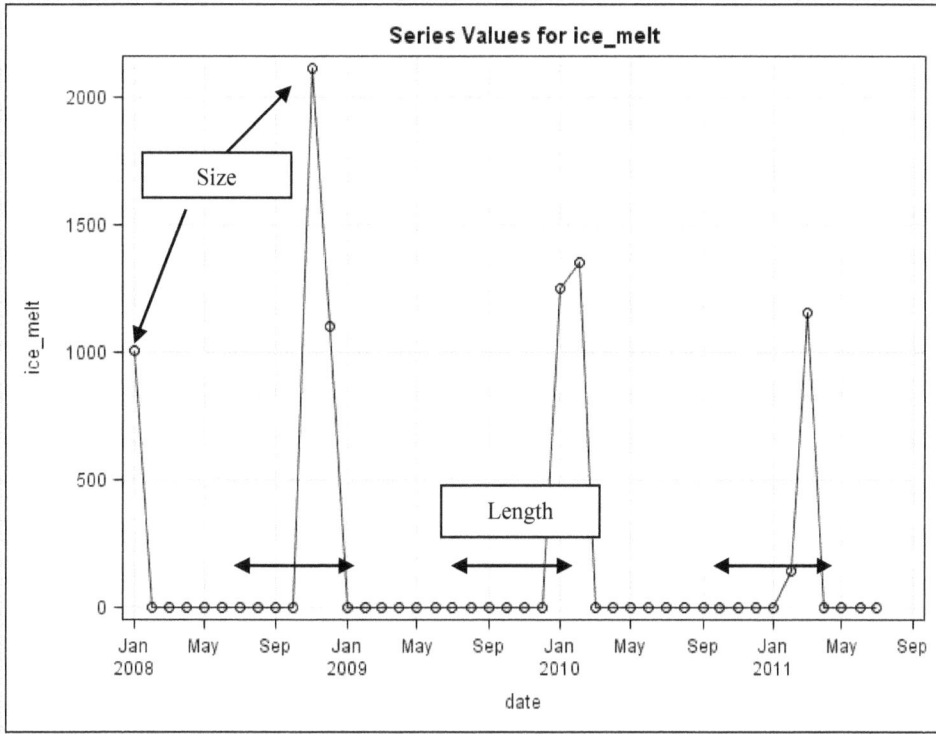

The plots show the two series that result from the decomposition.

Figure 10.5

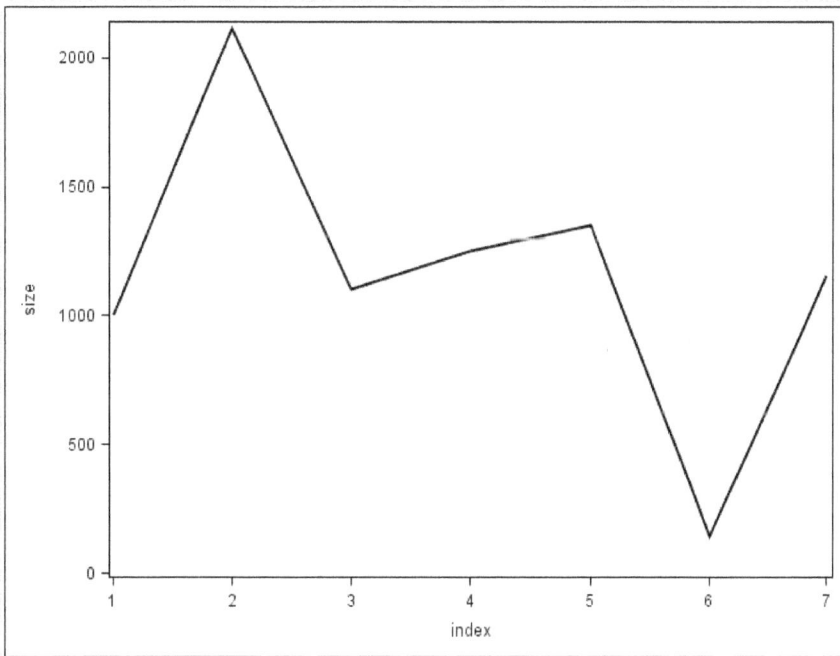

Figure 10.6

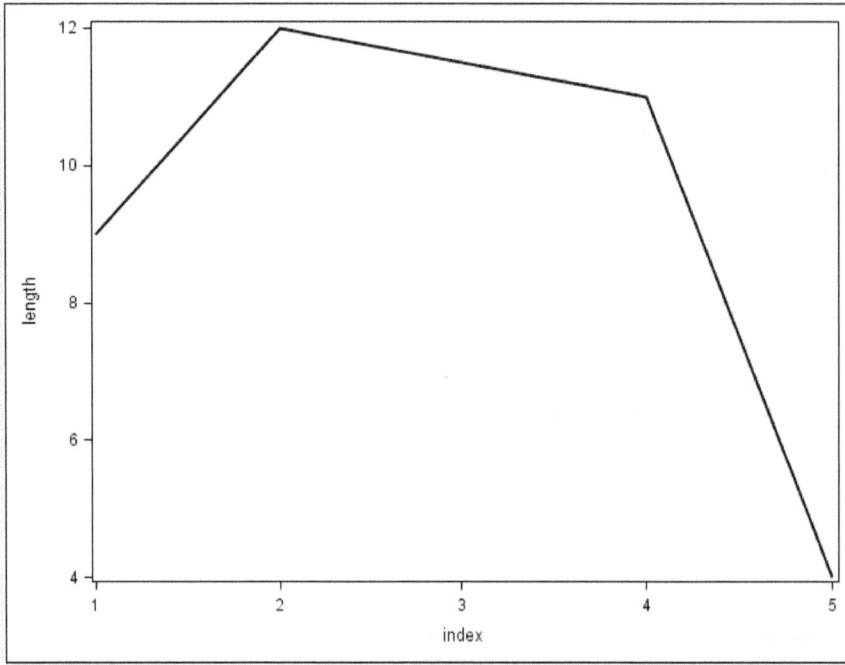

The time interval of the intermittent series no longer applies to the series that result from the decomposition. It has been replaced by an index.

In the modeling step, Croston's method uses exponential smoothing models to generate in-sample and lead forecasts for both the Size and Length series. The final step divides the Size series forecast by the Length series forecast. This ratio is the forecast for average demand or the recommended stocking level for the product. The following section uses SAS software that accomplishes the steps outlined for Croston's method automatically.

Methods

Two SAS Forecast Server procedures, PROC HPFDIAGNOSE and PROC HPFENGINE, were used to automatically accomplish the steps. These procedures and other SAS Forecast Server procedures run under the hood in the graphical user interface to SAS Forecast Server, which is SAS Forecast Studio. An example that illustrates other SAS Forecast Server functionality and that features SAS Forecast Studio is in a subsequent chapter.

A portion of the forecasts generated for the Ice Melt series using Croston's method is shown in the following data set. The lead forecast (the portion of the data where the actual values are missing) is a recommended stocking level. There is no trend, no ARMA produced, or no other variation in the lead forecast. This is an artifact of the intermittency of the data and the method used.

Figure 10.7

Variable Name	date	Actual Values	Predicted Values
ice_melt	OCT2010	0	881.67438272
ice_melt	NOV2010	0	881.67438272
ice_melt	DEC2010	0	881.67438272
ice_melt	JAN2011	0	881.67438272
ice_melt	FEB2011	145	881.67438272
ice_melt	MAR2011	1155	757.24163375
ice_melt	APR2011	0	814.15823492
ice_melt	MAY2011	0	814.15823492
ice_melt	JUN2011	0	814.15823492
ice_melt	JUL2011	0	814.15823492
ice_melt	AUG2011	.	814.15823492
ice_melt	SEP2011	.	814.15823492
ice_melt	OCT2011	.	814.15823492
ice_melt	NOV2011	.	814.15823492
ice_melt	DEC2011	.	814.15823492
ice_melt	JAN2012	.	814.15823492

The forecast plot for Ice Melt is shown next. The lead forecast in the plot appears to have variation in it, although the lead forecast in the data does not. The apparent discrepancy is explained by different indexes in the data set and the forecast plot. The horizontal axis in the plot is the index that was created in the data decomposition.

Output 10.16

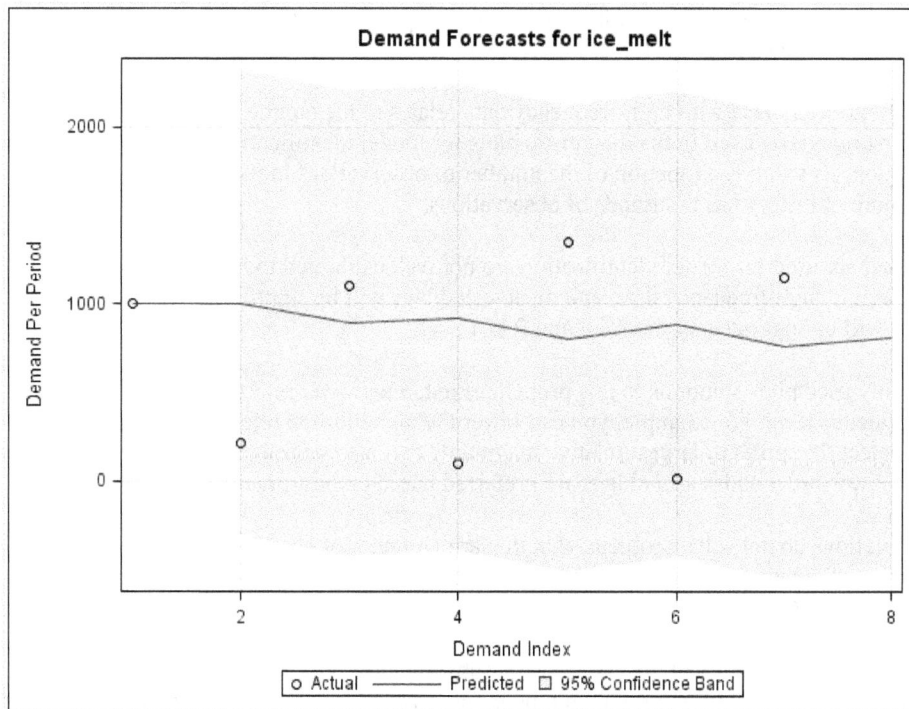

10.4 High-Frequency Data and Mixed-Frequency Forecasting

Introduction

Previous sections demonstrated that the selected time interval and aggregation method have an impact on the type and strength of signal contained in the data to be modeled. You might have gotten the impression that interval selection is a rigid proposition. That is, modelers are forced to choose one interval, and subsequent modeling proceeds exclusively on this frequency of the data.

This does not need to be the case. Often, the data contain multiple signals (primarily cycles) of interest. Model usefulness and accuracy can be greatly improved by accommodation of signals, such as day-of-the-week effects and an annual seasonal cycle that are strongest at different time intervals. Also, when the data for input variables is taken from outside sources, the interval of the input series is often different from the interval of the dependent series. This part of the book presents widely used ideas and techniques for modeling systematic variation across time intervals in the data.

These topics provide a reasonably complete introduction to the applied techniques in this area of time series analysis. However, mixed-interval or mixed-frequency forecasting is an active area in time series research. New techniques are emerging. Considering newer and more theoretical approaches to mixed-interval forecasting is beyond the scope of this section.

High-Frequency Data

Ideas

Frequency of the data refers to the rate of the data interval. For example, monthly data has 12 intervals per year and has a higher frequency than quarterly data, which has 4 intervals per year. High-frequency data usually denotes data with a daily or smaller interval (for example, hourly).

The first challenge that modelers face with high-frequency data relates to the number of observations in the series. The problem first manifests itself in the diagnostic plots for model identification. These plots are based on a 95% rejection region. Its width is a function of the number of observations in the series. An hourly interval series with a year or more of history has thousands of observations.

The diagnostic plots that are used for model identification are not well calibrated to handle the number of observations usually seen in high-frequency data, and most tested lags will be significant. Severely over-parameterized models will be suggested by the ACF and PACF.

There are two commonly used plots solutions to this problem. First, a narrower rejection region can be applied by expanding the confidence level. For example, you can lower the alpha for the rejection from 0.05 to 0.01. A second solution is to select the spikes of largest relative magnitude that are consistent with best practices and business knowledge. Lower-order and seasonal lags are preferred based on best practices.

However, these two solutions do not solve problems that the large number of observations cause in the estimated standard errors of the model parameters and in other areas. These solutions have proven to be unsatisfactory in our experience.

A better solution is to reduce the number of observations by subsetting the data. For example, business knowledge might suggest that history from more than nine months ago is not relevant in forecasting what is going to happen this afternoon. Data can be truncated to contain only the relevant history. Econometric literature provides more formal tests of structural changes in the series. These tests can be used to estimate truncation points. (See Capps' course notes for an excellent survey and explanation of these techniques.)

Another way to subset the data without truncation is to randomly sample the observations of the series over the range of the data (Woodfield 2009). This method has proven to be very useful, particularly when the relevant signals in the data might extend over the entire range of the data.

The second challenge that modelers face with high-frequency data is in the length of the forecast horizon. Something that new modelers often fail to appreciate is that the length of the lead forecast horizon is determined largely by the selected time interval of the data. For example, a year-ahead forecast might be short range if the data has a one-year interval. A forecast for tomorrow at noon is long range if the data has a one-minute interval.

The usefulness of the forecast model is often regulated by the number of intervals that it forecasts into the future. Consider how the models work.

- ARMA components die off or revert to the mean or trend quickly over a small number of intervals for lower-order lags. Seasonal effects and longer lags might be useful in longer horizons.

- Deterministic inputs like trends, events, seasonal dummies, and so on, can be extended into a long-range horizon, but their accuracy depends on data stability. Small changes going forward can cause a wide divergence between actual and predicted values or bias. The danger of biased forecasts grows with the length of the lead forecast horizon.

- Stochastic inputs must be forecast into the lead horizon. These forecasts must be built into the forecast for the dependent variable. Long-range forecasting with stochastic inputs is usually approached using a what-if approach instead of traditional forecasting approaches.

Long-range forecasting can be problematic. However, shortening the length of the forecast horizon by accumulating the data to a lower-frequency interval is not always feasible. It can be detrimental to the usefulness of the model. An example of this is a situation where an effect that is most evident in a high-frequency interval is important to the usefulness of the forecast model. However, forecasts must be provided for several months into the future. The next section outlines a widely used method to forecast across time intervals. This method accommodates multiple cycles of interest in the forecast.

Methods

A typical scenario that a modeler might face with high-frequency data might found in the case files of Margo X. Smythe, our fictional forecasting consultant. One of her contracts stipulated that she provide staffing forecasts for the number of magistrates, officers, and emergency medical technicians needed at local jails in more than three states in the southwestern United States. Staffing needs were determined using fixed proportions of counts of jail intake per hour.

Margo hypothesized that the data contained multiple cycles of interest that should be accommodated in a staffing forecast. Anecdotal evidence suggested that jail intake is higher from 18:00 to 3:00 than it is at other times of the day. Friday and Saturday see a higher rate of customers per hour than other days of the week. And, the number of arrests seem to peak in late summer and then cycle to a low point in mid-winter.

Margo needed to provide forecasts for the next year. Selecting an hourly interval to accommodate the time-of-day effect was problematic. Her forecast needed a lead horizon of approximately 8760 intervals. However, if she chose to use a lower-frequency interval for modeling, the time-of-day effect would be obliterated.

Margo's preliminary research indicated that the state- and county-level data were reasonably similar in terms of crime rates per capita and in the cycles that she was interested in estimating. Prior to data accumulation and aggregation activities, she decided to pool the data across states to estimate the cycles. PROC TIMESERIES syntax accomplished this. The INTAKE variable measures jail bookings per hour. Additional syntax includes the PRINT=(DECOMP) option.[6] Using this option, the series is decomposed into trend, seasonal, and irregular components. The hourly seasonal component is the most important.

Program 10.9

```
proc timeseries data=temp3a out=hfhr
        plots=(series) print=(decomp) outdecomp=hour;
    id date interval=hour accumulate=total;
    decomp / mode=pseudoadd;
    var intake;
run;
```

Output 10.17

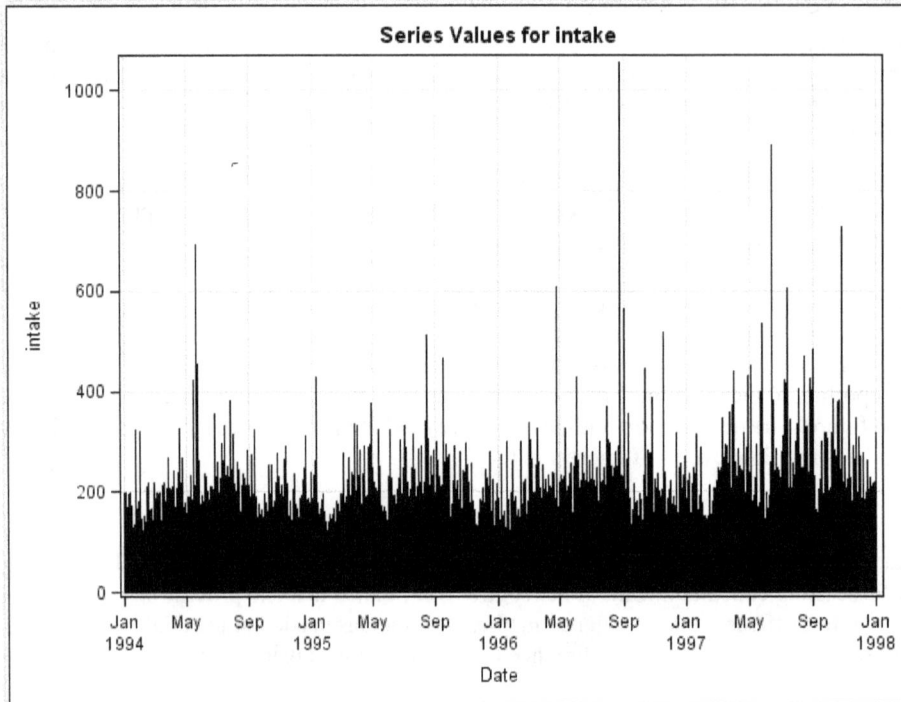

The number of observations makes it difficult to discern any features of interest in the plot of the hourly intake data shown above. However, the hourly seasonal decomposition, below, provides more insight. Note that, because the data contains zeros, a pseudo-additive decomposition (see the SAS/ETS documentation for the TIMESERIES procedure) is used to derive the component series. The largest seasonal components or factors occur between 19:00 and 2:00 (the seasonal index corresponds to the 24-hour daily cycle). This confirms the anecdotal evidence that Margo received.

The first 24-hour portion of the decomposition output is shown in the table below. This information is also in the output decomposition data set, Hour. It is useful to note two things about the output. First, the seasonal factors repeat. They represent the average hourly cycle effect over the range of the data, holding constant any effects from trend and irregular variation. Second, their average is one, and they can be thought of as proportions. For example, the seasonal factor for 1:00 is 1.355. This means that intake for any day at 1:00 is about 36% above the average hourly intake.

Output 10.18

Seasonal Index	Original Series	Trend-Cycle Component	Seasonal Component	Irregular Component	Seasonally Adjusted Series
1	199.5297155	.	1.3549033972	.	147.26490162
2	161.82192074	.	1.2014671553	.	134.68692842
3	154.59425647	.	1.0073482563	.	153.46654496
4	57.078436579	.	0.8077461286	.	70.663831816
5	55.062045128	.	0.808639509	.	68.092202414
6	154.67550144	.	0.7910594113	.	195.52956356
7	84.350436258	.	0.7489744096	.	112.62125272
8	0	.	0.752141331	.	0
9	37.602691614	.	0.7245244693	.	51.899822862
10	0	.	0.7595674474	.	0
11	0	.	0.7103201088	.	0
12	0	.	0.7897830587	.	0
13	0	58.030489203	0.7426705872	0.2573294128	14.93295171
14	20.740788377	53.402674008	0.786105674	0.6022791544	32.163317342
15	0	48.112457366	0.8663966195	0.1336033805	6.4279869503
16	46.387472463	45.239656805	0.9377467794	1.0876251119	49.203786797
17	55.779985501	44.464743551	0.8567140232	1.3977626935	62.151159712
18	39.263540016	40.909455321	0.8088048777	1.1509619955	47.085228331
19	0	36.330627345	1.3923771805	-0.39237718	-14.25530912
20	77.969369866	35.427427001	1.5330087509	1.6678106839	59.086241254
21	109.59037682	36.334882363	1.4431470685	2.572973246	93.488680217
22	19.433157148	37.265044998	1.2372327925	0.2842520776	10.592666462
23	109.13654871	37.795972216	1.4300763911	2.4574415721	92.881393382
24	52.920455287	37.819132074	1.5092445733	0.8900592283	33.66126751

Here is a plot of the seasonal factors. The plot summarizes the 24-hour intake cycle.

Output 10.19

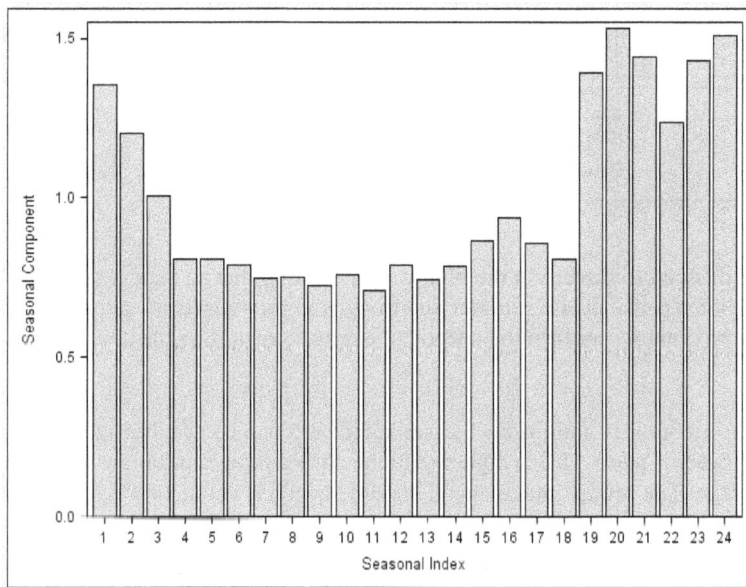

The day-of-the-week effect is explored. The data is accumulated to a daily interval.

Program 10.10

```
proc timeseries data=temp2 out=hfday
      plots=(series) print=(decomp) outdecomp=day;
   id date interval=day accumulate=total;
   decomp / mode=pseudoadd;
   var intake;
run;
```

Here is a plot of the daily interval intake series.

Output 10.20

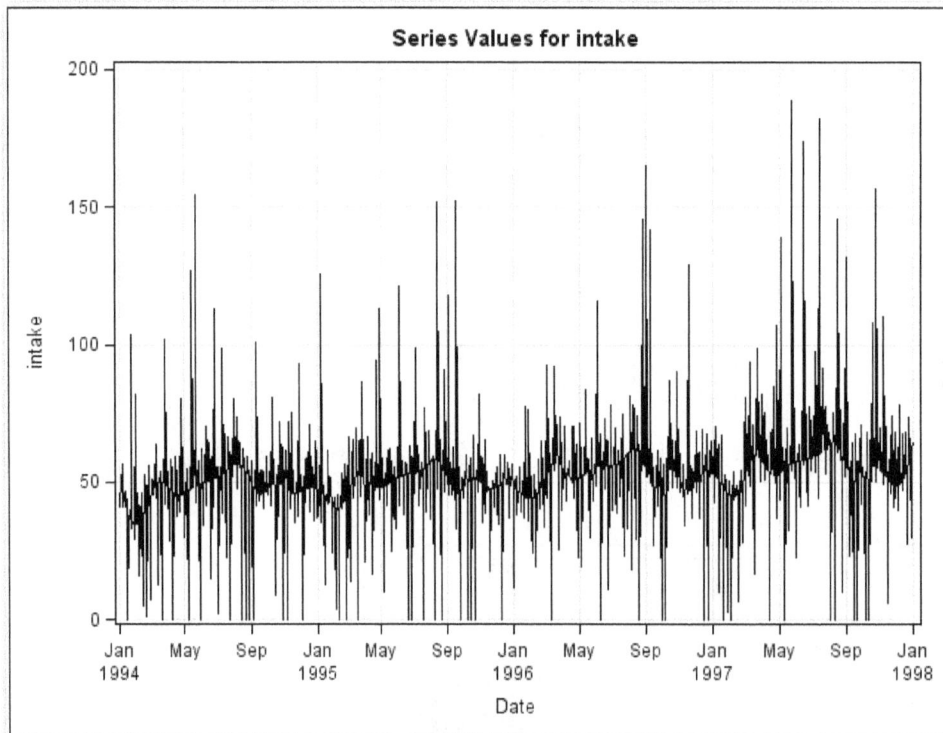

Although a day-of-the-week effect is difficult to discern in the plot, it is interesting that an annual seasonal pattern is emerging in the daily data. Some peaks in late summer and troughs in mid-winter are apparent. The seasonal component data and plot approximately confirm the anecdotal evidence regarding the day-of-the-week effect.

The day-of-the-week cycle is highest on Thursday and Friday (seasonal index=5 and 6). On average across the range of the daily data, intake on Thursday is about 18.7% higher than the daily average for the week. Friday is about 8.3% higher than the daily average. The lowest intake day is Sunday, which is about 14.4% lower than the daily average.

Output 10.21

Seasonal Index	Original Series	Trend-Cycle Component	Seasonal Component	Irregular Component	Seasonally Adjusted Series
7	46.249363157	.	0.893244053	.	51.776849787
1	40.91358675	.	0.8569150959	.	47.7452048
2	52.04691343	.	0.9928445794	.	52.422014995
3	50.559387625	49.195585774	0.9928705143	1.0348515227	50.91012685
4	49.134588454	48.433517238	0.9928282941	1.0216466244	49.481939394
5	56.515088602	49.197264335	1.1876975956	0.9610469406	47.280880375
6	48.950172403	47.929808132	1.0835998676	0.9376888525	44.94324679
7	40.914883405	46.734650442	0.893244053	0.982228065	45.90408527
1	46.259816423	45.612865047	0.8569150959	1.1572684327	52.786328844
2	43.174720009	44.231706265	0.9928445794	0.9832588463	43.49121647
3	42.193283798	43.13994756	0.9928705143	0.9851854682	42.500849437
4	41.282090691	41.833437207	0.9928282941	0.9939921406	41.582107798
5	46.846977123	35.224892004	1.1876975956	1.1422419573	40.235349588
6	41.307861473	34.440275715	1.0835998676	1.115805788	38.428658984
7	31.769310934	33.719754412	0.893244053	1.0489131924	35.369095248
1	0	33.060320244	0.8569150959	0.1430849041	4.7304327512

Here is the plot summarizing the day-of-the-week effect.

Output 10.22

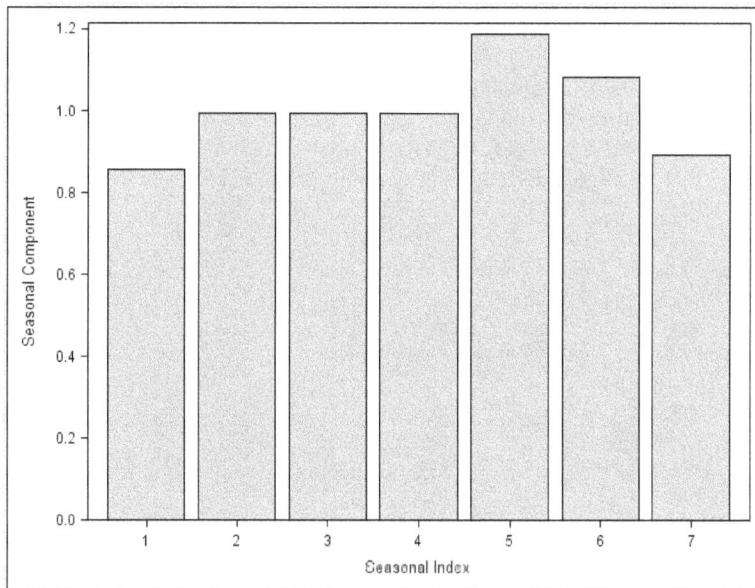

Finally, the monthly seasonal cycle is explored.

Program 10.11

```
proc timeseries data=temp2 out=hfmonth
      plots=(series) print=(decomp) outdecomp=month;
   id date interval=month accumulate=total;
   decomp / mode=pseudoadd;
   var intake;
run;
```

Output 10.23

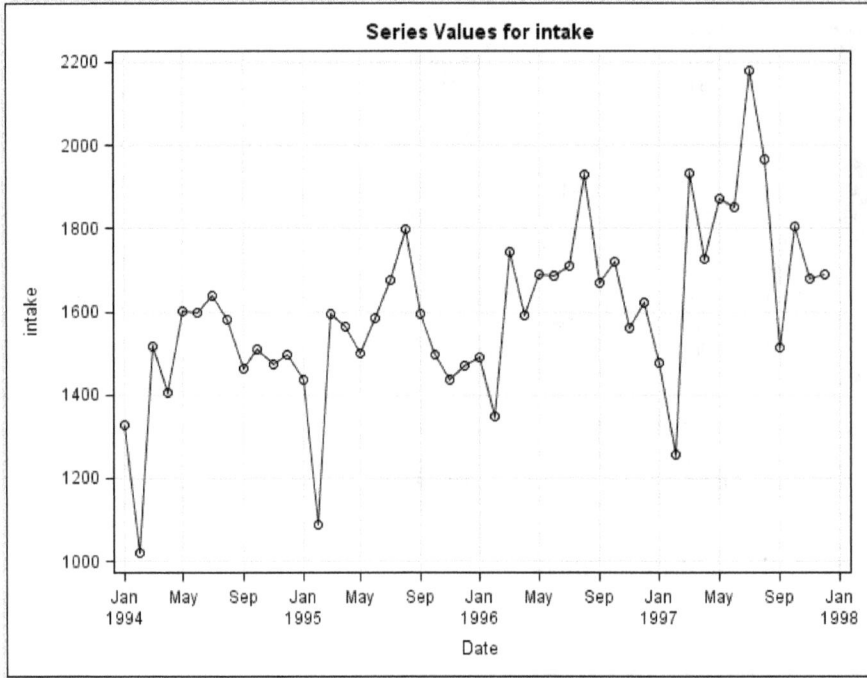

The monthly interval plot of the data confirms the late summer peaks in the intake hypothesis. It shows evidence of an increase in jail intake year over year that should be included in the forecast. The decomposition output quantifies the magnitude of the annual seasonal cycle. August (seasonal index=8) intake per month is the highest and is about 14% above the monthly average. February is lowest and is about 14.5% below the monthly average.

Output 10.24

Seasonal Index	Original Series	Trend-Cycle Component	Seasonal Component	Irregular Component	Seasonally Adjusted Series
1	1329.1535774	.	0.9210499541	.	1443.085222
2	1021.9537257	.	0.765545191	.	1334.9358571
3	1516.0173111	.	1.085895403	.	1396.0988388
4	1404.5559505	.	1.004464767	.	1398.3128096
5	1600.3103788	.	1.0347757842	.	1546.5286327
6	1597.4764164	.	1.0443534952	.	1529.6318955
7	1640.3601834	1474.1183559	1.084834674	1.0279390574	1515.3038332
8	1582.1076908	1481.3965774	1.1391556127	0.928828285	1375.9630424
9	1463.7968845	1487.4934035	1.0089265851	0.9751429113	1450.518648
10	1509.3691332	1497.457647	1.0019606282	1.0059938447	1506.4331755
11	1473.4568783	1500.0188029	0.944210073	1.0380821992	1557.1428178
12	1496.9867363	1495.3627585	0.9648278325	1.0362581767	1549.5818857

Output 10.25

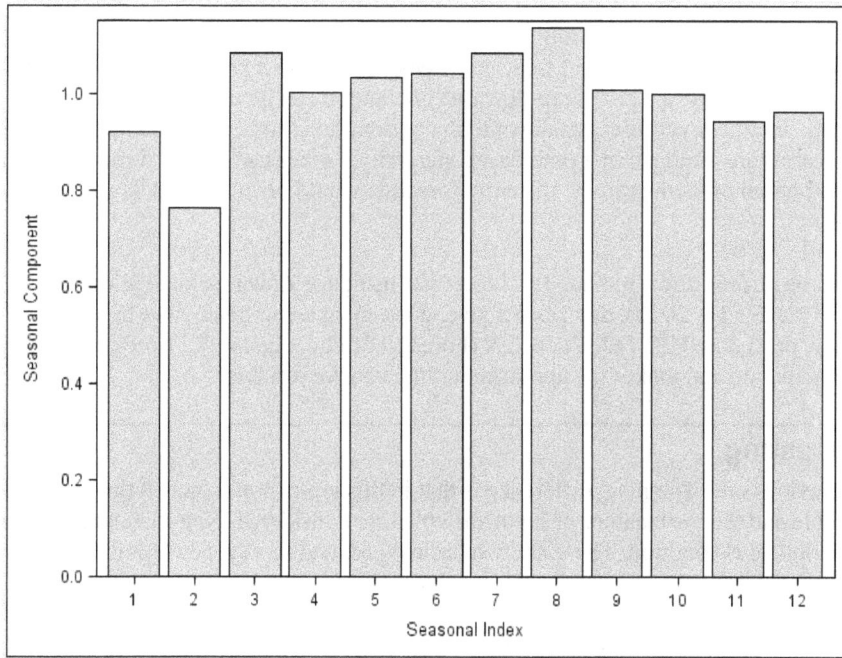

Using this information, Margo outlined a forecasting strategy. Accumulate and aggregate the data to a monthly interval to create a hierarchy by state to county. Estimate models and generate one-year ahead forecasts for all series.[7] The lead forecast horizon is only 12 intervals into the future.

This creates monthly intake forecasts that accommodate the monthly seasonal cycle and trend. Monthly forecasts are used as a basis for creating daily and hourly forecasts. Because the seasonal factors act like proportions, creating a forecast for the next Sunday, August 18, at 17:00, is straightforward. Scale the monthly forecast for August by the appropriate daily and hourly seasonal factors.

This approach effectively embeds the 24-hour and day-of-the-week cycles in the monthly forecasts. However, remember these facts:

- The strategy of estimating the hourly and daily cycle effects on the pooled data assumes that an average representation of these effects is adequate. If these cycles vary substantially by state or county, then they should be estimated separately by state or county.

- The accuracy of this approach depends on the cycles being uncorrelated. The cycle effects are estimated separately. If they are correlated, this could cause double-counting the cycle effects. A reasonable check for cycle correlation consists of the following steps.[8] Estimate cycle effects at the highest-frequency interval. In the previous example, a model is estimated that accommodates the hourly cycle on an hourly data interval. Take the residuals series of this model and accumulate it to a daily interval. If the day-of-the-week cycle exists in the accumulated residuals (at approximately the same magnitude as when estimated separately), then this is evidence that the cycles are uncorrelated. Similar steps should be performed to confirm that the monthly cycle is uncorrelated with the day-of-the-week and hourly cycles. If cycles are correlated, then the analysis could proceed by constructing the daily and monthly interval series from the residuals of models estimated on the hour interval data.

There are two newer techniques that can be used to accommodate multiple seasonal cycles in a forecast. The first technique is an unobserved components model (UCM). A UCM estimates the separate components of variation (for example, trend and cycles) in a series in separate models. A UCM is more general than an ARIMAX model and might do a better job accommodating multiple seasonal cycles and estimating them simultaneously. The UCM procedure (SAS/ETS) contains SEASONAL and BLOCKSEASONAL statements that are specifically designed to allow the model to accommodate multiple cycles.

The second technique is related to the approach of using forecasts in hierarchies and the reconciliation process to improve the usefulness and accuracy of lower-level forecasts. Recall that if the lower-level data consists of like items and the data is sparse and noisy, a form of top-down reconciliation can be used to push seasonality and trend accommodated in the middle-level forecasts down to the bottom-level forecasts. The idea here is the same, but in this case the forecasts are arranged in a time-based hierarchy. Forecasts generated from the hour interval data would be on the bottom of the hierarchy, the daily forecasts would be in the middle, and the monthly forecasts on the top.

Monthly and daily cycles can be pushed down to hour-level forecasts using top down reconciliation. The HPFRECONCILE Procedure (SAS/ETS, 2010), described above, does not accommodate time-based hierarchies. However, the new procedure HPFTEMPRECON (Forecast Studio, version 4.1) does accommodate time-based hierarchical reconciliation and makes the approach outlined above feasible.

Mixed-Interval Forecasting

This topic is similar to the previous one. The main difference is that in this topic, non-matched time intervals in the target or dependent variable and the explanatory or input variables are considered. Non-matching of time intervals between targets and inputs is common. There are two basic types, and these types provide the framework for the following discussion.

The first type occurs when candidate input variables have a lower-frequency interval than the target. This type is well known by forecasters that use data from external sources. Most businesses use a monthly or weekly interval for data collection, and they require forecasts on one of these intervals for efficient operation. However, much of the data generated outside the business (for example, data about macro-economic indicators or regional trends) is collected quarterly.

The usual solution for the first type of non-matching is to disaggregate the quarter interval inputs to a higher frequency (for example, monthly).[9] There are several ways to do this. The most straightforward way is to divide the quarter-interval observations by 3 to create monthly observations. This most straightforward method might prove unsatisfactory, however. A different and perhaps more satisfactory approach is to use PROC EXPAND in SAS/ETS. PROC EXPAND is a tool for interval conversion, and it provides methods for converting series to a higher-frequency interval. These methods include the use of splines to incorporate seasonal effects into the converted series. PROC EXPAND documentation is very good and provides additional details.

A word of caution: in these straightforward methods, the converted series is a scaled proportion of the original lower-frequency series. The noise in this representation of the series might be overly simplified. Consider the description of the most straightforward method. Monthly observations within each quarter are repeating scalars. The variance of monthly observations within each quarter is zero. A recent generalization of the straightforward approach is the use of a Brownian Bridge (Glasserman 2004). The Brownian Bridge would use a white noise series that effectively incorporates a more representative intra-quarter variance into the converted series. This bridges the variance information gap between observations of the original low-frequency series.

We do not want to imply that any of the straightforward methods should not be used or that applied forecasters need to immediately pursue an advanced degree in financial mathematics. However, it is prudent to remember that oversimplification is an issue when implementing the straightforward methods for series conversion and when interpreting the significance of estimated effects based on them.

The second basic type of interval non-matching occurs when candidate inputs are collected at a higher-frequency interval than the target variable. This situation is common and can be an opportunity for improving the accuracy and usefulness of generated forecasts. This type of non-matching might occur in businesses when the dependent variable (for example, sales) is accumulated at a monthly interval, and variables such as sales on book (finalized sales) are collected weekly.

In this example, large shifts in sales on book could be used to adjust the month's end forecasted stocking levels intra-month and to refine supply chain operations. Other examples include intra-day server hit counts and monthly server capacity utilization, weekly accrued expenses and monthly budget allocations, and so on. There are widely used methods for building this information directly into the model for forecasting the dependent series. The next section provides an example (Lucas 2008) of one of the more useful ones.

Methods

To discuss an example of the second type of interval non-matching in forecasting, let's imagine another modeler, Dr. Johan Slick, a professional forecaster who has spent most of his career at a large research firm. Representatives of the finance department met with Dr. Slick and described some problems that poor budget allocation forecasts were causing in the company. Under-forecasting last month meant that staffing and other needs for projects would be unmet for a time. Also, over-forecasting costs four months before had hurt the projected bottom line and angered shareholders.

Dr. Slick asked the accountants what method they were using to produce their forecasts, and the accountants told him that they were fitting a regression line to a rolling two-year window of total cost history. The accountants added that about a third of the business costs was tied to infrastructure and was effectively fixed. Remaining costs were tied to the number of revenue-producing projects that the company performed for its clients at any given time. Dr. Slick requested and received samples of company cost data and data on other variables related to project development.

In scrutinizing the data, Dr. Slick noticed a variable called CONTRACT that was denominated on a weekly interval. In a second meeting with the subject matter experts, he was informed that as contracts in negotiation reached a certain milestone, they were denoted "Pending Contracts," or "Contracts" for short. The CONTRACT variable was produced by the legal department, and it was a weekly count of new pending contracts. "What else can you tell me about contracts?" Dr. Slick asked. The experts answered, "As you know, it can be feast or famine around here. Some months the number of contracts seems to build and build, some months that it remains flat, and so on."

Dr. Slick asked, "Would you say that a reasonable hypothesis might be that rate of change or momentum in contracts might be a leading indicator for variation costs?"[10] The experts answered, "Well, you know that it is the total number of finalized contracts that drives costs. Also, not all pending contracts are finalized, but it might give us an indication." What follows is an outline of Johan Slick's momentum approach to building a model to forecast variable costs.

Plotting, cleaning, and screening his data, Dr. Slick came up with three years of history and the following data sets for analysis. Portions of the data sets are shown.

Figure 10.8

wk_dt	contract	indicator	t
06JAN08	7.2232421533	1	1
13JAN08	7.2269059507	1	2
20JAN08	7.2217653875	1	3
27JAN08	7.2084271842	1	4
03FEB08	7.2122084371	2	1
10FEB08	7.2099056746	2	2
17FEB08	7.2224509871	2	3
24FEB08	7.2252031789	2	4
02MAR08	7.2126062534	3	1
09MAR08	7.2144109302	3	2
16MAR08	7.2130687282	3	3
23MAR08	7.1984282059	3	4
30MAR08	7.2229732067	3	5
06APR08	7.2336716929	4	1
13APR08	7.2188794397	4	2
20APR08	7.1969991343	4	3
27APR08	7.2247767403	4	4
04MAY08	7.2209858542	5	1
11MAY08	7.2189452962	5	2
18MAY08	7.2163112242	5	3
25MAY08	7.2263631763	5	4

Consider the following facts about the data.

- The CONTRACT variable has been transformed and made anonymous.
- The INDICATOR variable flags weekly observations that fall within the same month and year. For example, observations flagged with a 1 fall in January 2008. The reason for this is below.
- The T variable creates a linear trend for weekly observations within each month.

Here is a portion of the data set containing monthly costs. Costs have been transformed to make them anonymous.

Figure 10.9

mo_date	costs
01JAN2008	79810.826554
01FEB2008	72771.463591
01MAR2008	74791.896864
01APR2008	80123.197418
01MAY2008	87820.109889
01JUN2008	85632.008423
01JUL2008	87962.378938
01AUG2008	70748.171092
01SEP2008	73825.255447
01OCT2008	70299.915827
01NOV2008	70784.803981
01DEC2008	62653.295152
01JAN2009	54749.461096
01FEB2009	64584.673235
01MAR2009	59396.833592
01APR2009	60140.1932
01MAY2009	51650.415172

Following a careful analysis of his data, Dr. Slick noted the following:

- The company was mainly interested in changes in variable costs from month to month, so he would model costs as percentage changes from one month to the next.[11]

- Because the CONTRACT variable was denominated weekly, it could not be incorporated in a model for percentage changes in costs in a straightforward way.[12] He would calculate a monthly indicator of the rate of change or slope in the number of weekly contracts and use this in his model.

The first step is creating the monthly slope or momentum of weekly contracts. The following PROC REG syntax in SAS/STAT creates the slopes.

Program 10.12

```
proc reg data=e noprint outest=est;
   model contract = t;
   weight t;
   by indicator;
run;
```

The REG procedure is a widely used SAS tool for fitting linear regression models. It is used here because the BY statement allows fitting individual slopes to observations within or by each month. The T variable is the within-month linear trend variable, and the INDICATOR variable identifies observations that fall within the same month and year.

A portion of the output data set, Est, is shown. This data set contains the within-month intercept and slopes associated with weekly contract observations. The variables have been renamed and a month date has been added that matches the time ID variable in the data set that contains costs.

Output 10.26

contract_slope	contract_int	mo_date
-0.006658747	7.2375821368	01JAN08
0.0056584492	7.2030431987	01FEB08
0.0018543608	7.2058148873	01MAR08
-0.00059953	7.2179520848	01APR08
0.0025590405	7.213649161	01MAY08
-0.000120785	7.2162960701	01JUN08
0.000233665	7.2267357164	01JUL08
0.0096880781	7.1882359676	01AUG08
-0.00496611	7.2389357903	01SEP08
-0.000388995	7.2271941642	01OCT08
-0.007526163	7.2377381951	01NOV08
0.0026866397	7.2115390265	01DEC08
0.0020962787	7.2013451736	01JAN09
-0.006681949	7.2264219722	01FEB09
-0.000302837	7.190546398	01MAR09
0.0096795403	7.1651694234	01APR09

This data set is merged with the data containing costs to create the Momentum data set for analysis. The dependent variable COSTS has been transformed to represent percentage changes in costs from one month to the next.

Output 10.27

mo_date	costs	contract_slope	contract_int	pc_costs
01JAN2008	79810.826554	-0.006658747	7.2375821368	.
01FEB2008	72771.463591	0.0056584492	7.2030431987	-0.088200602
01MAR2008	74791.896864	0.0018543608	7.2058148873	0.0277640874
01APR2008	80123.197418	-0.00059953	7.2179520848	0.0712817936
01MAY2008	87820.109889	0.0025590405	7.213649161	0.0960634712
01JUN2008	85632.008423	-0.000120785	7.2162960701	-0.024915722
01JUL2008	87962.378938	0.000233665	7.2267357164	0.0272137786
01AUG2008	70748.171092	0.0096880781	7.1882359676	-0.195699662
01SEP2008	73825.255447	-0.00496611	7.2389357903	0.0434934827
01OCT2008	70299.915827	-0.000388995	7.2271941642	-0.047752488
01NOV2008	70784.803981	-0.007526163	7.2377381951	0.0068974216
01DEC2008	62653.295152	0.0026866397	7.2115390265	-0.114876476
01JAN2009	54749.461096	0.0020962787	7.2013451736	-0.126151929
01FEB2009	64584.673235	-0.006681949	7.2264219722	0.1796403461
01MAR2009	59396.833592	-0.000302837	7.190546398	-0.080326173
01APR2009	60140.1932	0.0096795403	7.1651694234	0.0125151387
01MAY2009	51650.415172	-0.000282428	7.2050135552	-0.141166458
01JUN2009	51597.442231	0.0103383893	7.1559715323	-0.001025605
01JUL2009	51962.673715	-0.006896615	7.212443741	0.0070784804
01AUG2009	40050.908232	0.0050125732	7.1626994918	-0.229236962
01SEP2009	35115.647765	-0.004648868	7.1698808425	-0.123224683

The ARIMA procedure in SAS/ETS is used to test Dr. Slick's hypothesis that the rate of change in contracts provides a leading indicator of changes in monthly costs.

Program 10.13

```
proc arima data=momentum;
identify var=pc_costs crosscorr=(contract_int contract_slope);
run;
```

The cross-correlation function plot for the CONTRACT_SLOPE variable provides evidence to support Dr. Slick's hypothesis.

Output 10.28

Correlation of pc_costs and contract_slope	
Variance of input =	0.000029
Number of Observations	35

Lag	Covariance	Correlation	-1 9 8 7 6 5 4 3 2 1 0 1 2 3 4 5 6 7 8 9 1
-8	-0.0001252	-.19505	. ****	.	
-7	5.11854E-6	0.00797	.	.	
-6	0.00011322	0.17636	.	**** .	
-5	-0.0000623	-.09704	. **	.	
-4	-0.0000389	-.06063	. *	.	
-3	-0.0000539	-.08398	. **	.	
-2	0.00001597	0.02488	.	.	
-1	0.00012238	0.19062	.	**** .	
0	-7.2519E-7	-.00113	.	.	
1	0.00023149	0.36057	.	******* .	
2	-0.0000384	-.05988	. *	.	
3	-0.0000394	-.06133	. *	.	
4	-0.0000922	-.14364	. ***	.	
5	-0.0001133	-.17653	. ****	.	
6	0.00002436	0.03794	.	* .	
7	0.00003728	0.05806	.	* .	
8	0.00003913	0.06094	.	* .	

"The rate of change or momentum in contracts seems to be significantly and positively correlated with variation in changes in costs. Further, momentum in contracts leads variation in changes in costs by one month. That is, it provides a one-month leading indicator for changes in variable costs," Dr. Slick informed the experts in an interim report. "This makes sense because it takes approximately 30 days to implement a pending contract," the experts answered.

Based on this information, Dr. Slick built a forecasting model around the MOMENTUM variable. This model was used by the finance department in two ways. First, weekly projections were created as new contracts data came in to update monthly cost predictions. Second, Dr. Slick modified the linear trend forecasting model used by the subject matter experts to include momentum. The precision of cost projections was significantly improved.

10.5 Holdout Samples and Forecast Model Selection in Time Series

Introduction

In Chapters 8 and 9, when we discussed the fundamentals of model building, AIC is used as the primary model selection diagnostic. AIC features a built-in penalty term that is designed to prevent overfitting. An alternative and preferred method for avoiding overfitting is the use of holdout samples and diagnostics based on them in the model development process.

You might be familiar with the use of holdout samples in static models. This section begins with a review of ideas in this context. Although holdout samples are the preferred method for forecast model selection, implementation of this method is not always feasible in time series data. Large-scale forecasting problems have unique features that can exacerbate problems associated with the use of holdout samples. This section concludes with guidelines for when and how to implement holdouts for model selection in time series and when to use AIC instead.

This section does not have a methods or applications discussion. If you use automated forecasting tools like SAS Forecast Server, the successful implementation of the ideas discussed in this section consists of changing two default settings before running the project. If you are using more manual tools like PROC ARIMA in SAS/ETS, implementation depends on data manipulation that might be involved for less-experienced programmers. (The course notes for "Forecasting Using SAS Software: A Programming Approach" provide a step-by-step example for implementing holdout sample selection using PROC ARIMA.)

Holdout samples are not complicated. A basic assumption is that practitioners are working with a representative sample of data that is drawn from the population. The variation in the sample can be split into two components—signal and noise. Any representative sample has approximately the same systematic variation or signal as the population that the modeler is trying to make inferences about. For example, an analyst that is interested in the relationship between changes in price and sales might draw a sample from his company's products. If the sample is representative, the estimated relationships should provide a good estimate of the true average-price effect across all products.

Population and sample signals are very similar if samples are drawn correctly. A key fact is that the noise or non-systematic variation that each sample contains is unique to the sample. This is an important fact for applied forecasters. If the model is over-fit or accommodates the signal *and* noise that is unique to the sample, then model accuracy and usefulness will likely be degraded. Consider a naïve modeler named Mr. N. N's favorite model selection diagnostic is R-squared. He is provided with a sample of his company's sales data and data on 300 candidate input variables, some of which were purchased from sources outside his company.

N fits a reasonable model using guidelines provided by an experienced modeler, Mrs. E. N generates an R-squared value, .65. This is approximately equal to the fit of previous models. In an effort to improve his model, N explores other candidate regressor variables and finds two that have a marginally significant relationship with his sample of sales data. These two are PET (price of eggs in Tasmania) and AIT (Arctic ice cap thickness). Including these variables in the model improves his generated R-squared value to .67. N is happy. His diligence has improved the fit of the model. However, N's happiness is short-lived. His model is put into production to estimate sales associated with new customers. His model's performance is found to be substantially worse than models used in the previous three quarters.

"How could this happen," N asked E, "when the overall fit of the model was better than its predecessors?" E explains that the two new variables, PET and AIT, were probably correlated with his sample by chance. In other words, they were correlated with the unique noise in his sample and not with the systematic variation. Because they were in the model, predictions included effects of variation in PET and AIT. Extrapolating these effects onto a new sample that has a completely different and unique set of noise (that did not happen to be correlated with PET or AIT) degraded prediction performance.

The moral of the story is to try to build a model that captures all of the systematic variation or signal in the sample of data that you have and that also ignores the noise. Holdout sampling helps modelers do this because this approach mimics what is done in practice. The model is fit or trained on one sample of data (with its own unique partition of noise), but its performance is measured or validated based on a diagnostic that is calculated on a separate data set (with a different and unique partition of noise).

An outline of the holdout sample approach for static models is as follows:

- Break the sample into two samples. These are usually called the train sample and validation sample.
- Fit candidate models (see Chapters 8 and 9) on the train sample.
- Apply these fitted models to the validation data.
- Use a fit statistic that is based on the residuals of the application of models to the validation data to select the best model.[13]

This process should give the modeler a good idea of which model will have the best performance in application. This is the model that does the best job at capturing the signal and ignoring the noise in the training sample.

The main problem associated with the use of holdout samples in the static context is lack of data. The training sample must have enough observations to provide reasonably good parameter estimates. The validation sample must have enough observations to provide a good idea of future model performance.[14] If this is not the case, the holdout sample approach should not be used. A widely used and recommended alternative is the use of AIC for model selection. Other alternatives include resampling or simulation-based techniques for creating the subsamples.

In forecasting time series data, the basic ideas are exactly the same, but the nature of the data creates additional restrictions on the holdout sampling approach for model selection. Also, modelers like Mr. N might fail to see the usefulness of holdout samples for model selection in time series data. N's approach can be summarized by his statement, "My company has been around for 10 years and I have all the data. Since I have the population, I don't need to worry about holdout samples, and I can go back to using good old R-squared for selecting my models!" This approach is faulty.

In time series data, the history can be thought of as one sample, and the future as another sample. It is hoped that the history contains the same signal as the future. The future has a unique and different set of noise than the noise that is contained in the historical sample.

Additional restrictions associated with time series data in implementing holdout samples for model selection come from the following idea—the most relevant data for predicting what is going to happen in the next time interval is the most recent data. Therefore, the holdout or validation sample to pick the best model should always be drawn from the most recent or last data in the historical sample.

This modification to the static approach seems reasonable, but it presents a problem. If the most recent data is the most important, modelers will want this information in the parameter estimates of their forecast model.

An outline of the time series approach to implementing holdout samples is as follows:

- Break the sample into two samples. The most recent data is used for the validation sample. Older historical data is used for the training sample.
- Fit candidate models on the training sample.
- Apply these fitted models to the validation data.
- Use a fit statistic that is based on the residuals of the application of models to the validation data to select the best model.
- After the best model is selected, refit the champion model to the entire range of history (training and validation) before forecasting or putting the model into production.

A diagnostic calculated on a holdout sample is the preferred measure for selecting a forecast model. However, holdout sampling is not always feasible in time series. As previously noted, the problem with scarce or short data is relevant for time series data.[15] In addition, the process of creating the holdout or validation data for time series can present problems.

Recent structural changes can make the use of holdout samples infeasible. For example, many firms whose activities were impacted by September 2001 or December 2008 found using holdout samples infeasible for several months following the date.[16] This is because the data in the holdout sample was very different from the preceding data on which the models were fit. No model validated reliably. In this circumstance, a penalized fit statistic like AIC is probably the best alternative measure for model assessment. The penalized statistic can be used until the history contains enough information past the event to re-implement the out-of-sample approach.

Out-of-sample validation can be problematic in large-scale forecasting applications. The main characteristic of these applications is that the analyst does not have enough time to visually explore and validate each time series to be modeled. A substantial number of series might be short or contain late structural changes. Prudent modelers in large-scale forecasting applications spend a lot of their time prescreening and filtering the data that flow to the forecasting algorithms to remove problematic series and to enable the use of holdout samples on the series that remain. Problematic series are usually best dealt with separately using modeling and selection techniques that do the best job possible.

This section concludes with some rules for implementing holdout sample selection.

- A standard rule is to put 25% of the data into validation and 75% into training. Another popular partition is 50/50. The tradeoff for all rules for partitioning is that more data in the training sample means more information on which to base parameter estimates, and probably more precise parameter estimates. More data in the validation sample means more information for model selection, and probably more reliable validation results. Modelers should create their own partitions based on a value-weighting of each side of the tradeoff and on the characteristics of their data.

- If the data has a seasonal pattern, the holdout or validation sample should include at least one full seasonal cycle. For example, if the data has three years of history and seasonality, a reasonable split would be 66% in training and 33% in validation.

- Holdout sample validation should not be used for short or intermittent series or for series that have undergone recent structural changes.

10.6 Planning Versus Forecasting and Manual Overrides

Manual overrides are adjustments made to statistical forecasts. Forecasts are usually adjusted to more closely align them with stakeholder group plans or goals over the lead forecast horizon. This topic has already received excellent coverage in several texts. The purpose of this section is to summarize how planning activities can impact the accuracy and usefulness of generated forecasts. A method for implementing manual overrides in a way that increases the likelihood of enhancing forecast usefulness is discussed.

Statistical forecasts are not usually implemented as provided in the supply chain or planning processes that they feed. Forecast users include C-level executives, non-statistician forecasters, production managers, planners, sales managers, and so on. Each of these people usually has an opinion about the accuracy of generated statistical forecasts and adjusts or ignores them accordingly. In many businesses, each of these people might be allowed to adjust (manually override) generated forecasts in a haphazard way.

Some of the ideas covered in reconciliation apply to this topic, too. The statistical forecast is an objective best guess, based on historical patterns, about what will happen in the future. Manually changing or overriding these forecasts has the potential to substantially degrade their accuracy and usefulness. However, this does not mean that forecasts should never be adjusted. Forecast users might have information about competitor activities or forthcoming restrictions on business processes that is not in the historical data. Under these circumstances, forecast accuracy can be enhanced by embedding the hypothesized effect of future events into lead forecasts.

Forecasting best practices include the systematic and supervised adjustment of generated forecasts (Chase 2010). Forecasts should only be adjusted based on reliable information that is not contained in the historical data. Forecasting worst practices (see Gilliland 2010) includes haphazard forecast adjustments from stakeholder groups that might have different and conflicting objectives and small adjustments based on gut instincts (Gilliland 2010).

Some interesting studies provide evidence to confirm this categorization. Defining a small adjustment as around 5%, Fildes and Goodwin (2007) found that small manual overrides tend to diminish the accuracy of forecasts. (See also Armstrong 2001.) Manual adjustments in the range of 25% or more are likely to have the opposite effect. The authors (cited above) conclude that small adjustments tend to be based on factors like a stakeholder's need to appear to be contributing something to the planning process and not on evidence of future changes. Larger changes are more likely to be based on plausible evidence of future changes because an explanation for the override will likely be required by other members of the firm. This leads to another forecasting best practice—the requirement that all manual overrides be tracked and reasons for changes to the baseline forecast be documented (Chase 2009).

Systematic overrides that are based on plausible evidence of future changes tend to enhance forecast accuracy and usefulness. Haphazard overrides that are based on other factors usually lead to worse forecasts. Chase (2009) provides a framework for conducting overrides in a systematic, supervised, and documented planning environment. His approach is called demand-driven forecasting and his book is recommended if you are interested in this area of applied forecasting.[17]

10.7 Scenario-Based Forecasting

Scenario-based forecasting or what-if analysis describes a process where the fitted model is used to answer questions about what happens to the target or forecast variable after hypothesized changes in input variables. This section provides background on this approach and an illustration of scenario-based analysis using SAS Forecast Studio.

Ideas

From an analyst's point of view, scenario-based forecasting or what-if analysis occurs at the end of the forecasting cycle. Models capturing the systematic variation in each series have been fit and forecasts have been generated. At this point, it is common for forecast users to ask questions like, "What would happen to projected sales if we increase price by 10%?" or "What is the lift associated with running a buy-one-get-one-free promotion next November?" The model is not refit following a change to the underlying input data. The data is modified and then run through the model that was fitted on the original data.

The ability to answer these types of questions depends on the forecast specifications implemented. To use the model as a basis for answering questions about price effects, the model should have a significant price term in it. There are less obvious challenges to providing useful scenario-based answers. If the price term enters the model as a contemporaneous-only (has an effect only at lag 0) term, then providing an answer is relatively easy.

If a model has a significant term that can be used as a basis for the hypothesized change, and if the relationship of this input to the target variable is static (contemporaneous), then what-if analysis can be implemented using spreadsheet software. However, if the relationship between the input and target is dynamic, the situation becomes more complex. Anytime the specification embeds a scenario where what happens to an input today has an impact on what happens to the target in the future, the scenario-based analysis becomes more difficult to implement.

A common request from analysts using manual forecasting software is the ability to use fitted specifications in spreadsheet software to answer scenario-based questions. This is possible when the relationship between the input and target is dynamic, but implementing it is a cumbersome process that has a large potential for error. Another alternative for SAS/ETS users is to do the scenario analysis in the software. The ARIMA procedure has a NOEST option. Users can manipulate the input variable data and run it back through the fitted specification without estimating parameters to produce scenario results.[18]

A preferred method for doing scenario analysis, particularly for analysts who work in a large-scale forecasting environment, is to use software that is designed to do scenario analysis using models that have been generated and fitted based on characteristics of the data. SAS Forecast Studio functionality includes a Scenario Analysis View that is designed for this purpose. The remainder of this section presents an example of doing scenario-based forecasting using this tool.

Methods

A project is created in SAS Forecast Studio that is based on the CITIMON data set found in the SASHELP library. The data contains history beginning January 1980 to December 1991. The interval of the data is month. The dependent variable is CCIUAC and measures aggregate consumer demand for credit. Candidate independent variables are various macro-economic indicators hypothesized to impact the demand for credit.

The initial aim of the project is to quantify the effect of variation in macro-economic indicators on the target. The only indicator selected in the forecast model is FM1D82, a measure of the money stock. The project is set up and run. Here is the fitted model for consumer demand for credit.

Output 10.29

Component	Parameter	Estimate	Standard Error	t Value	Approx Pr > |t|
CCIUAC	AR1_1	0.28846	0.08753	3.30	0.0013
CCIUAC	AR1_2	0.14127	0.08904	1.59	0.1152
CCIUAC	AR1_3	0.32408	0.08655	3.74	0.0003
CCIUAC	AR2_12	0.26446	0.09225	2.87	0.0049
FM1D82	SCALE	38.57784	18.97052	2.03	0.0442
FM1D82	NUM1_1	31.24726	19.41954	1.61	0.1102
FM1D82	NUM1_2	-6.73975	18.49195	-0.36	0.7161

Additional details provide information about shift (delay) effects.

Output 10.30

Name: LEAF_0
Description: "ARIMA: CCIUAC ~ P = ((1,2,3)(12)) D = (1) NOINT + INPUT: Lag(11)Dif(1) FM1D82 NUM = 2"
Details: "ARIMA: CCIUAC ~ P = ((1,2,3)(12)) D = (1) NOINT + INPUT: Lag(11)Dif(1) FM1D82 NUM = 2"
Model family: ARIMA
Model type: GENERALARIMA
Source: Generated by HPFDIAGNOSE

Intercept: None
Forecast variable: CCIUAC
Delay: 0
Differencing: (1)
P: (1,2,3)(12)

Input: FM1D82
  Delay: 11
  Differencing: (1)
  Numerator polynomial: (1,2)

It appears that the demand for consumer credit is highly autoregressive and nonstationary (differencing=1). Also, shocks to the money supply seem to have short-lived persistence (numerator=2). They take, on average, 11 months to impact a consumer's demand for credit after they are initiated (delay=11).

The phone rings on your desk in Washington, DC, and the chairman of the Federal Reserve Board is on the line. He is considering expanding the money supply in an effort to boost the demand for consumer credit specifically and to influence overall economic growth. "If the money supply is expanded by 2% starting next month (December1991), I need to see the estimated future impact on consumer credit demand, and I need it on my desk today."

You switch to the Scenario Analysis View in SAS Forecast Studio. You navigate the software and create scenarios once the models are generated. You create two scenarios. The first scenario, Baseline, is the forecast based on the original data. The second scenario, M1iall, is based on the modified future values of the money stock variable. The output below shows the estimated impact of a permanent 2% increase in the money supply starting in December 1991 on the demand for future consumer credit using the forecast model as a basis.

Steps to create the M1iall scenario follow. After clicking **New**, future values of the input variable can be changed in the table. The method highlights all cells to be changed and adjusts them using the Input Calculator.

Output 10.31

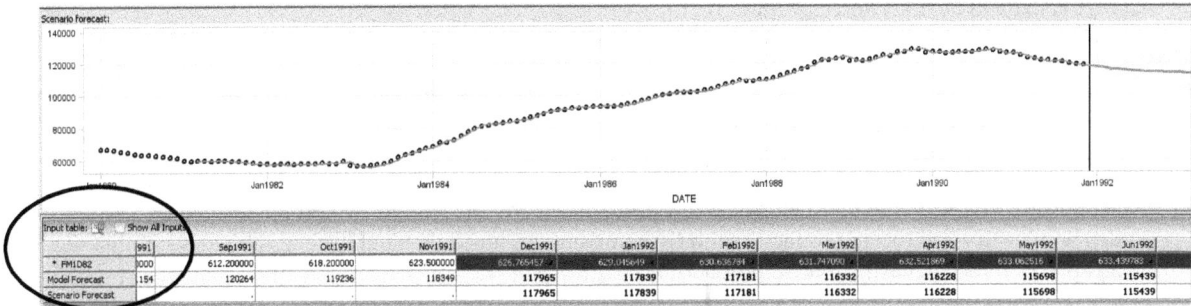

Output 10.32

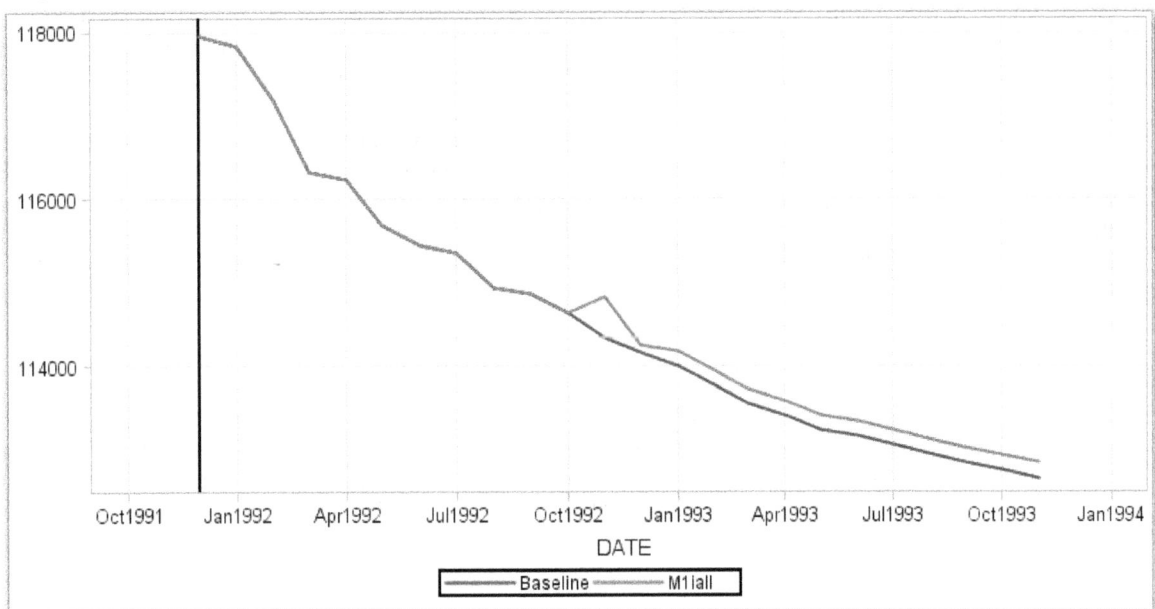

After the scenario is set up, it is run by clicking **Run Scenario**. The output shows the compare table that overlays the Baseline and M1iall scenarios. Using the model as a basis, the change in the money stock has no impact on a consumer's demand for credit until about 11 months after the change is initiated. The consumer's demand for credit is increased by the amount shown below.

Output 10.33

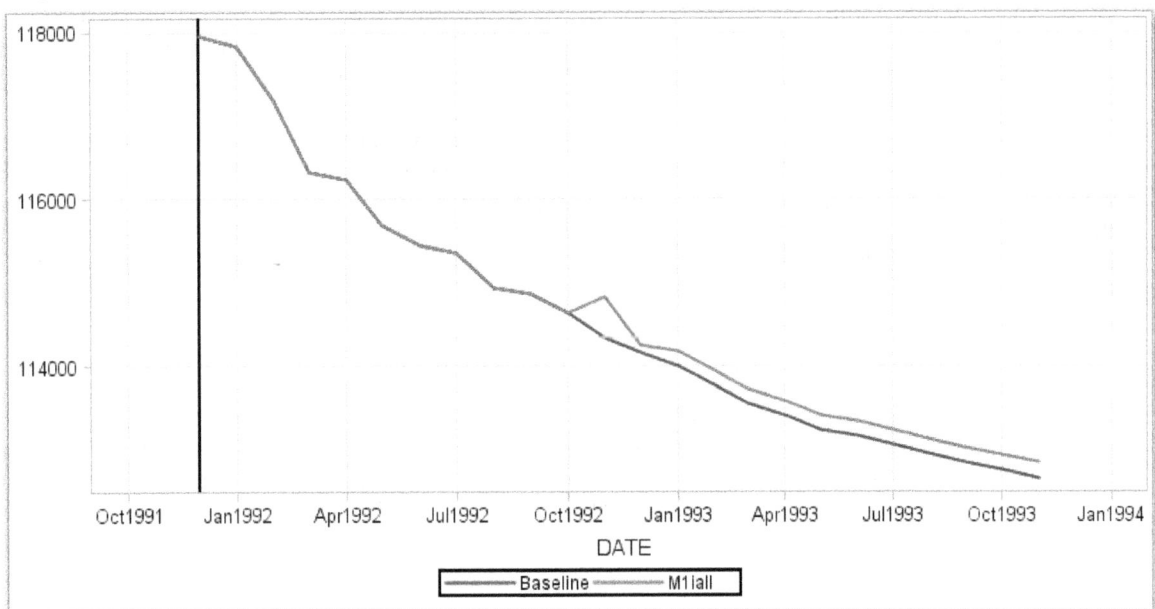

Scenario-adjusted target forecasts are also available in a table in the Scenario Analysis View and in an output data set generated by SAS Forecast Studio.

10.8 New Product Forecasting

Providing forecasts for new products is challenging because, by definition, new products have little or no demand history on which to build models. New products can be subset into two groups—pending products with no demand history and recent products that have some but not much demand history. The focus of this section is on an approach for forecasting recent products.

New product forecasting is an active area in applied forecasting. There is a wealth of information about the topic. This section discusses the basic ideas, provides references if you want to learn more about this topic, and presents an example using the SIMILARITY procedure in SAS/ETS.

The first type of new product is a pending product. These products are goods that are about to be introduced into the market. They have no accumulated history. Movie DVD releases and new drugs are two examples. A recent and useful approach to forecasting pending products is forecasting by structured analogy. (See Gilliland 2010.) The basic idea is that the pending product's future demand will behave similarly to the demands of related and existing products. The challenge is to find which existing products are most like the product to be released. The most analogous products can be used to create a proxy history for the pending product. Models can be fit to the proxy history and forecasts can be generated. Forecasting new products, pending products in particular, has been a very active area of research.[19] Armstrong (2001) provides a survey of work done on forecasting by analogy and alternative approaches.

The second type of new product and the focus of this section is the recent product. Some data on product demand does exist but, as the name suggests, there is not much of it. The approach to forecasting recent products is similar to the approach for pending products. Find similar or analogous series that can be used as proxy history. The analyst is faced with several challenges in this framework. Analyst judgment plays a larger role than in traditional forecasting approaches.

The first challenge that an analyst might face is to determine which of the thousands of candidate similar series to start with. It is not a good idea to introduce thousands of candidate series into an algorithm and let it crank away to select the best (most similar) ones. Subsetting rules do not need to be complex. They should be based on hypothesized (not hoped for) demand characteristics. For example, the analyst could make a first cut by considering only series with positive trend or seasonality if this seems appropriate. The best information at this stage generally comes from the market and from the knowledge of stakeholders. This information should be used to derive a reasonably sized subset of candidate series with which to proceed.

Recent product history is usually much shorter than the history of the candidate or analogous products in the analysis. This leads to another challenge. What is the most useful definition of similarity? If the recent product is fairly unique, and the analyst is interested in forecasting the rate of product penetration or market acceptance, then the focus will be on finding series with early demand patterns that are similar. If the recent product will compete with several other similar products in the market, or if the market environment has undergone significant and recent structural change, then the most useful focus might be on finding series with similar and recent demand patterns.

The most useful similarity might not occur at the beginning or end of longer candidate series. For example, forecasters might be interested in the candidate series with the seasonal cycle that is most similar to the one displayed in the recent product, regardless of where it occurs in the candidate series. There are also other, more technical considerations that need to be assessed after the candidate series have been selected and the definition of usefulness has been agreed on. "How is similarity to be measured?" "Should the series be transformed before calculating similarity statistics?" "Are there missing values in series?" All of the definitions of useful similarity and the technical considerations can be accommodated in the tool that is used in the next section, PROC SIMILARITY in SAS/ETS.

Applications

WhyFi is a manufacturer of electronics. It introduced a new model of the flat-screen TV, New_p, 15 weeks ago. Among New_p's features are its 3-D-cubed and new Sonic Boom technologies. WhyFi also has 32 existing models of flat-screen TVs that have characteristics that are somewhat similar to New_p. Although sales of the New_p have begun at a good pace, supply chain and distribution processes associated with it have been inefficient because an established historical demand series does not exist on which to formulate reliable forecasts.

WhyFi's upper management has agreed to a forecasting by analogy approach for New_p sales. Project stakeholders met to decide on likely candidate products and to discuss other relevant topics. Marketing experts agreed that many New_p features were revolutionary. However, they conceded, no new features were unique relative to competitors' flat-screen TVs. The firm's economist determined that an increase in unemployment and diminishing consumer disposable income were the causes of the recent trend of flat to diminishing sales for all TVs. The engineering department provided a list of 10 WhyFi TVs most similar to New_p based on picture size, audio and visual features, and so on.

Based on this information, the WhyFi forecasting team determined that the most recent year of data would be used to implement the forecasting by analogy approach. New_p has 15 weeks of history. A similarity analysis of the 10 TVs identified by the engineers would be conducted. If one candidate TV's history proved to be a clear winner, then its associated forecast model would be used as a basis to provide forecasts for New_p until sufficient history was obtained. If no clear winner among the candidates emerged, the history from the three most similar brands would be combined, and a new forecasting model would be created based on the combined data.

The PROC SIMILARITY syntax that the WhyFi forecasters used is listed below. What follows is a brief description of the syntax. More information can be found in SAS/ETS documentation for PROC SIMILARITY.

- The PROC SIMILARITY statement lists the data set that contains the historical demand data for the 10 candidate TVs and for New_p. The OUTSUM option creates a data set named SIMMTX that contains the similarity measures on each series and stores it in the WORK library.

- The ID statement lists the time ID variable in the input data set (DATE), the interval of the data (WEEK), and the method of accumulation used to create the weekly time interval. The START and END options create a time subset (the most recent year) of the historical candidate series data for the analysis.

- The INPUT statement lists the 10 candidate input series, A through J. It specifies that the inputs be transformed using ABSOLUTE normalization before the similarity analysis.

- The TARGET statement lists the recent product series, New_p, and specifies that it be transformed. The similarity measure used to do the analysis is mean absolute deviation (MABSDEV). The EXPAND and COMPRESS options list restrictions on the time-warping process that is part of the similarity analysis.[20]

Program 10.14

```
proc similarity data=simin outsum=simmtx;

   id date interval=week accumulate=total
      start='25JAN2010'd end='31jul2011'd;
   input a b c d e f g h i j / normalize=absolute;
   target new_p /  normalize=absolute measure=mabsdev expand=(localabs=3 globalabs=3)
                   compress=(localabs=3 globalabs=3);

run;
```

The output shows the generated and sorted similarity measures. Series G is the most similar based on the measure used.

Output 10.34

		Input Variable Name	Analysis Status	new_p
1	G		0	0.1169036185
2	E		0	0.1229897843
3	B		0	0.1359030067
4	A		0	0.1374352234
5	I		0	0.1560733143
6	C		0	0.1587176998
7	J		0	0.1595137813
8	D		0	0.1775833294
9	H		0	0.1803397623
10	F		0	0.1996958587

Further analysis of the New_p and G pair indicates that the G series might be a good candidate for use as proxy history. The TIMESERIES procedure is used to produce overlay and CCF plots of New_p and the top three candidate series. ODS graphics options are not shown.

Program 10.15

```
proc timeseries data=simin plots=(series) crossplots=(ccf series);
id date interval=week;
var new_p;
crossvar G E B;
run;
```

Although the New_p series history is more volatile, the patterns seem to map fairly well to the most recent G series data.

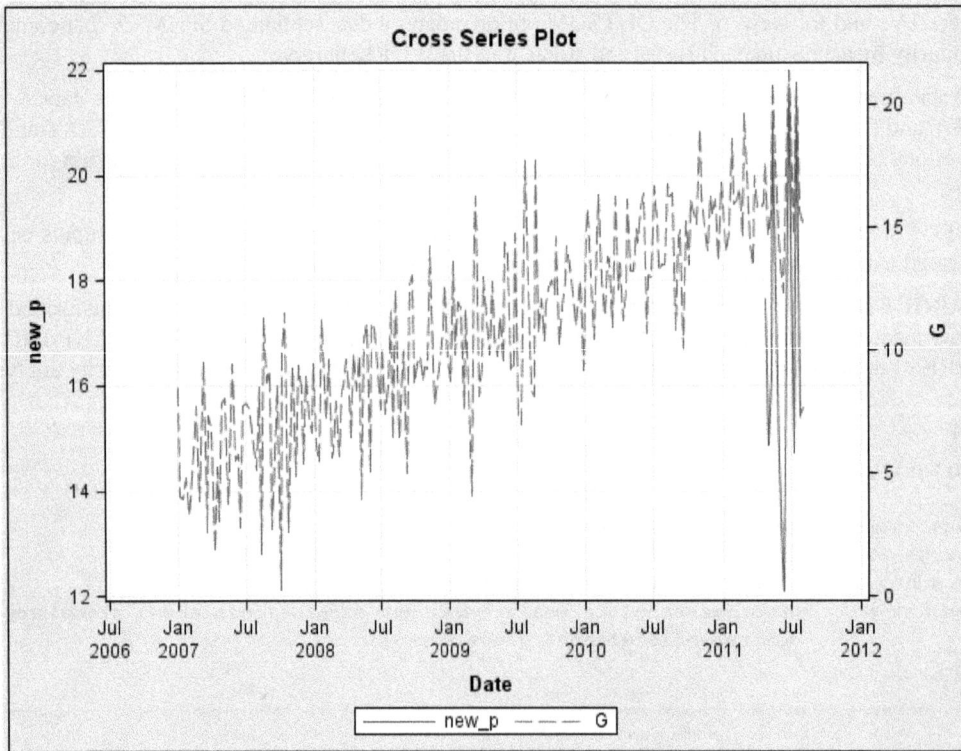

The CCF plot indicates significant correlation between the New_p and G series at lag 0. They seem to move together contemporaneously.

Output 10.36

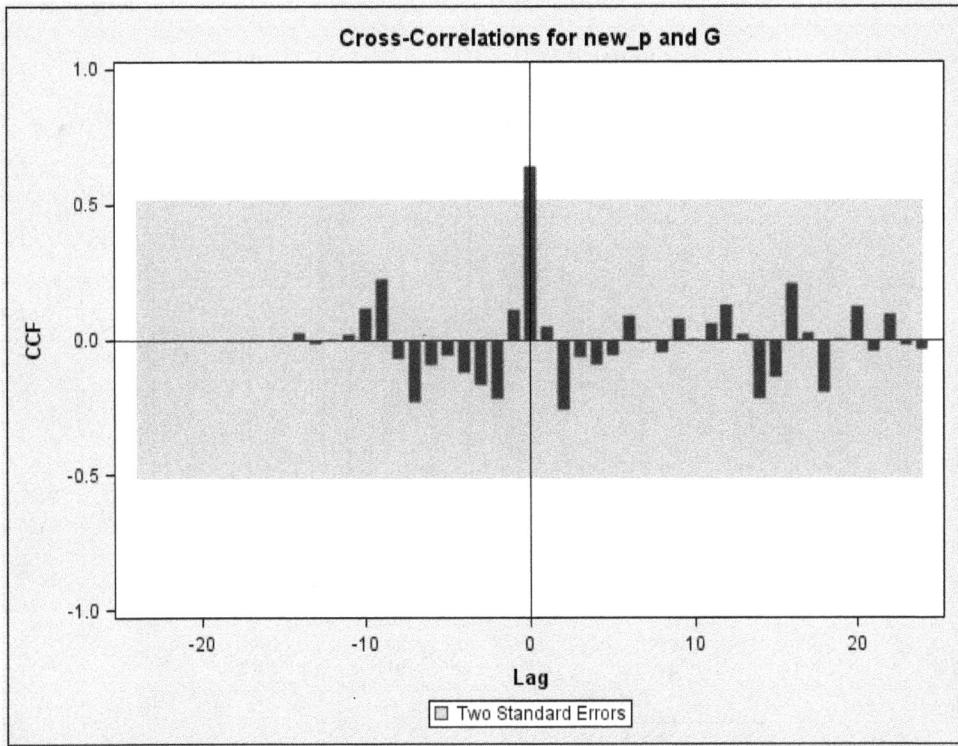

Here are overlay plots of New_p and the second and third place candidates.

Output 10.37

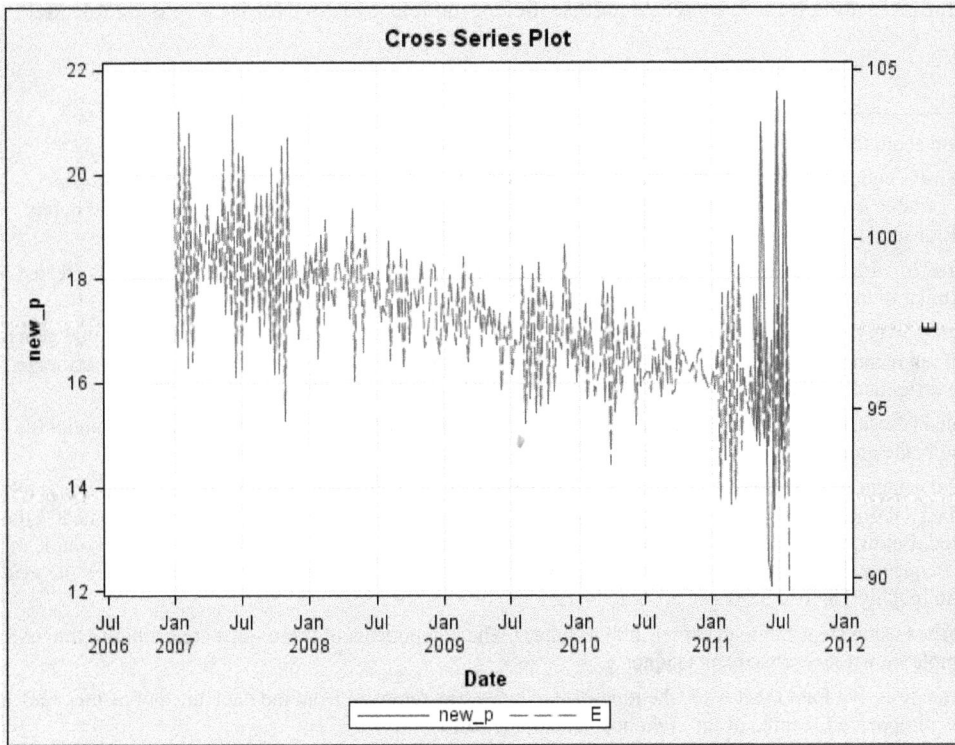

Output 10.38

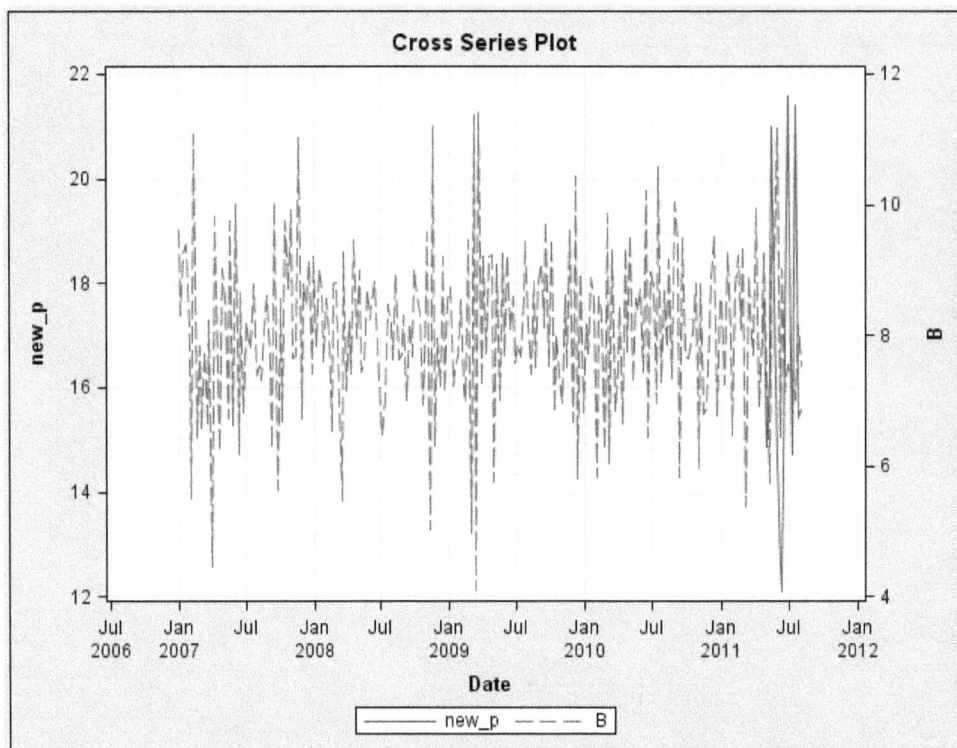

Neither the E or the B series had significant correlation with the New_p series at what was considered a reasonable lag. The G series was selected as proxy history for the New_p series. Its forecasting model will be used as a basis for generating New_p forecasts until sufficient modeling history for New_p is established.

---

[1] Additional information about this topic can be found in Dagum and Cholette 2006.

[2] PROC TIMESERIES runs under the hood in SAS Forecast Server to create data hierarchies in SAS High-Performance Forecasting. The ID variable in the input data set has a datetime format. This results in the datetime (DT) interval being specified. Date (D) formatted intervals are also feasible and common.

[3] The overlay plot is created with PROC SGPLOT in SAS/GRAPH.The middle or regional level of the hierarchy is created using syntax that is very similar to the code. The only changes are in the output data set and in the BY statement.

[4] Each series in the hierarchy is forecast one at a time. Each series has its own model. The cost of this approach is that generated forecasts usually will not reconcile for each time interval. The benefits of this approach include marginal effects estimated individually for each series and legitimate error variance estimates for all generated models in each level.

[5] SAS High-Performance Forecasting procedures are a component of SAS Forecast Server. These procedures run under the hood of SAS Forecast Studio, the graphical user interface to SAS Forecast Server.

[6] The absence of a BY statement indicates that the generated series contained in HFHR represents intake per hour across all states and counties in the data. ODS graphics statements are not shown to eliminate clutter. MODE=PSEUDOADD indicates the decomposition method. Below, the pooled jail intake data is shown at an hourly interval. The problems that the volume of observatons in high-frequency data can cause are apparent. However, the seasonal decomposition output confirms the evidence that intake peaks from roughly 18:00 to 3:00.

[7] This can be accomplished using steps outlined earlier in this chapter. The components of these steps are similar to the SAS Forecast Studio example shown in a subsequent chapter.

[8] In a hospital staffing analysis, we found that when the time-of-day cycle was removed from the data, the day-of-the-week and monthly cycle effects disappeared in three of the eight hospitals in the study.

[9] "Disaggregate" is the widely used term for moving series from a lower- to a higher-frequency interval. It is different from the "disaggregation" term used in the data hierarchy section.

[10] See, for example, Lehman 2010.

[11] The COST variable is nonstationary. The percentage change transformation creates a stationary version of the dependent series. See Chapter 8 for more information.

[12] The relatively new and more theoretical mixed-interval or MIDAS approach provides a way to accommodate mixed-interval data in the forecast specification. (See Ghysels 2010.)

[13] Popular fit statistics for this approach are MAPE (mean absolute percent error) and RMSE (root mean squared error).

[14] A barely enough rule for how much data is needed for parameter estimates is five observations per estimated parameter.

[15] An alternative method for choosing a holdout sample size is to select the last n percent of the series. Percentage-based holdout size can mitigate the problems associated with holdout sampling on series of different lengths to a limited extent.

[16] The dates of terrorist attacks in the United States and India, respectively.

[17] SAS has developed a planning tool called SAS Demand-Driven Forecasting that is based on Chase's approach.

[18] Woodfield (2008) provides an example.

[19] SAS has developed patented algorithms to handle forecasts associated with new products based on the forecasting by structured analogy approach.

[20] Because the target and input series are different lengths, the series will be warped to facilitate comparison. Warping is a process by which the shorter series are expanded and the longer series are compressed so that similarity statistics can be calculated in a more useful way than simply restricting the calculation to matching date intervals in two series. Other functionality includes the use of sliding scales to compare the target to subsequences of candidate input variables. Additional information about similarity analysis is found in the procedure documentation and in Sankoff and Kruskal 1999.

Chapter 11: Model Building: Alternative Modeling Approaches

11.1 Nonlinear Forecasting Models

The objective of this section is to give you an overview of some of the most popular nonlinear forecasting methods. Because most of these methods are still in the research domain, the focus is on describing their key principles without examples. The section includes a generic summary of the nonlinear modeling features and short descriptions of selected approaches for nonlinear forecasting. These include neural networks, support vector machines, and genetic programming.

11.1.1 Nonlinear Modeling Features

There is a growing interest toward developing and implementing nonlinear models in forecasting. (For more information, see the current surveys of Tsay 2002 and Clements et al. 2004.) Established application areas for nonlinear models are macro-economics and finance. For example, most real business cycle models are highly nonlinear. Applications include bond pricing models, product diffusion processes, and almost all other continuous time finance models. The key idea of the nonlinear nature of economic time series is that major economic phenomena, such as output growth in a market economy, are represented by the presence of two or more regimes (for example, recessions and expansions). This is true of financial variables with periods of high and low volatility. Other types of nonlinearity might include the possibility that the effects of shocks accumulate until a process reaches a catastrophic event. (This was the case of the financial crisis in 2008.)

A general nonlinear dynamic model with an additive noise component can be defined as follows:

$$y_t = f(\mathbf{z}_t; \boldsymbol{\theta}) + \varepsilon_t \qquad\qquad 1$$

In this equation, $\mathbf{z}_t = (\mathbf{w}_t', \mathbf{x}_t')'$ is a vector of explanatory variables. $\mathbf{w}_t = (1, y_{t-1}, \dots, y_{t-p})'$ is the vector of strongly exogenous variables $\mathbf{x}_t = (x_{1t}, \dots, x_{kt})'$. $\boldsymbol{\theta}$ is the corresponding model parameters vector. It is assumed that y_t is a stationary process and the noise ε_t has zero mean and constant standard deviation. The nonlinear function f could have a different representation such as a switch function (in the case of switch models), a combination of sigmoid functions (in the case of neural networks), a kernel function (in the case of support vector machines), or any arbitrary function (in the case of genetic programming). The overview of nonlinear models begins with threshold models based on different types of switch functions. The smooth transition regression model, self-exciting threshold autoregressive model, and Markov switching autoregressive model are discussed. (For more detailed information, see Clements et al. 2004.)

The smooth transition regression (STR) model assumes two regression lines and a smooth transition from one line to the other. The function to characterize the transition is the hyperbolic tangent function, which is also a bounded function of a continuous transition variable s_t. The STR model is capable of characterizing asymmetric behavior. For example, suppose that s_t measures the phase of the business cycle. The STR model can describe processes whose dynamic properties are different in expansions from what they are in recessions, and the transition from one distinct regime to the other is smooth.

Another nonlinear model is the self-exciting threshold autoregressive (SETAR) model, which is a piecewise linear autoregressive model in the threshold space. It is similar to the usual piecewise linear models in regression analysis. Piecewise linear models are piecewise linear in time, but the SETAR model is nonlinear. The SETAR model might be used to characterize processes in which the asymmetry lies in growth rates. For example, the growth of a series when it occurs might be rapid, but the return to a lower level is slow.

The third model is the Markov switching autoregressive (MSA) model. The MSA model uses a hidden Markov chain to govern the transition from one conditional mean function to another. This is different from the SETAR model, where the transition is determined by a lagged variable. Consequently, a SETAR model uses a deterministic scheme to govern the model transition whereas an MSA model uses a stochastic scheme. This difference has important practical implications in forecasting. For example, forecasts from an MSA model are always a linear combination of forecasts produced by sub-models of individual states. But, forecasts from a SETAR model come from a single regime only.

One of the fundamental issues of nonlinear forecasting models is whether the in-sample model estimation criterion and out-of-sample forecast accuracy criterion should match (Clements et al. 2004). It is often difficult to choose asymptotically valid inferential strategies using standard estimation procedures that minimize one-step-ahead errors for many nonlinear models. These include smooth transition models to neural networks to support vector machines. Currently, estimation and in-sample inference of nonlinear forecasting models remain a potentially difficult task with much work remaining to be done.

Another issue with nonlinear models is that they fit the data better than the corresponding linear models, but they give less accurate forecasts. Reasons for this are discussed in the documentation (Tsay 2002 and Clements et al. 2004). One reason could be that nonlinear models might explain features in the data that do not occur very frequently. If these features are not present in the series during the period to be forecast, then there is no gain from using nonlinear models for generating the forecasts. This could be the case when the number of out-of-sample forecasts is relatively small. It appears that nonlinear models can generate accurate forecasts when the number of observations for specifying the model and estimating its parameters is large. In other words, short time series are not good candidates for nonlinear forecasting modeling.

The consensus about the contributions of nonlinear models in forecasting is that expectations should be reduced. Several factors contribute to this lack of progress. First, no commercially available statistical packages exist that can handle various nonlinear models. Second, there is no general agreement about how to judge the real contributions of nonlinear models over linear ones. This is particularly true in forecasting because evaluation of forecasting accuracy depends on forecasting origin, forecasting horizon, and how forecasts are used.

If you believe that the underlying phenomenon is nonlinear, it is worth considering a nonlinear model. However, expectations that these models will always do well and over-perform the linear models should be low. There are too many unknowns. And, the economic system is too complex to support the belief that generalizing a linear model in one (simple!) direction, such as adding another regime, will improve matters. It is a good idea to seek alternative nonlinear methods derived from different assumptions. Three examples—neural networks (derived from machine learning), support vector machines (derived from statistical learning theory), and genetic programming (derived from evolutionary computation)—are discussed in the next sections.

11.1.2 Forecasting Models Based on Neural Networks

An artificial neural network is a computer system originally modeled after the learning process of the human brain. (You can find detailed information in Simon Haykin's *Neural Networks: A Comprehensive Foundation*.) At the basis of neural networks is the artificial neuron or processing element, which emulates the axons and dendrites of its biological counterpart by connections and emulates the synapses by assigning a certain weight or strength to these connections. A processing element has many inputs and one output. Artificial neurons use an activation function to calculate their activation level as a function of total inputs. The main popular activation function options are based on a sigmoid function and radial basis function (RBF).

An artificial neural network consists of processing elements organized in a network structure. The output of one processing element can be the input of another. Numerous ways to connect the artificial neurons to different network architectures exist. The most popular and widely applied neural network structure is the multilayer perceptron. It consists of three types of layers (input, hidden, and output) and is shown in Figure 11.1.

Figure 11.1 Typical Three-Layer Structure of Multilayer Perceptron

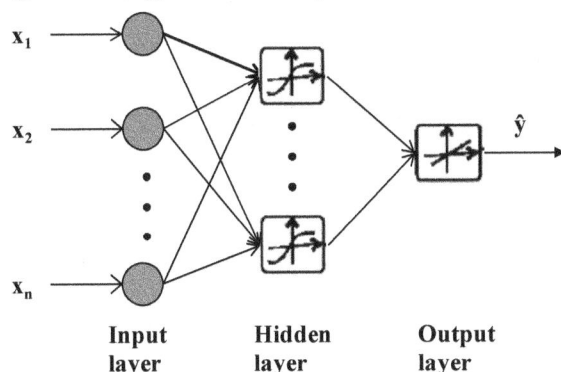

The input layer connects the incoming patterns and distributes these signals to the hidden layer. The neurons in the hidden layer are with the sigmoid activation function. The hidden layer is the key part of the neural network because its neurons capture the features hidden in the input patterns. The learned patterns are represented by the weights of the neurons. These weights are used by the output layer to calculate the prediction. Artificial neural networks with this architecture might have several hidden layers with different numbers of processing elements and outputs in the output layer. However, the majority of applied neural networks use only three layers, which are similar to those shown in Figure 11.1.

The functional relationship represented by a neural network is a special case of the generic nonlinear equation with the following form of the nonlinear function f:

$$f = \beta_0' \, \mathbf{z_t} + \sum_{j=1}^{q} \beta_j \, G(\gamma_j' \, \mathbf{z_t}) \qquad\qquad 2$$

In this equation, $\mathbf{z_t} = (1, \, y_{t-1}, \dots, y_{t-p} \, , x_{1t}, \dots \dots, x_{kt})'$ is the vector of inputs including the intercept and lagged values of the output. $\beta_0' \, \mathbf{z_t}$ is a linear unit. $\beta_j, j = 1, \dots, q$ are parameters called connection strengths or weights. $G(.)$ is the activation function. And, $\gamma_j, j = 1, \dots, q$ are parameter vectors.

The parameters of the neural network are obtained by a variety of learning algorithms, the most popular of which is back-propagation.[1] Unfortunately, the derived parameters are locally optimal because the algorithm can get stuck in local minima. The total available data is usually divided into a training set (in-sample data) and test set (out-of-sample data). The training set is used for estimating the parameters (weights), and the test set is used to estimate the generalization ability of the neural network model.

The key advantages of neural networks can be summarized as follows (Kordon 2009):

- Nonlinear representation. Neural networks deliver a variety of flexible nonlinear models between input and output variables.

- Adaptive learning. Neural networks are data-driven, self-adaptive models with very few a priori assumptions. They learn the underlying relationships from the training data without theoretical knowledge.

- Universal approximators. Neural networks can approximate any continuous function to any degree of accuracy.

- Fault tolerance. Because the information in the neural network is distributed in many process elements (neurons), the overall performance of the network does not degrade drastically when the information of a node is lost or connections are damaged. The neural network repairs itself by adjusting the connection weights based on the new data.

- Wide range of architectures. Neural networks are one of the few research approaches that offer high flexibility in the design of different architectures with its basic component, the neuron. In combination with the variety of learning algorithms such as back-propagation, competitive learning, Hebbian learning, and Hopfield learning, there are many potential solutions. The key architectures from an application point of view are multilayer perceptrons (delivering nonlinear functional relationships), self-organizing maps (generating new patterns), and recurrent networks (capturing system dynamics).

The key disadvantages of neural networks can be summarized as follows (Kordon 2009):

- The purely mathematical description of an even simple neural network is not easy to understand. A black box links the input parameters with the outputs, and it does not give sufficient insight on the nature of the relationships. As a result, black boxes are not well accepted by the majority of users.

- The excellent approximation capabilities of neural networks within the range of model development data are not valid when the model operates in unknown process conditions. Unfortunately, neural networks are very sensitive to unknown process changes and demonstrate poor generalization capabilities.

- Large amounts of training, validation, and test data are required for model development. It is necessary to collect data with the broadest possible ranges to overcome the poor generalization capability of neural networks.

- Results are sensitive to the choice of initial connection weights.

- There is no guarantee of optimal results because of the possibility of being stuck in local minima. As a result, the training data might bias the neural network toward some operating conditions and deteriorate the performance of others.

- Model development requires substantial intervention of the modeler and can be time-consuming. A specialized knowledge of neural networks is assumed for model development and support.

An interesting type of neural network structure applicable to forecasting is a recurrent neural network. A recurrent neural network is able to represent dynamic systems by learning time sequences. A good survey of neural networks in forecasting is in the paper of Zhang et al. entitled "Forecasting with artificial neural networks: The state of the art" (1998). One key requirement for capturing process dynamics is the availability of short-term and long-term memory. Long-term memory is captured in the neural network by the weights of the static neural network. Short-term memory is represented by using time lags in the input layer of the neural network.

A common feature in most recurrent neural networks is the context layer, which provides the recurrence. At current time t, the context units receive signals resulting from the state of the network at the previous time sample, $t-1$. The state of the neural network at a moment in time depends on previous states because of the ability of the context units to remember aspects of the past. As a result, the network can recognize time sequences and capture process dynamics. Examples of the most popular architectures of the recurrent neural network are the Hopfield, Elman, and Jordan networks (Haykin 1998).

11.1.3 Forecasting Models Based on Support Vector Machines

An implicit assumption in developing and using neural networks is the availability of sufficient data to satisfy the hunger of the back-propagation learning algorithm. Unfortunately, in many practical situations and especially in forecasting, you do not have sufficient data. The machine learning method, support vector machines (SVM), can be used to address this issue. It is based on the solid theoretical ground of the statistical learning theory, which can effectively handle statistical estimation with small samples with short time series.

The basis of SVM is the statistical learning theory, which focuses on defining and estimating the capacity of the machine to learn from finite data. Unlike most of the traditional learning machines that adopt the empirical risk minimization principle, an SVM implements the structural risk minimization principle, which seeks to minimize an upper bound of the generalization error rather than minimize the training error. This results in better generalization than conventional techniques.[2]

The key idea of the SVM algorithm is to map the input data x_i into a high-dimensional feature space through nonlinear mapping $\phi(x_i)$. You then perform a linear regression in the feature space. Usually, nonlinear mapping is performed by an a-priori-defined kernel function, $K(xi, xj)$. The value of the kernel is equal to the inner product of two vectors, Xi and Xj , in the feature space, $\phi(xi)$ and $\phi(xj)$. For example, $K(xi, xj)=\phi (xi)*\phi (xj)$. Common examples of appropriate kernels are the polynomial kernel, sigmoid kernel, radial basis function kernel, and the Gaussian kernel. The final nonlinear solution includes a linear combination of the kernels weighted by the support vectors. Mathematically, these optimal weights are associated with the nonzero Lagrange multipliers from the convex optimization. An interesting practical interpretation of SVM is that because the nonzero Lagrange multipliers are typically less than the entire data set, you do not need the entire data set. You just need the support vectors to define the nonlinear relationship. The typical SVM architecture (shown in Figure 11.2) can be represented like the architecture of the three-layer neural network. Instead of the sigmoid transfer function in the hidden layer, you have the kernel function. The output layer includes the linear sum of the kernels with the support vectors.[3]

Figure 11.2 Typical Structure of SVM

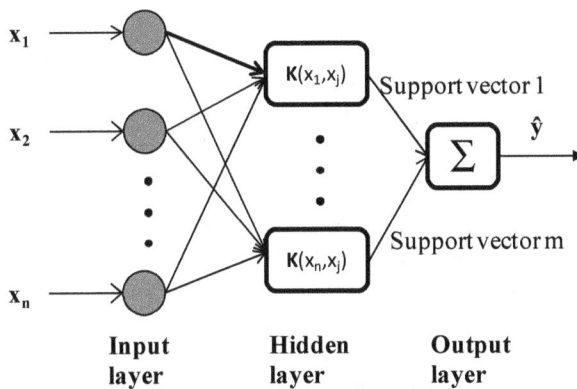

The key advantages of SVMs can be summarized as follows (Kordon 2009):

- Solid theory. One of the important benefits of SVMs is that the solid theoretical basis of the statistical learning theory is closer to the reality of practical applications.

- Explicit model complexity control. SVMs provide a method to control complexity independently of the dimensionality of the problem. In the same way, the modeler has full control over model complexity by selecting the relative number of support vectors. The best solution is based on the minimal number of support vectors.

- Maximization of generalization ability. In contrast to neural networks, where the weights are adjusted only by the empirical risk (in other words, by the capability to interpolate training data), SVMs keep the empirical risk fixed and maximize the generalization ability. As a result, the performance of the derived models involves the best possible generalization abilities. Derived models do not deteriorate dangerously under new operating conditions like the neural network models.

- No local minima. SVMs are derived from global optimization methods such as linear programming (LP) or quadratic programming (QP). They are not stuck in local minima such as the typical back-propagation-based neural networks.

- Repeatable results. SVM development is not based on random initialization as is the case with most neural network learning algorithms. This guarantees repeatable results.

The key disadvantages of SVMs can be summarized as follows (Kordon 2009):

- For a non-expert, interpreting SVMs is a bigger challenge than understanding neural networks. The main flaw is the black-box nature of the models. The better generalization capability of SVMs relative to neural networks cannot change this fact.

- There are very demanding requirements for practitioners who must understand the extremely mathematical and complex nature of the approach. Knowledge of both statistics and optimization is required.

- SVM model development is very sensitive to the selection of a good kernel function. Because the kernel is problem-dependent, the final user must have some knowledge of SVMs.

- The application record of SVMs is relatively short. The experience in large-scale industrial applications and in model support is very limited.

SVMs are still knocking at the industrial door. Most of the applications in time series forecasting are in an early research phase. (A good survey of the current state is in the paper by Sapankevych and Sankar [2009]). Some comparative studies demonstrated a superior performance of SVM relative to neural networks. Of all practical applications using SVM for time series forecasting, financial time series prediction and electrical load forecasting are the most studied.

The application of SVM to time series forecasting is relatively new. However, initial results from several research studies have demonstrated the potential of this approach in predicting time series data. Of special importance is the capability to build models based on short time series, which is one of the key issues of the other nonlinear methods. Further research is needed on the performance of SVM with different kernel functions and optimal tuning parameter selection.

11.1.4 Forecasting Models Based on Evolutionary Computation

Genetic programming (GP), developed by John Koza, is a subcategory of evolutionary computation in which solutions are represented as tree structures (Koza 1992). Internal nodes of solution trees represent appropriate operators, and leaf nodes represent input variables or constants. For time series forecasting applications, the operators are mathematical functions, and the inputs are lagged time series values and explanatory variables. An example of a solution tree for time series forecasting, based on one of the GP-generated nonlinear functions, is shown in Figure 11.3.

Figure 11.3 Tree Representation of Nonlinear Function

$$y_t = \log(y_{t-2}) + \sqrt{\frac{y_{t-1}}{y_{t-3}}}$$

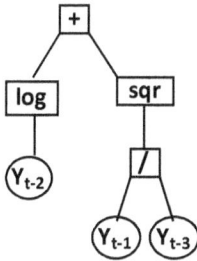

Candidate solution trees are randomly constructed to create the initial population. Each solution is ranked based on its prediction error in a set of training (in-sample) data. A new population of solutions is generated by selecting fitter solutions and applying a crossover and mutation operation. Crossover is performed by exchanging subtrees from two parent solutions. Mutation is performed by selecting a subtree of a single solution and replacing it with a randomly constructed subtree.

The task of time series prediction is accomplished by using previous values of a time series to predict future values. Therefore, the GP terminal set will consist of previous values of the time series up to a certain maximal lag limit. To measure the predictive performance of a solution, the following method is used. First, a vector of past time series values for the last (most recent) test case is applied to the solution (function), and the predicted future value, $y(t + 1)$, is calculated. This predicted value is appended to the past values to produce a new vector, which is applied to the solution to generate the next predicted future value, $y(t + 2)$. This process is continued until the appropriate number of future values to predict is generated. The mean of the squared errors between the actual time series values and the predicted values is calculated.

Usually, the average fitness of the population increases from generation to generation during the GP runs. Often, the final high-fitness models are inappropriate for practical applications because they are very complex, difficult to interpret, and have terrible out-of-sample performance. In practice, simplicity of the applied solution is as important as the accuracy of the in-sample model predictions. A special version of GP called Pareto-front GP offers a solution to this problem by using multi-objective optimization to direct the simulated evolution toward simple models with sufficient accuracy (Smits and Kotanchek 2004). An example of Pareto-front GP results is in Figure 11.4.

Figure 11.4 Pareto-Front GP Model Distribution

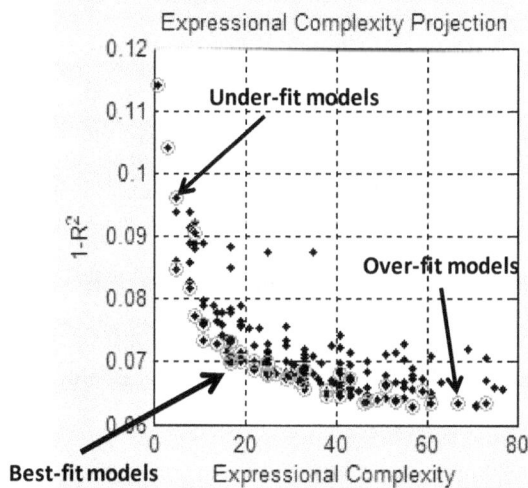

In Pareto-front GP, the simulated evolution is based on two criteria—prediction error (for example, based on 1-R^2) and complexity (for example, based on the number of nodes in the equations). In Figure 11.4, each point corresponds to a certain model with the X coordinate referring to model complexity and the Y coordinate referring to model error. The points marked with circles form the Pareto front of the set of models. Models at the Pareto front are non-dominated by other models in both criteria simultaneously. In other words, all models on the Pareto front are chosen as best members of a population. The Pareto front itself is divided into three areas. The first area contains the simple under-fit models that occupy the upper left of the Pareto front. The second area contains complex over-fit models and lies in the bottom right of the Pareto front. The third and most interesting area contains best-fit models and is around the tipping point of the Pareto front, where the biggest gain in model accuracy for the corresponding model complexity is. Model selection is reduced to a small number of candidate models from this area.

Another interesting approach for applying GP in time series forecasting is the Dynamic Forecasting Genetic Programming (DyFor GP). The key difference in this method relative to the standard GP technique is that it adds features that are customized for forecasting time series whose underlying data-generating processes are non-static. These time series are often used for real-world forecasting in which environmental conditions are constantly changing. The distinctive feature of DyFor GP is its adaptive data window adjustment. This feedback-driven window adjustment is designed to automatically home in on the currently active process in an environment where the generating process varies over time. (For more information, see Wagner et al. 2007.)

When applied to time series forecasting, GP has four advantages over classical methods. First, the functional form of the forecasting model does not need to be prescribed. It is automatically discovered. Second, the functional form is not restricted to the linear model. Third, it has the ability to produce several high-quality solutions. Fourth, it has a built-in nonlinear variable selection.

There are three key disadvantages of GP. First, it has very slow model generation relative to other nonlinear approaches. Second, the derived nonlinear models have no classical statistical interpretation. Third, the whole approach is new and lacks solid theoretical basis.

Numerous studies have applied GP to time series forecasting with favorable results. (See many references in the survey paper by Barbazon et al. 2008.) The main innovation of GP is that it automatically selects and self-adjusts its functional form and the time period on which its forecasts are based. GP has been used for the US demand of natural gas forecast, daily exchange rate forecast, real estate prices forecast of residential single family homes in southern California, oil revenue in Kuwait forecast, stock trading rules generation, and so on. From all presented nonlinear forecasting methods, GP is the least developed, with almost nonexistent theoretical basis for forecasting and few options for software. However, even in its early phase, it has big potential that could be explored in future applications.

11.2 More Modeling Alternatives

11.2.1 Multivariate Models

Chapters 7, 8, and 9 focus on building univariate forecasting models. A main assumption in this approach is that target or dependent variables can be modeled one at a time. For example, variation in red tennis shoe sales is modeled as a function of only deterministic red tennis shoe price, lagged past values, and so on. It is implicitly assumed that red tennis shoe sales are independent of the sales of other tennis shoe types.

If variation in a set of dependent variables is related, then these variables can be modeled as a system of equations. To understand what an interrelated system of equations might look like, consider a company selling peanut butter. The amount sold is a function of the price that the company charges. The company has an interesting pricing policy. If the monthly number of cases sold falls above or below certain thresholds, then the price is automatically adjusted by a preset increment.

In this case, there are really two dependent variables—the units sold or demand is a function of price, and price is a function of demand. Looking at a CCF plot of sales and price, you would expect to see significant positive and negative spikes. Lead *and* lag relationships exist between sales and price. Another name for this type of relationship is a feedback effect. Univariate models cannot accommodate them.

Common sense tells you that modeling either peanut butter price or demand in isolation will yield sub-optimal results. However, it is difficult to imagine how an adequate specification can be written to accommodate the relationships. The solution is to specify an equation for each related dependent variable and to estimate them simultaneously as a system of equations. Relationships between dependent (in this setting, endogenous) variables are specified in the system. This is shown in the simple example below.

Much has been written about multivariate or simultaneous equation modeling, and it has a wide range of applications, primarily in economics.[4] Further information about this approach is beyond the scope of this section.

Applications

Joe Hoops, lead forecasting analyst at HiTops shoes, wants to increase the accuracy of his shoe forecasts. He is considering a multivariate approach to accommodate possible sales interrelationships between brands of shoes. To begin, Joe builds on his previous univariate analyses of the two leading shoe brands at HiTops, red tennis shoes (RTS) and black tennis shoes (BTS). His analysis is shown below. (See Chapter 8 and the ARIMA procedure in SAS/ETS documentation for information about the syntax.)

Program 11.1

```
proc arima data=varmax_sim plots=forecasts(forecast);

identify var=RTS;
estimate q=1 p=(1 4) plot method=ml;
forecast lead=12 id=date printall;

identify var=BTS;
estimate q=1 p=(1 6) plot method=ml;
forecast lead=12 id=date printall;

run; quit;
```

The TIMESERIES procedure in SAS/ETS is an excellent tool for investigating the time series components of individual and multiple series.

Program 11.2

```
options helpbrowser=sas;
ods html;
ods graphics on;

proc timeseries data=varmax_sim plot=(series corr) crossplots=(series ccf) vectorplot=(series);
    id date interval=month;
    var RTS;
    crossvar BTS;
run;
```

The cross series plot seems to confirm Joe's suspicions about correlation between the two dependent variables, RTS and BTS.

Output 11.1

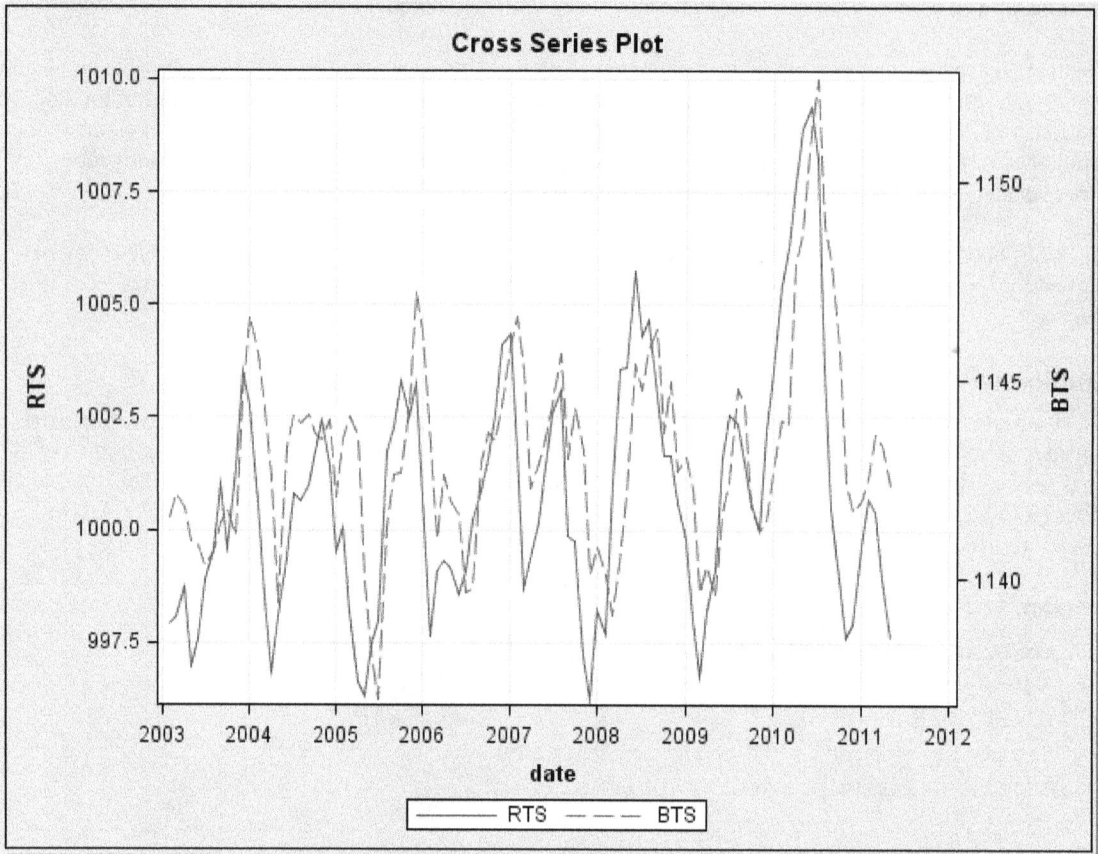

Investigating further, the CCF plot indicates that sales of BTS and RTS are significantly correlated. It shows that sales in BTS seem to both lead and lag sales in RTS. (See Chapter 9 for additional information about CCF plots.)

Output 11.2

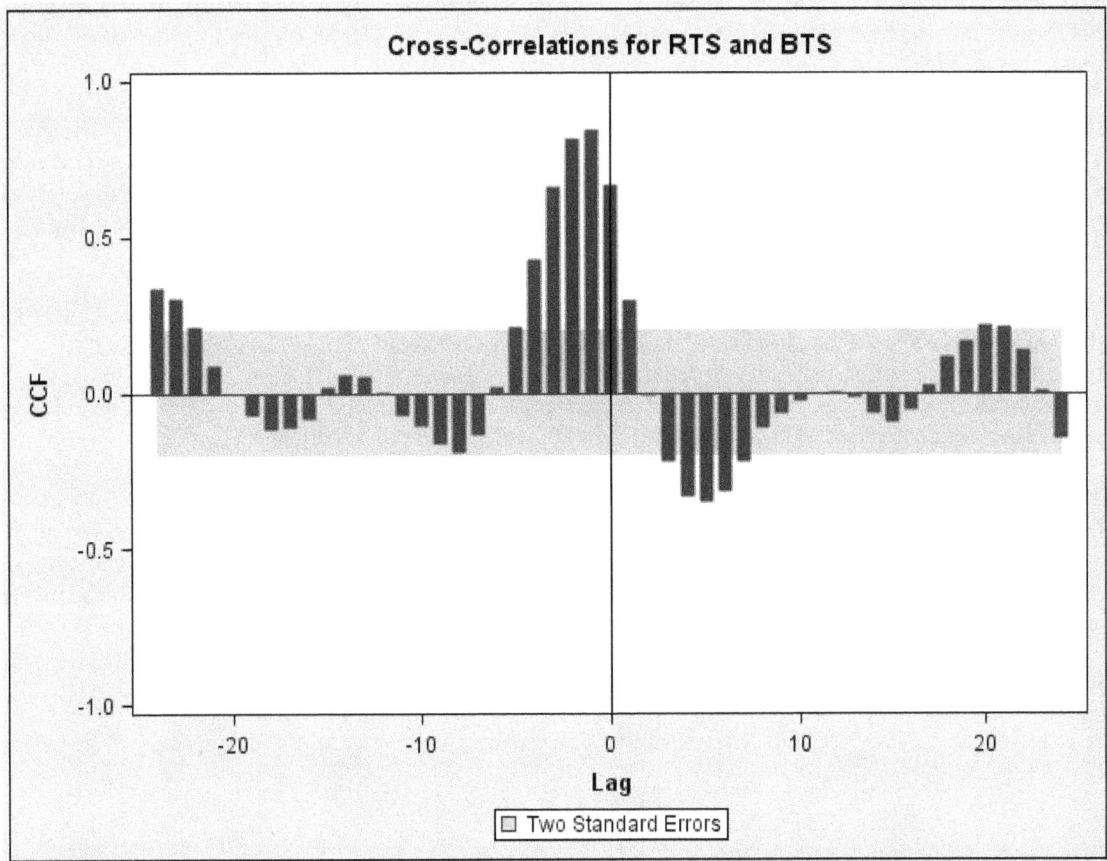

Cross-Correlations for RTS and BTS

Joe's univariate analysis indicates that the RTS and BTS variables are significantly autocorrelated. Furthermore, the CCF analysis shows that the dependent variables are correlated with each other and might have a dynamic (lead, lag) relationship. (The RTS and BTS series have not been prewhitened. For more information, see Chapter 9.)

To adequately accommodate the indicated relationships, Joe needs a more general modeling framework. Using SAS/ETS software, a straightforward way to model a system of equations is by using PROC VARMAX.[5]

Joe writes the following syntax to accommodate the relationships.

Program 11.3

```
proc varmax data=varmax_sim;
    id date interval=month;
    model RTS BTS / p=2 noint lagmax=3;
    output out=for lead=5;
run;
```

This syntax specifies a vector autoregression with lag length of 2 or VAR 2 (Hamilton 1994.) Joe has effectively specified a two-equation system that looks like the following:

$$RTS_t = \delta_1 RTS_{t-1} + \delta_2 RTS_{t-2} + \varphi_1 BTS_{t-1} + \varphi_2 BTS_{t-2} + \varepsilon_{1t} \qquad 3$$

$$BTS_t = \chi_t BTS_{t-1} + \chi_2 BTS_{t-2} + \gamma_1 RTS_{t-1} + \gamma_2 RTS_{t-2} + \varepsilon_{2t} \qquad 4$$

Notice the interrelationships between the sales of the two variables in the equations. In equation 3, RTS is a function of autoregressive terms and lagged values of BTS. Equation 3 captures the leading effect of variation in BTS on RTS. Equation 4 captures the leading effect of variation in RTS on BTS and the autoregressive component in BTS.

How did Joe come up with a VAR lag length of 2 when the univariate analysis contained significant lag lengths as long as 6? This is actually Joe's final model. He began with a lag length of 6 and reduced it until the majority of the insignificant terms were omitted. The VARMAX procedure in SAS/ETS contains several diagnostics that are analogous to the diagnostics for univariate modeling (white noise tests, AIC, and so on). The modeling approach of identify, estimate, and forecast is still intact in this more general framework.

The VAR approach is one of several approaches that can be used to model a system of equations. One of its advantages is flexibility in handling lead and lag relationships between endogenous variables and the autoregressive structure of individual dependent variables. However, the VAR approach is not very parsimonious. You might see an insignificant parameter estimate in the table below. If a system includes more than a few dependent variables or longer significant lag lengths, a lot of history is needed.

The VARMAX procedure has generalized the traditional VAR framework, and it can help augment parsimony. If a variable is hypothesized to only lead, it can be specified as an exogenous-only variable in the system. For more information, see PROC VARMAX in the SAS/ETS documentation. The OUTPUT statement in the syntax generates the data set FOR, which contains the five-step lead forecasts for each dependent variable generated by the system. The output shows the results of the estimation.

Output 11.3

Model Parameter Estimates						
Equation	Parameter	Estimate	Standard Error	t Value	Pr > |t|	Variable
RTS	AR1_1_1	1.58080	0.10076	15.69	0.0001	RTS(t-1)
	AR1_1_2	-0.53975	0.12151	-4.44	0.0001	BTS(t-1)
	AR2_1_1	-0.42948	0.13492	-3.18	0.0020	RTS(t-2)
	AR2_1_2	0.40726	0.09020	4.52	0.0001	BTS(t-2)
BTS	AR1_2_1	0.72293	0.08990	8.04	0.0001	RTS(t-1)
	AR1_2_2	0.30861	0.10840	2.85	0.0054	BTS(t-1)
	AR2_2_1	-0.14371	0.12037	-1.19	0.2355	RTS(t-2)
	AR2_2_2	0.18430	0.08047	2.29	0.0242	BTS(t-2)

Joe's intuition told him that this new approach should result in better forecasts for the RTS and BTS series because it captures systematic variation in the series that the univariate approach could not accommodate. However, the multivariate approach is computationally expensive and relatively manual. This is one reason why the automated forecasting systems that run in Joe's production shop are exclusively univariate. The multivariate forecasting approach is usually only feasible for selected high-value series.

Joe wondered whether this approach was necessary. What evidence did he have that these variables were correlated in how the diagnostics indicated? The first indication is that lagged values of BTS are significant in the RTS equation. There seems to be a significant lead relationship between BTS and RTS. The same can be seen for lagged values of RTS in the BTS equation. Because these relationships are captured by the system, their effects are embedded in the lead forecasts for each variable.[6]

Looking at the output further reveals the estimated variance-covariance matrix of the system. The diagonal elements are the estimated variances of the target variables. The off-diagonal elements represent their estimated covariance. The estimated covariance is not close to zero. This is an indication that the independence assumption is violated, and that the univariate approach is inadequate for modeling the RTS and BTS variables.

Output 11.4

Covariances of Innovations		
Variable	RTS	BTS
RTS	1.80752	0.53925
BTS	0.53925	1.43867

11.2.2 Unobserved Component Models (UCM)

Unobserved components models (UCMs) are univariate or one-series-at-a-time forecasting models. However, they share some characteristics with the multivariate approach outlined above. Recall equations 1 and 2 above. Each dependent variable has its own equation, and each equation has its own source of error or error term. The UCM approach is univariate, but it decomposes any series into its constituent components. Each component has its own equation and error term. For example, if a series contains trend, seasonality, and irregular (ARMA) variation, a UCM representation of the series could consist of a three-equation system—a trend equation, a seasonal equation, and an irregular equation.

A primary distinction between UCMs and other univariate models is multiple sources of error. Another difference is that the component relationships to the dependent variable or correlation structure of the model can change over time in UCMs. No time subscripts exist on any parameter in the ARIMAX specifications shown in this book. This implies that the correlation structure is fixed or not a function of time. In UCMs, the parameters regulating the correlation between the dependent variable and components can evolve or be a function of time.

These two differences or generalizations from the ARIMAX framework provide UCMs with greater flexibility in modeling phenomena that include trend slopes that steepen or dampen over time, seasonality with periods of relatively greater or diminished variance, and so on. This is the main advantage of UCMs. They provide the analyst with more visibility into and relatively greater information about features of interest in the series. Also, because the component equations are additive, the analyst could obtain a forecast for the seasonal effect for example, and then add up forecasts from other components to obtain an overall or total forecast for the series.

This additional flexibility adds computational costs. Some automated forecasting systems, notably SAS Forecast Server, create and fit UCM specifications based on series components. However, many analysts prefer the speed and robustness of simpler univariate models when fitting the majority of their series in a large-scale forecasting scenario. Like multivariate models, UCMs are primarily useful for investigating phenomena in selected high-value series.

A simple example of how a UCM works follows. If you want to learn more about this approach to modeling, start with PROC UCM documentation for SAS/ETS. It is written by Dr. Rajesh Selukar, the lead developer of PROC UCM, and it provides an excellent introduction to the topic. A more comprehensive and technical reference is found in Harvey (1989). This section concludes with a slightly more realistic example of building a forecast model using the UCM approach.

Consider the simple random walk model:

$$y_t = y_{t-1} + \varepsilon_t \qquad\qquad 5$$

The error term in this equation is assumed to have the usual properties—normality and independence with mean 0 and constant variance.

If a plot of the dependent series looks like the following, can the random walk model be used to accommodate the variation in it?

Output 11.5

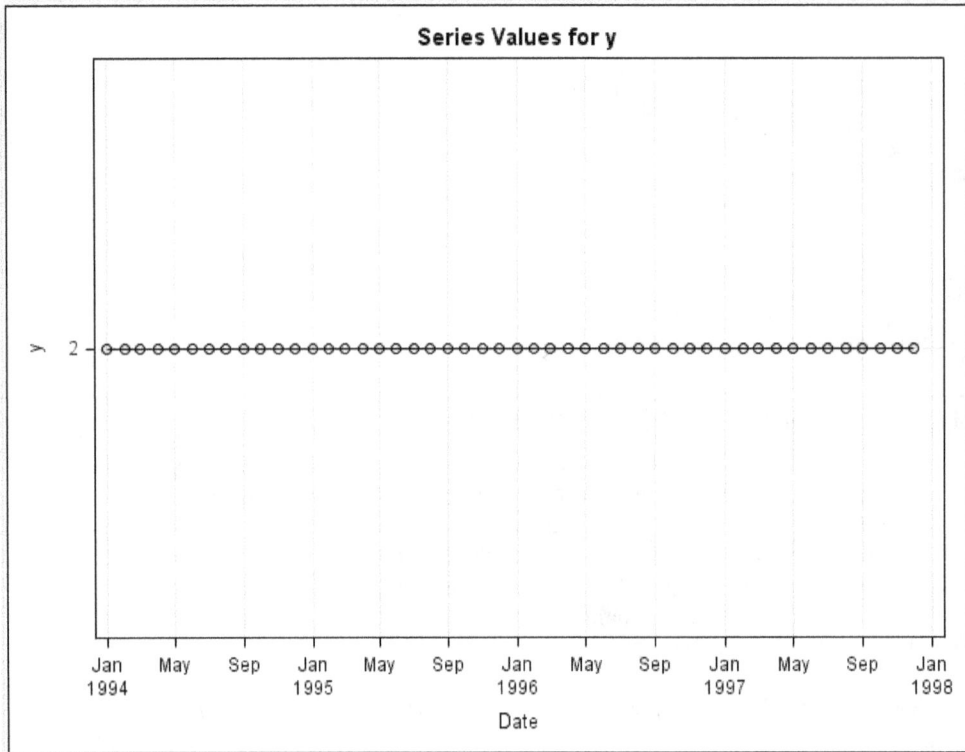

Yes, an excellent fit to the series can be obtained using the random walk model by setting the variance of the error term to zero in this equation (assuming an initial value for y=2).

Consider the series z. Again, variation in the series z could be accommodated using the random walk model by setting the variance to an appropriate nonzero value.

Output 11.6

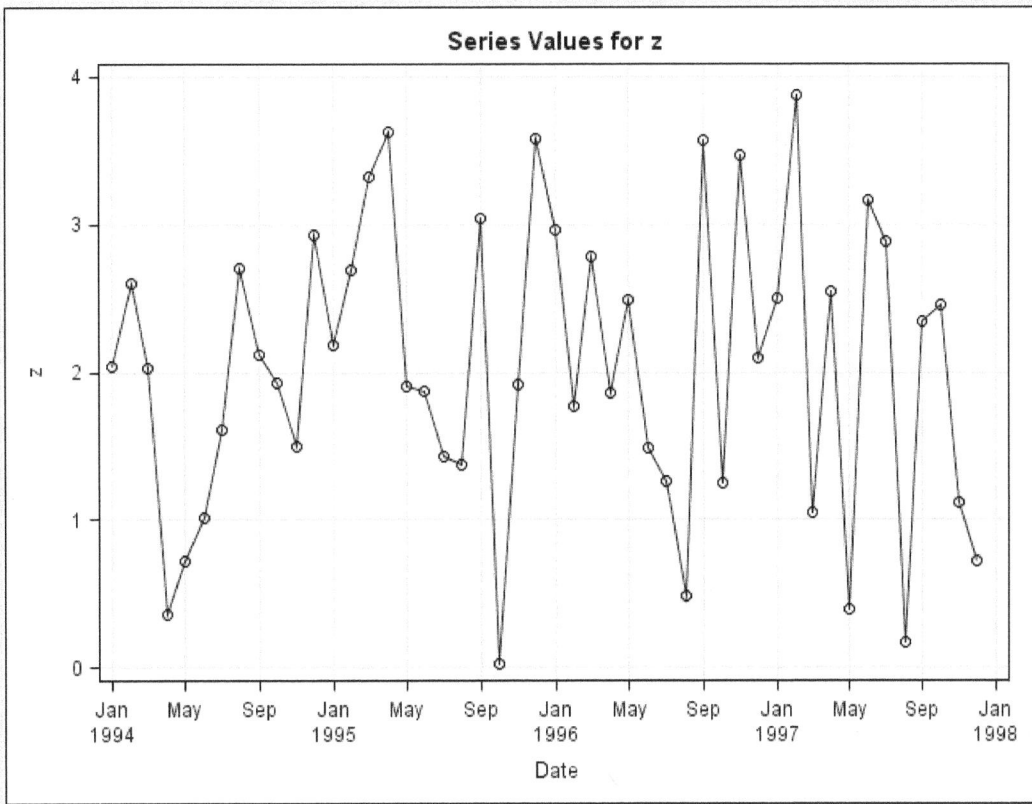

Series Values for z

This simple example outlines how the LEVEL component of a UCM is obtained. The variance of an equation that represents the component is estimated in a way that minimizes the difference between the series and the fitted values. In practice, the fitting is actually done in an updating or filtering process that begins at the first observation. It updates the best fit variance estimate as the history is iterated. The variance of equations representing seasonality, slope, and so on, are fit in a similar way.

To illustrate a more realistic example of the UCM approach, start with the monthly interval, jail intake data introduced in a previous chapter. A plot of the data indicates that the series contains seasonal and trend variation. Irregular variation is also possible. You can find descriptions of the various statements and options in the PROC UCM documentation for SAS/ETS.

Output 11.7

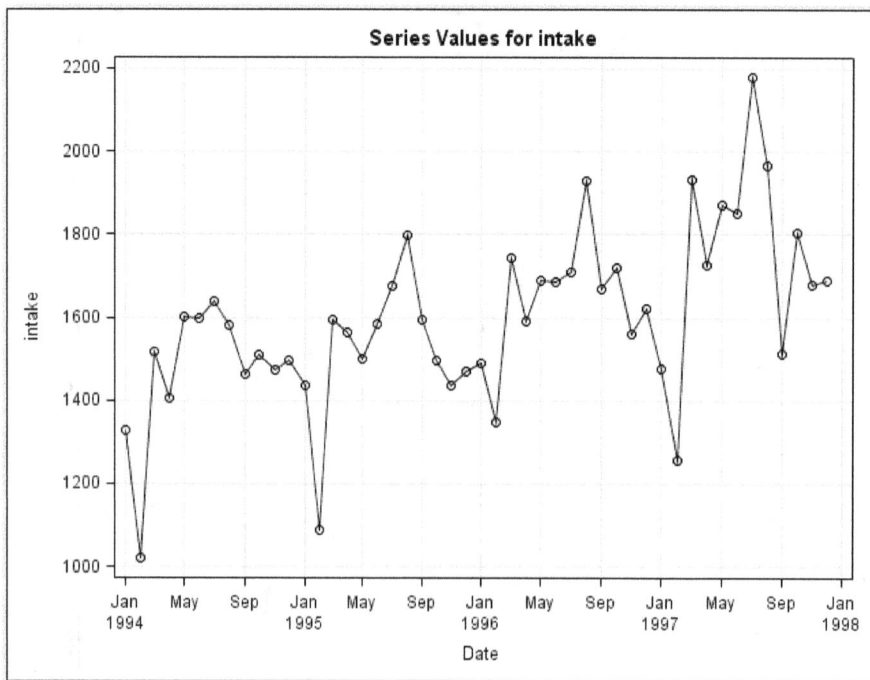

To focus ideas, UCM modeling will begin by accommodating the trend component seen in the plot of the data as follows:

Program 11.4

```
ods graphics on;

proc ucm data=hfmonth plots=all;
    id date interval=month;
    model intake;
    level;
    slope;
    estimate;
    forecast lead=12 print=decomp;
run;
```

The initial model contains two components. The LEVEL component is described above. The SLOPE component combines with the LEVEL component to model the trend of the variation in the intake series.

The table below provides estimates of the error variance in each component equation. The LEVEL variance is significantly different from zero. This indicates that the LEVEL component is stochastic. The estimated variance of the SLOPE component is very close to zero. This does not indicate that the series trend has a zero slope. It indicates that this effect is essentially regular or deterministic.

Output 11.8

Final Estimates of the Free Parameters					
Component	Parameter	Estimate	Approx Std Error	t Value	Approx Pr > |t|
Level	Error Variance	49745	10372.6	4.80	<.0001
Slope	Error Variance	0.00105	0.56004	0.00	0.9985

This information is used to refine the model (set the slope variance equal to zero and specify the NOEST option). A component to accommodate the seasonality in the series is added.

Program 11.5

```
ods graphics on;

proc ucm data=hfmonth plots=all;
    id date interval=month;
    model intake;
    level;
    slope variance=0 noest;
    season length=12 type=trig print=smooth;
    estimate;
    forecast lead=12 print=decomp;
run;
```

The variance estimates indicate that the SEASON component is stochastic.[7] The LEVEL component seems to have become fairly deterministic in the presence of the SEASON component.

Output 11.9

Final Estimates of the Free Parameters					
Component	Parameter	Estimate	Approx Std Error	t Value	Approx Pr > |t|
Level	Error Variance	61.77700	87.38634	0.71	0.4796
Season	Error Variance	183.03269	45.62425	4.01	<.0001

The significance analysis table gives information about the importance of the components in the model. All three components seem to be prominent in the data and contribute significantly to the fit of the model.

Output 11.10

Significance Analysis of Components (Based on the Final State)			
Component	DF	Chi-Square	Pr > ChiSq
Level	1	6829.08	<.0001
Slope	1	33.13	<.0001
Season	11	126.98	<.0001

A term to accommodate possible irregular variation is added to the UCM specification.

Program 11.6

```
ods graphics on;

proc ucm data=hfmonth plots=all;
    id date interval=month;
    model intake;
    level;
    slope variance=0 noest;
    season length=12 type=trig print=smooth;
    irregular;
    estimate;
    forecast lead=12 print=decomp;
run;
```

The variance estimates indicate that the IRREGULAR component is stochastic. However, the Significance Analysis of Components table indicates that there is not a prominent IRREGULAR component in the intake data (in the presence of the other components in the model).

Output 11.11

Final Estimates of the Free Parameters					
Component	Parameter	Estimate	Approx Std Error	t Value	Approx Pr > |t|
Irregular	Error Variance	7302.54626	1745.6	4.18	<.0001
Level	Error Variance	0.00000367	0.03281	0.00	0.9999
Season	Error Variance	4.007016E-8	0.0052338	0.00	1.0000

Output 11.12

Significance Analysis of Components (Based on the Final State)			
Component	DF	Chi-Square	Pr > ChiSq
Irregular	1	0.25	0.6179
Level	1	5136.20	<.0001
Slope	1	75.07	<.0001
Season	11	159.84	<.0001

The IRREGULAR component is dropped and the results are generated using the previous model. The following output shows the in-sample and lead forecasts for each component in the model and the combined intake forecast.

The output shows the smoothed evolution of the LEVEL component and its lead forecast.[8] It uses all data.

Output 11.13

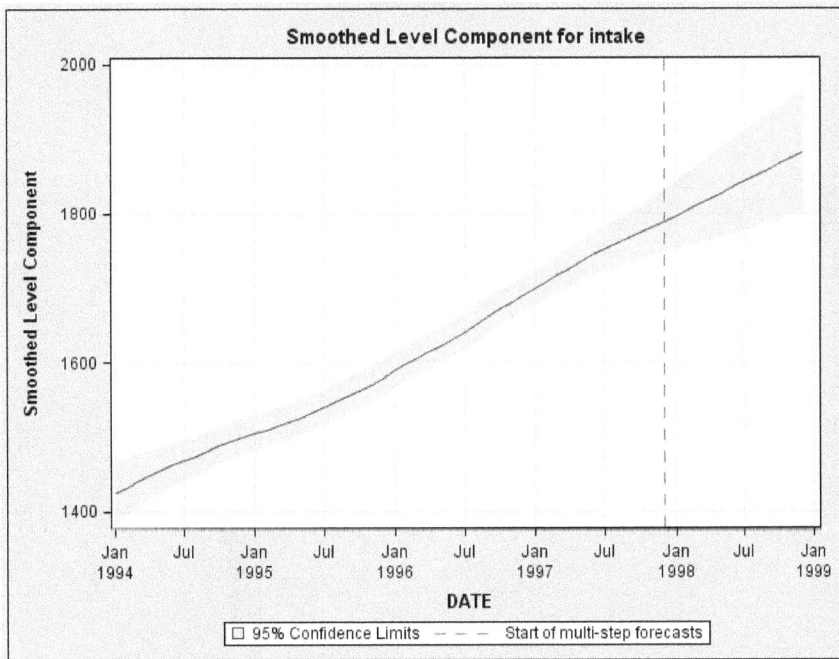

The output shows the deterministic slope.

Output 11.14

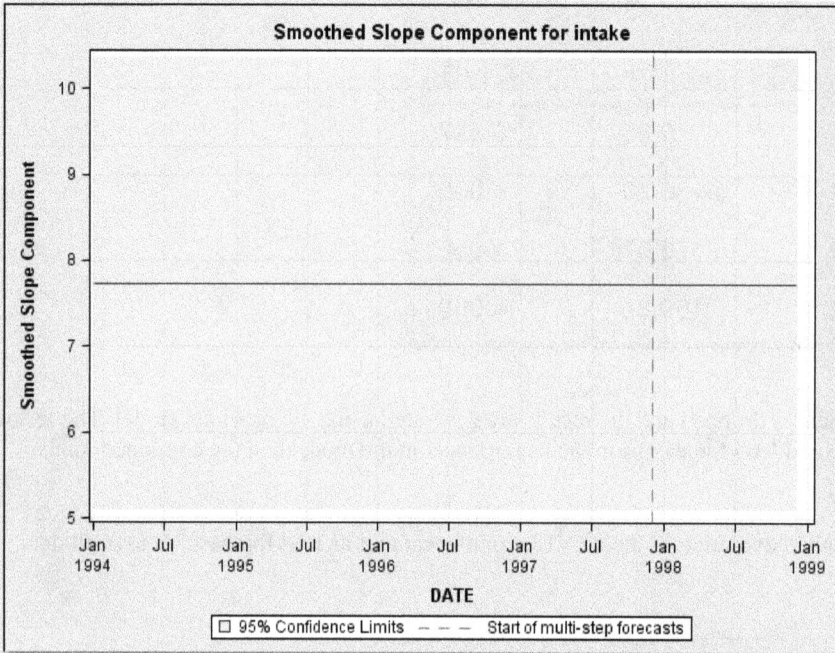

The output shows the trend forecast. This forecast combines the LEVEL and SLOPE components.

Output 11.15

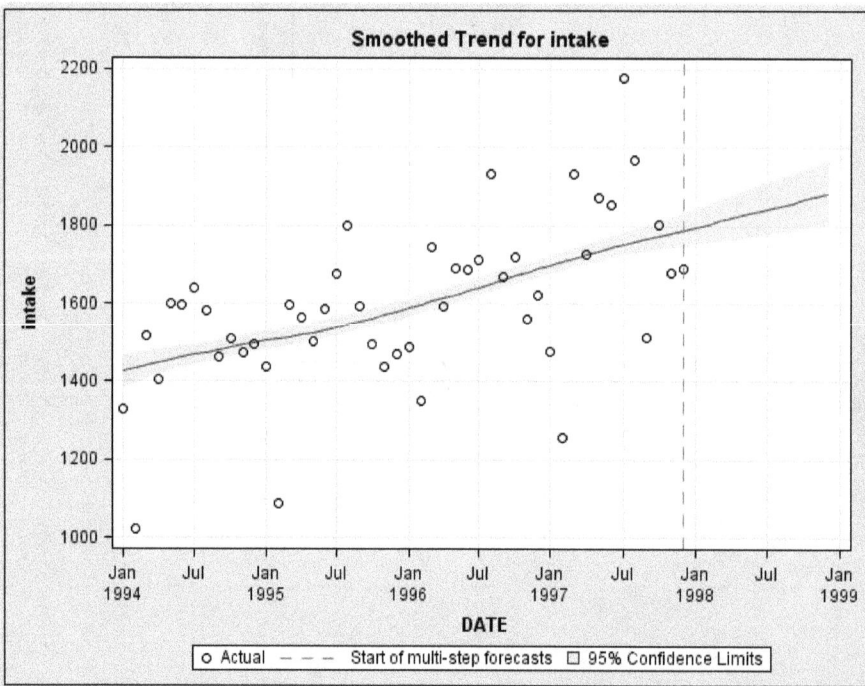

The output shows the SEASON component. As the stochastic SEASON component evolves, the variance seems to increase.

Output 11.16

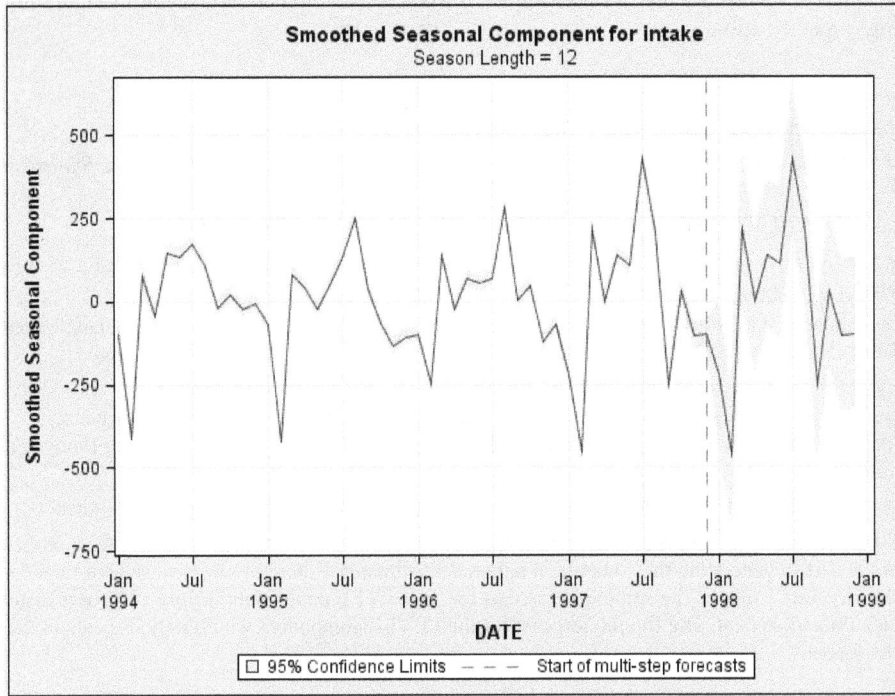

Smoothed Seasonal Component for intake
Season Length = 12

The final UCM forecast for intake is the sum of the component forecasts (trend and season) shown above.

Output 11.17

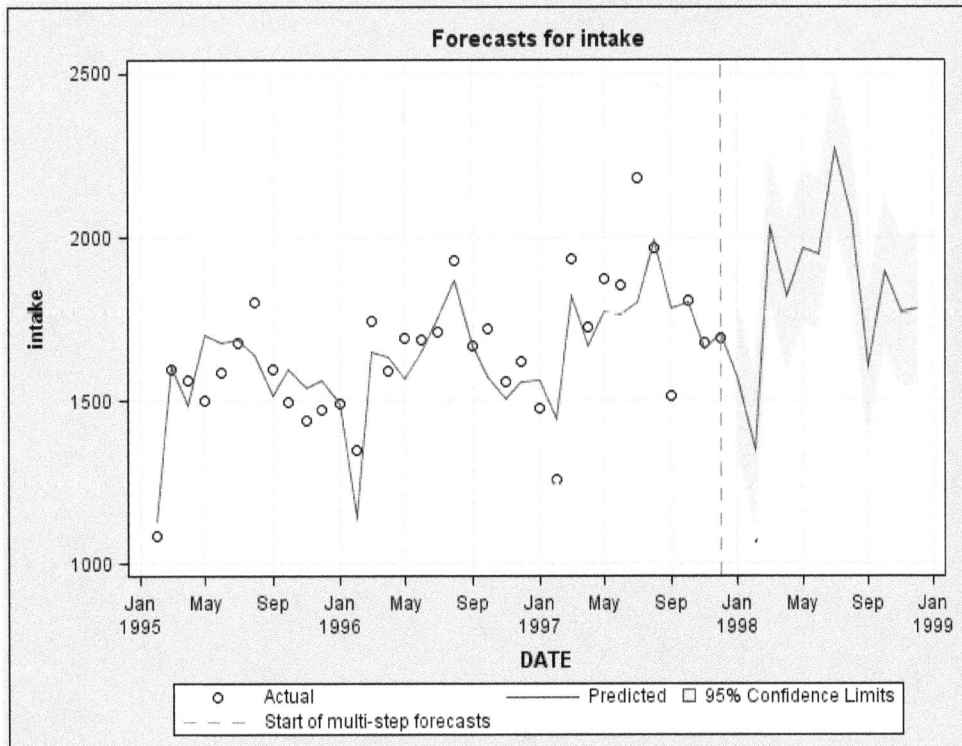

Forecasts for intake

This simple example focuses on the greater visibility into data components that UCMs provide. However, much of the functionality was abstracted from PROC UCM in SAS/ETS. For example, PROC UCM features outlier detection functionality, and it is one of the few forecasting tools that can simultaneously accommodate multiple seasonal cycles in a single specification.

---

[1] Information about back-propagation and other learning algorithms is in Haykin (1998).

[2] For complete descriptions of SVM and the statistical learning theory, see the essential book by Vladimir Vapnik, *Statistical Learning Theory* (1998).

[3] For a more detailed introduction to SVM, see the book by Cherkassky and Mulier (2007).

[4] If you want to augment your forecasting expertise in this area, start with Pindyck and Rubinfeld (1991). To look for additional, more technical details, see Hamilton (1994).

[5] Other SAS/ETS procedures accommodate a multivariate analysis. The MODEL procedure is a tool designed to handle nonlinear systems of equations. The SYSLIN and SYMLIN procedures are designed to handle linear systems of equations.

[6] You might be wondering why Joe did not set up the specifications in equations 3 and 4 in a univariate modeling tool and estimate them independently. This would have avoided problems with manual processing and computational expenses. However, severe statistical problems can arise when correlated dependent variables are handled in this way. See Pindyck and Rubinfeld (1991) for more information.

[7] The SLOPE component has been set to deterministic. It is present in the model, but its variance is no longer estimated.

[8] The UCM procedure in SAS/ETS produces two types of forecasts for any component specified in the model. The smoothed forecasts use all of the historical data in generating the component parameter estimates. Filtered forecasts use historical data up to a given time point in the history. For example, the smoothed forecast for the LEVEL component in July 1996 uses historical data to the left and right of this time increment. The filtered forecast for the LEVEL component would only use data to the left of July 1996 in generating the forecast.

Chapter 12: An Example of Data Mining for Forecasting

12.1 The Business Problem

For many companies, in order to manage resources ranging from people to brick-and-mortar assets, it is important to have a sales or demand forecast. Many large corporations use a weekly, monthly, or quarterly process called Executive Sales and Operations Planning (ES&OP) (Wallace 2004). A key component to this process is a good forecast. Even though a corporation can buy marketing research reports that indicate where a market is going, it is often useful to be able to build their own holistic forecast. The following example demonstrates the vast majority of the steps involved in the data mining for forecasting problem. The emphasis here is on showing the methods approach more than the particulars of the business problem.

This problem focuses on the North American paint and coatings market. All of the Y or dependent variable data is publicly available though somewhat derived for purposes of the example. All of the X data in this problem was graciously donated by IHS, Global Insights, Inc. The business problem is to develop a demand forecast (millions of gallons) for the US paint market. The hope is that the model has no more than 10% error measured by SMAPE on the holdout set and outperforms a baseline model (Gilliland 2010). The emphasis is more on prediction and forecast accuracy rather than on structure, although an understanding of the structure would be beneficial. The markets included in the study are all those that would contribute to the use of paint in architectural facilities. The value proposition is to ensure, through the ES&OP process, that resources are put into place to provide the right product for the right customer at the right time in the right quantity. This is not to say that this model is developed for detailed supply chain operations.

12.2 The Charter

As mentioned in the business problem statement, the scope is the United States architectural coatings market. This problem is restricted to the paint market. The business team is hoping to have a model for use in ES&OP that has less than 10% error and out performs a baseline model. The model will be used on a quarterly basis to plan for a 12-quarter time horizon. The value to the business is to ensure that resources are in place to eventually do more detailed supply chain planning. The team involved in defining the potential drivers is the marketing manager, sales manager, supply chain manager, financial analyst, and the TS&D representative.

Concerning data and modeling, the business data subject matter expert along with the data and modeling expert round out the team. The modeling expert is the resource that drives the project once the charter is finished.

12.3 The Mind Map

A mind-mapping session was conducted to hypothesize drivers for this market. Focus was on drivers for coatings. Applications are for both an industrial as well as a consumer setting. An example, though not complete, is given below.

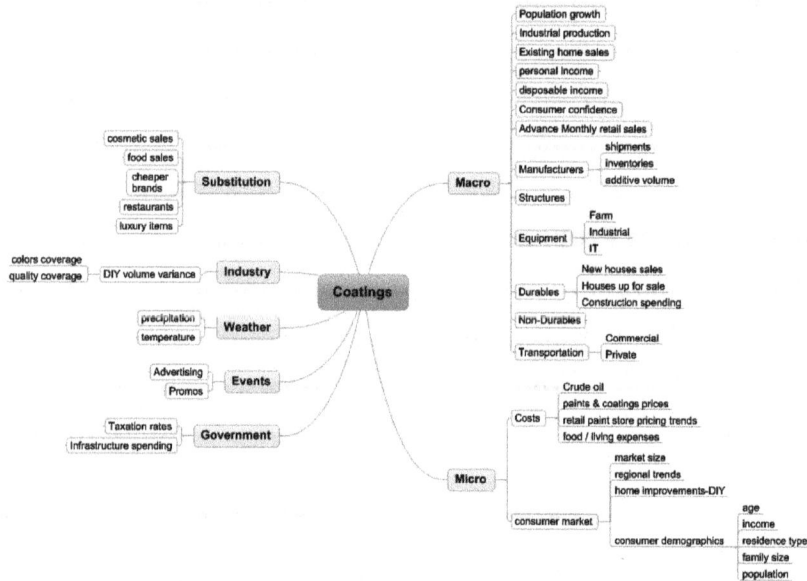

12.4 Data Sources

In this particular problem all of the data was readily available in IHS, Global Insights, Inc. More attention could have been paid to the reduction of the X variables a priori, but, for the purposes of this example, more variables were included rather than less. The initial data set had 1433 X variables in it. This data was delivered by IHS, Global Insight, Inc. in Excel format (complete with GI's short variable descriptions), which was then converted to a raw data file and a separate metadata file. All data was delivered in a quarterly format, given in Figure 12.1, and was available through 4th quarter 2010 with forecasts out to 2020. But the charter stated a need for only 12 quarters or 3 years of forecasts.

Figure 12.1: Original Raw Data from Global Insights for Xs

Formula bar: Full-employment federal NIPA budget surplus, Source: IHS Global Insight, Units: billions of dollars- annual rate, Last updated: 12/30/10 - 10:32

GLOBAL INSIGHT

Created on Tue 1 Feb 2011, 7:50 AM EST (13:50 GMT)

Short Label	1981 Q1	1981 Q2	1981 Q3	1981 Q4	1982 Q1	1982 Q2	1982 Q3	1982 Q4	1983 Q1	1983 Q2	1983 Q3	1983 Q4	1984 Q1	1984 Q2	1984 Q3
Full-employment federal NI	#N/A	#N/A	#N/A	#N/A	#N/A	#N/A	#N/A	#N/A	#N/A	#N/A	#N/A	#N/A	#N/A	#N/A	#N/A
proportion of depreciation	0.17	0.17	0.17	0.17	0.17	0.17	0.17	0.17	0.17	0.17	0.17	0.17	0.17	0.17	0.17
Average tax lifetime of oth	8.48	8.48	4.48	4.48	4.48	4.48	4.48	4.48	4.48	4.48	4.48	4.48	4.48	4.48	4.48
Averate tax lifetime of tele	14.50	14.50	14.50	14.50	14.50	14.50	14.50	14.50	14.50	14.50	14.50	14.50	14.50	14.50	14.50
proportion of depreciation	0.36	0.36	0.43	0.43	0.43	0.43	0.43	0.43	0.43	0.43	0.43	0.43	0.43	0.43	0.48
Long-term government bon	#N/A	11.96	12.74	12.35	11.98	11.68	11.44	10.56	10.21	10.05	10.32	10.12	10.14	10.37	10.24
Wage accruals less disburs	0.00	0.00	0.00	0.00	0.00	0.00	0.00	0.00	0.00	0.00	0.00	0.00	0.00	0.00	0.00
Industrial production--Text	84.15	85.45	82.41	78.68	79.18	79.49	81.78	83.05	85.70	92.01	98.19	95.87	101.38	104.52	102.08
Wage accruals less disburs	0.00	0.00	0.00	0.00	0.00	0.00	0.00	0.00	0.00	0.00	0.00	0.00	0.00	0.00	0.00
proportion of depreciation	0.17	0.17	0.17	0.17	0.17	0.17	0.17	0.17	0.17	0.17	0.17	0.17	0.17	0.17	0.17
Rate on 3-month negotiabl	15.95	16.53	17.52	13.48	14.20	14.25	11.67	9.10	8.51	8.78	9.58	9.39	9.66	10.93	11.46
Diffrence between income	58.40	35.90	26.00	26.00	-3.00	-0.60	2.10	20.60	23.20	44.10	65.10	66.40	41.20	36.30	30.30
proportion of depreciation	0.54	0.54	0.17	0.17	0.17	0.17	0.17	0.17	0.17	0.17	0.17	0.17	0.17	0.17	0.17
Demand for all fuels, Sourc	74.36	73.82	73.22	72.56	71.24	70.69	70.31	70.11	69.94	70.14	70.57	71.23	73.28	73.93	74.35
Average tax lifetime of com	4.50	4.50	4.50	4.50	4.50	4.50	4.50	4.50	4.50	4.50	4.50	4.50	4.50	4.50	4.50
Average tax lifetime of ligh	2.50	2.50	2.50	2.50	2.50	2.50	2.50	2.50	2.50	2.50	2.50	2.50	2.50	2.50	2.50
Ratio of state & local health	#N/A	#N/A	#N/A	#N/A	#N/A	#N/A	#N/A	#N/A	#N/A	#N/A	#N/A	#N/A	#N/A	#N/A	#N/A
Real Consumer Spending -S	#N/A	#N/A	#N/A	#N/A	#N/A	#N/A	#N/A	#N/A	#N/A	#N/A	#N/A	#N/A	#N/A	#N/A	#N/A
Consumer Spending -Thera	5.72	5.59	5.61	5.73	5.93	6.05	6.40	6.59	6.76	6.76	7.47	7.56	7.75	8.29	8.31
Industrial production--Carp	72.28	72.50	66.36	61.52	65.77	63.61	65.45	67.44	72.07	78.11	85.61	81.00	87.65	91.89	88.36
Chained Price Index--Compu	26,385.49	26,182.44	25,515.84	24,833.87	23,755.43	21,981.23	20,293.94	18,744.87	16,692.00	15,210.68	14,075.66	13,410.85	13,056.45	12,880.79	12,689.43
Chained Price Index--Therap	48.77	49.81	50.54	51.45	52.29	53.24	53.97	54.58	55.21	55.95	56.45	56.95	57.41	58.07	58.54
Consumer Spending -Pharm	20.38	20.98	21.64	22.14	22.56	23.27	23.76	24.39	25.52	26.41	27.52	28.04	29.07	30.04	30.58
Consumer Spending -Food	131.64	133.86	134.59	135.57	137.13	141.02	144.57	147.32	150.10	151.97	155.08	157.05	161.62	162.69	167.01
Real Consumer Spending -H	759.74	770.78	772.38	766.06	761.88	766.58	770.54	778.27	785.33	793.59	798.97	801.84	807.88	810.08	813.14
Real Consumer Spending -I	#N/A	#N/A	#N/A	#N/A	#N/A	#N/A	#N/A	#N/A	#N/A	#N/A	#N/A	#N/A	#N/A	#N/A	#N/A
Real Consumer Spending -T	36.59	37.25	36.83	37.16	37.18	37.87	38.43	39.13	39.11	39.76	39.30	39.73	40.11	40.04	39.24
Consumer Spending -Consu	1,828.72	1,867.69	1,899.70	1,925.32	1,953.65	1,980.64	2,028.47	2,082.10	2,123.32	2,173.46	2,240.65	2,285.41	2,327.23	2,384.20	2,426.41
Consumer Spending -Water	9.95	10.26	10.71	10.99	11.18	11.51	11.80	12.14	12.58	12.86	13.24	13.75	14.26	14.69	15.19
Real Consumer Spending -P	#N/A	#N/A	#N/A	#N/A	#N/A	#N/A	#N/A	#N/A	#N/A	#N/A	#N/A	#N/A	#N/A	#N/A	#N/A
Consumer Spending -Other	37.01	36.43	38.23	38.13	37.58	38.13	39.09	40.26	41.49	42.69	44.44	46.52	48.43	49.85	51.90
Consumer Spending -Teleco	29.40	30.25	31.19	32.71	33.23	34.68	35.74	36.91	37.96	38.82	38.64	39.14	41.60	41.84	41.37

Note that the data is in a horizontal format rather than vertical and that there are missing data designated by "#N/A". Also note that the headers for date are not necessarily in a proper SAS DATETIME format.

12.5 Data Prep

This data has to be converted to a proper time series data set format. The first thing to do is to get rid of the "#N/A" since SAS interprets this as a character variable if we try to simply read this data directly. At this point it is personal preference regarding how to do some of the following basic data cleaning manipulations. These data sets are not that large, so a simple Find and Replace in Excel can easily convert these items to a ".", which SAS does recognize as numeric. Also, adding a column with short variable names (in this case X1-X1433) for purposes of the example was done as well. Then, the short descriptors from GI and the new variable names are copied to another work sheet and serve as the metadata for later use, as shown in Figure 12.2 The new vertical format, ready to be transposed, is shown in Figure 12.3.

Figure 12.2: Metadata

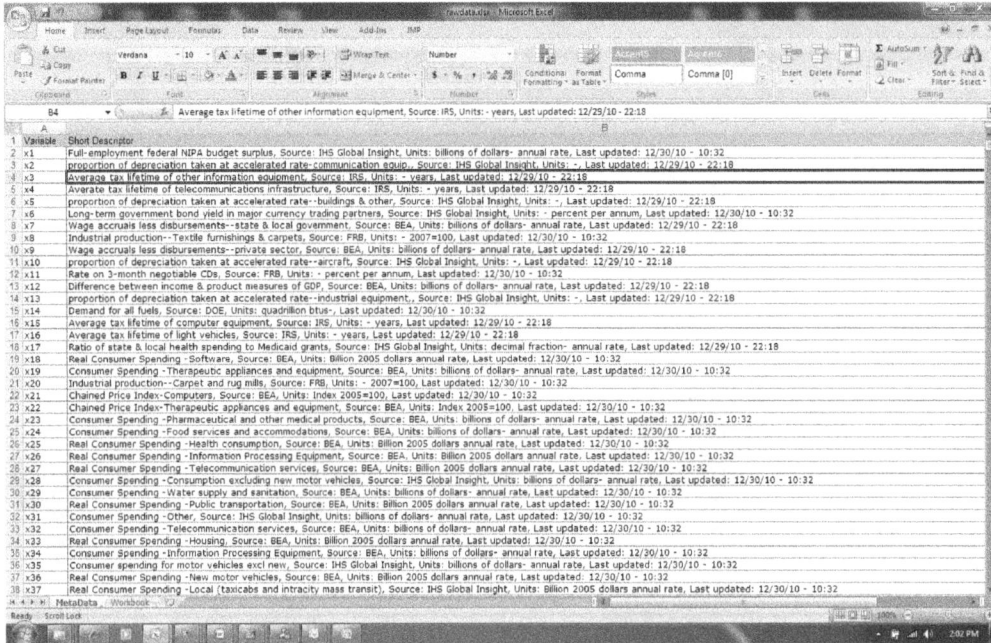

Figure 12.3: New Vertical Data

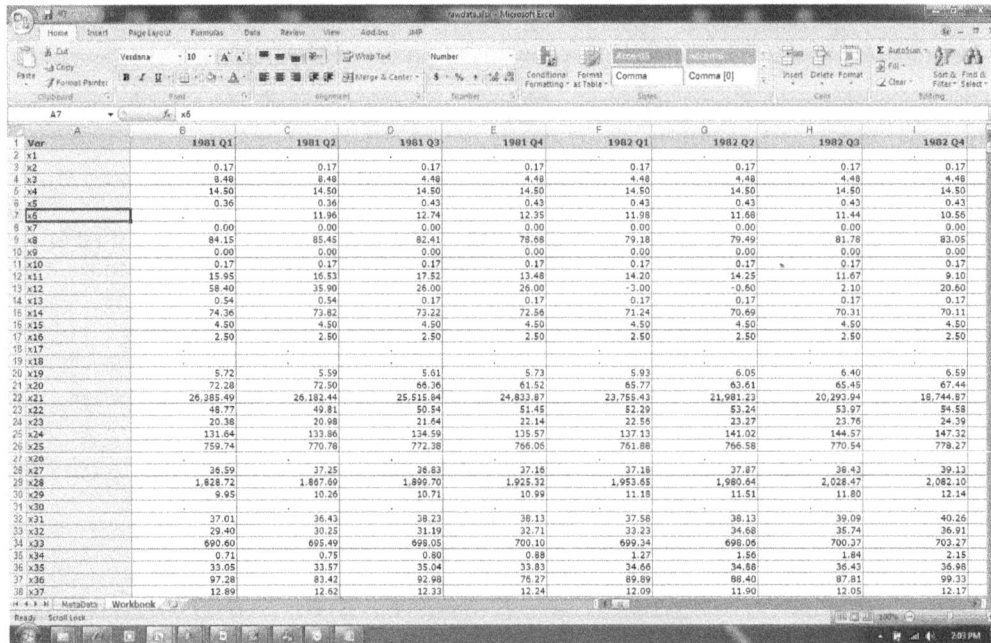

The last two steps necessary to make this a Time Series data set is to make the rows the columns and then have a proper DATE variable. Figure 12.5 shows the transposed data set using the TRANSPOSE wizard in the SAS Enterprise Guide flow in Figure 12.4.

Figure 12.4: SAS Enterprise Guide Flow for Final Processing

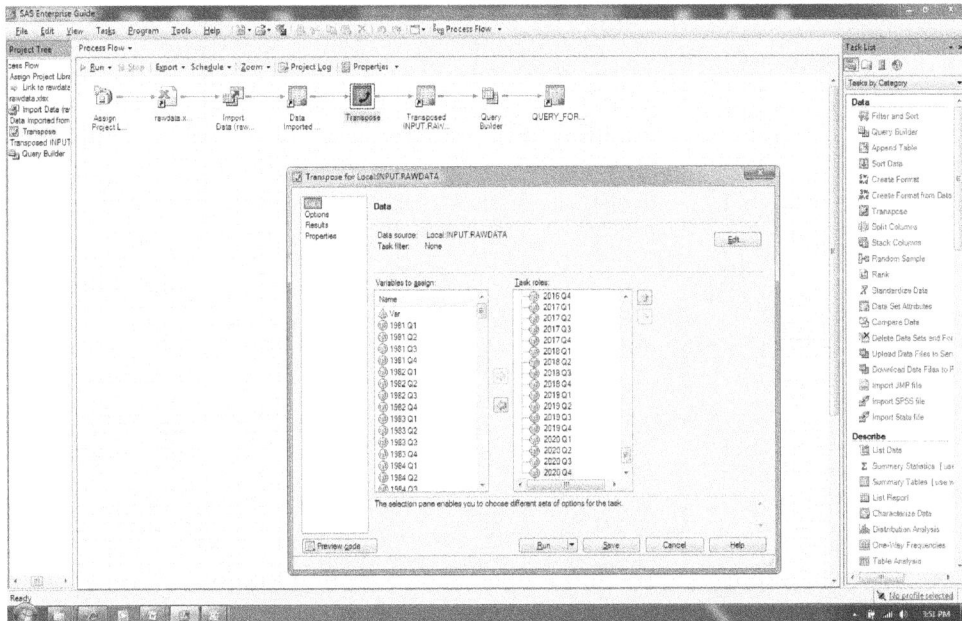

Figure 12.5: Transposed Raw Data

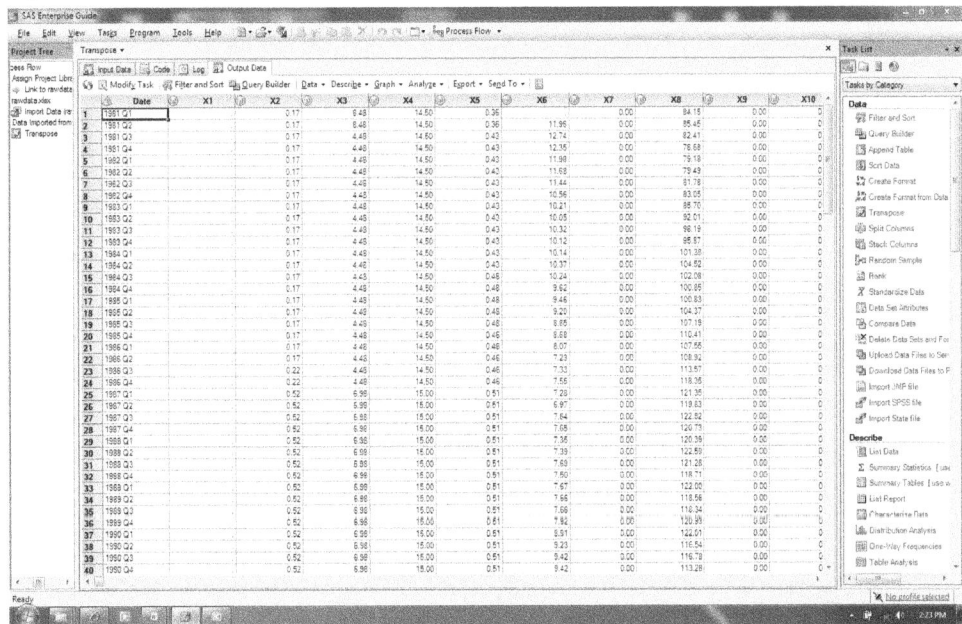

The last step in the process of generating a true Time Series data set, Figure 12.6, is to get the date variable in a proper SAS DATE format. Here the QUERY wizard is used to convert the character variable Date_OLD to two new variables Quarter and YearFinal using the SUBSTR and INPUT functions and then using the MDY function to form a proper DATE format. This flow produces the final Time Series data set for the Xs given in Figure 12.7.

Figure 12.6: Date Conversion Approach

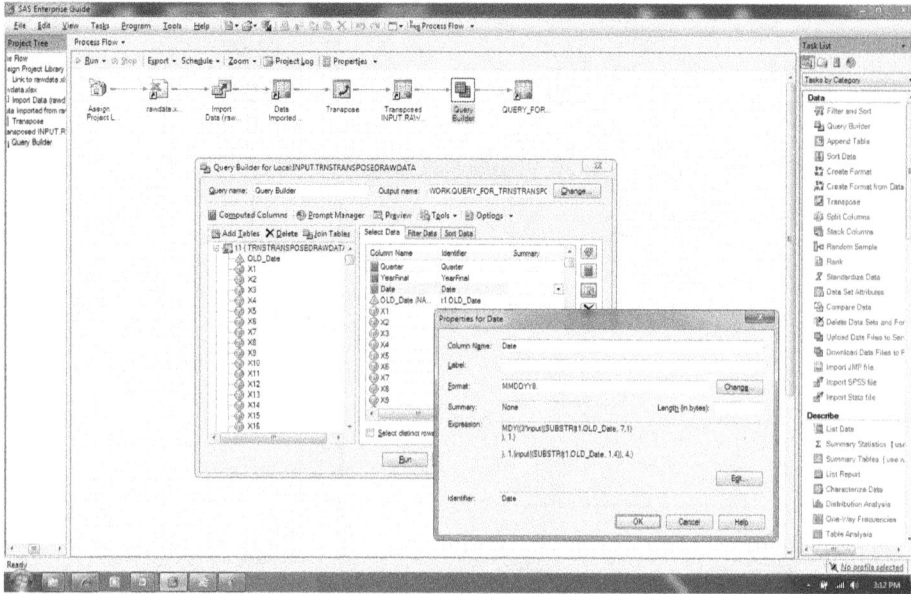

Figure 12.7: Final Time Series Data Set for Xs

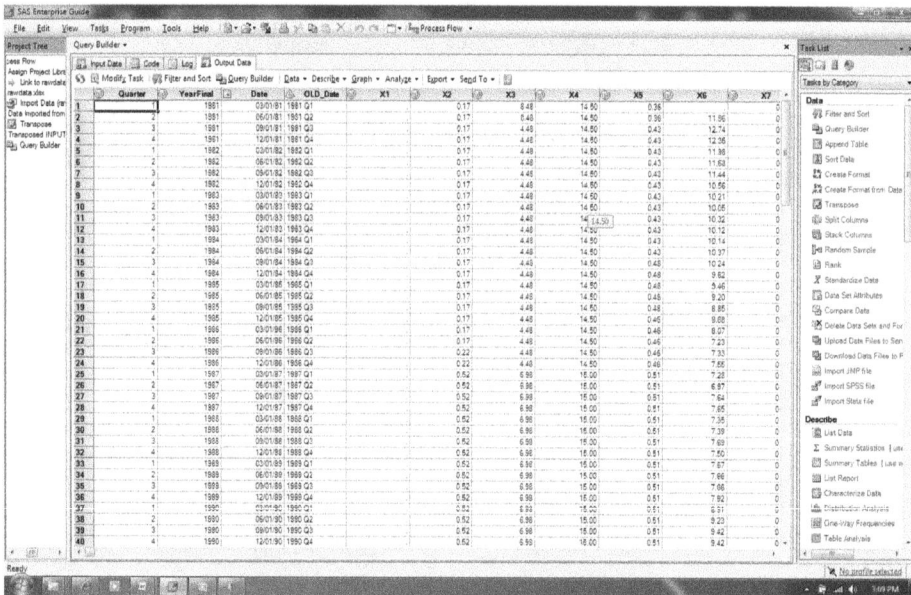

Before moving to the exploratory phase of the project, the Y data has to be merged with the X data. Both data sets have to have the same quarterly based MDY Date variable, as they do. Figure 12.8 contains the view of the Y data. Note that Y data starts in 1981 as does the X data. This is not always the case. Generally, the Y data is the controlling data set for the data range. We also note that the X data goes through the end of 2020. Given the study was done in the 1st quarter of 2011, the Y data only extends to 4th quarter 2010. This is actually 20 years of data or 80 quarterly data points for the Y. It is a common practice to limit the forecast range to 5 to 1. That is, 20% of the historical data. So, we would be limiting the forecast range to 16 quarters, which is 4 years, one year longer than that necessary for the Charter (36 months or 12 quarters for ES&OP). A simple merge is all that is necessary to bring the Y and X data together. This is the first step in the exploratory SAS Enterprise Guide flow in the next section.

Figure 12.8: Raw Y Data

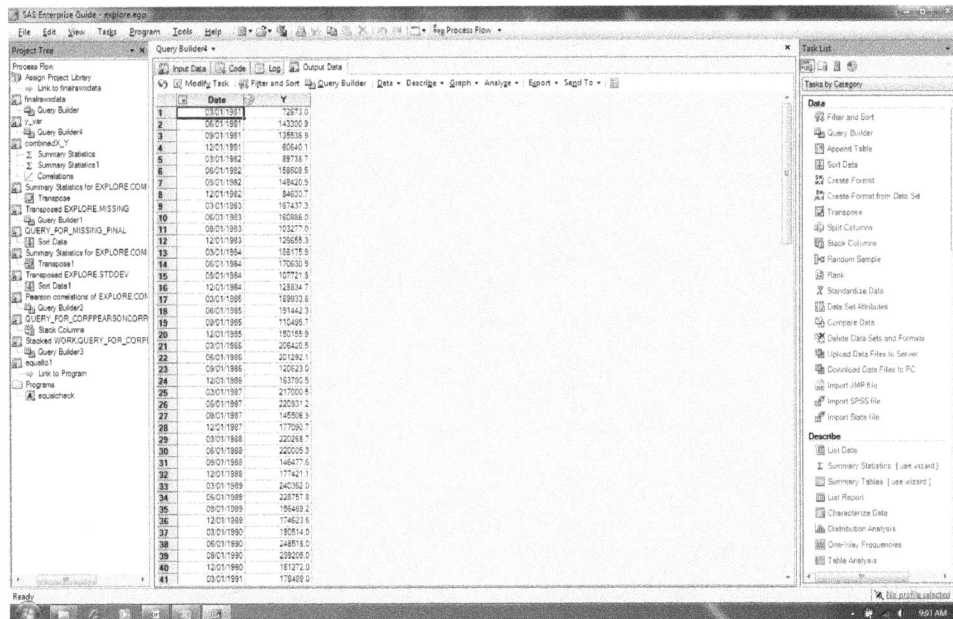

12.6 Exploratory Analysis and Data Preprocessing

Now the fun begins. This analysis is separated in three main phases. First, we explore the Xs for redundancy, lack of information, and missing information. For those X variables that stay in the data set but have intermittent or groups of missing data, various imputation or model based approaches are used to fill that data in. Next we determine how many of the Xs actually have forecasts out at least three years and, for those that do not, develop them.Regardless of the length of the X forecasts, all Xs with forecasts longer than 12 quarters are truncated to 12 quarters in accordance with the Charter requirements.

Figure 12.10 is the data set formed after running the merge sequence in the SAS Enterprise Guide flow in Figure 12.9. Here are some general observations. The X data starts in the 1st quarter of 1981 and goes through the 4th quarter 2020. On the other hand, the Y data start 3rd quarter 1981 and goes through 4th quarter 2010. Thus, the analysis is restricted to the time frame for which we actually have Y data. Notice that we can already see missing values for the Xs in the early quarters. We can also see that some of the X variables are really not continuous and that some of them seem to have many 0s. These are just some of the anomalies that are detected in the exploratory phase of the project and thus show that not all Xs offer useful information for forecasting. Various analyses will show where the problems actually are.

Figure 12.9: Merge Sequence for Merging Y and X Data

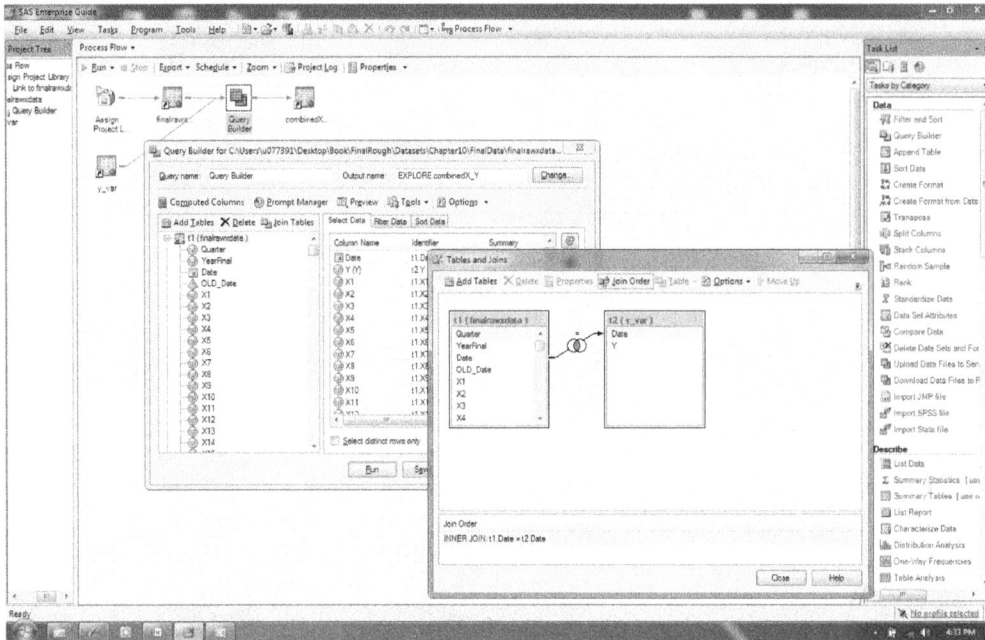

Figure 12.10: Final Merge Y and X Time Series Data

Figure 12.11 contains the SAS Enterprise Guide flow used to do the initial exploratory analysis, which includes checking for missing data, variables that are constant (that is, do not change over time) and variables that have exact correlations to others. We might ask, why were these variables put into the data set in the first place if they did not offer pertinent information? Often the data service provider does not give enough information in the metadata on the variables of interest to discern these issues ahead of time.

Figure 12.11: Exploratory Enterprise Guide Flow

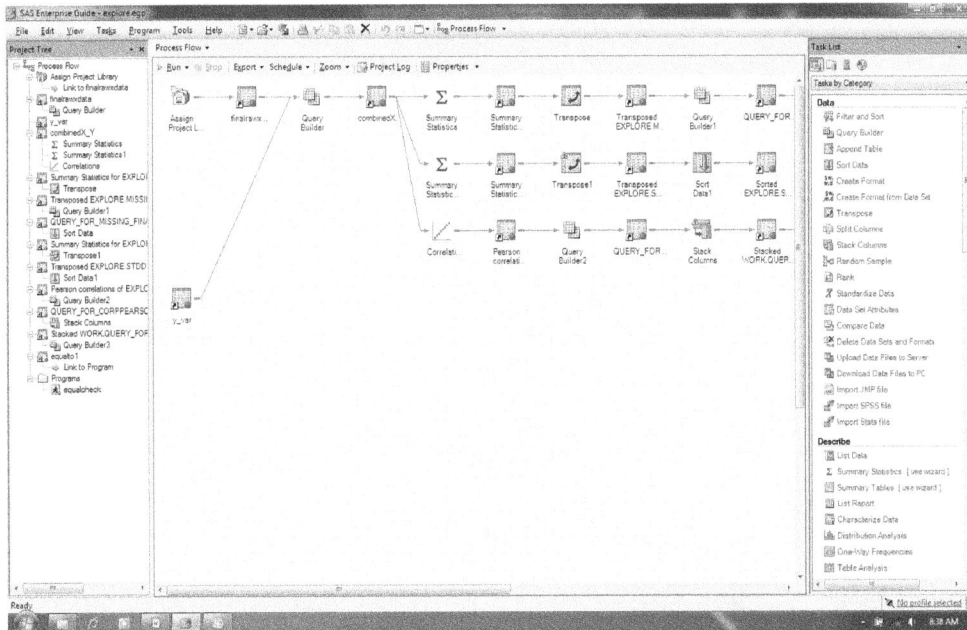

Figure 12.12 contains the fraction of missing data. We can see that some variables are completely missing and therefore are removed from the analysis. We normally use the .25 fraction rule in that for Xs, we are willing to impute (generally backcast) the missing data whereas all other Xs with more than .25 missing are dropped from the initial analysis. SAS Forecast Server has its own imputation methods for minor individual missing data points within the series.

Figure 12.12: Fraction Missing

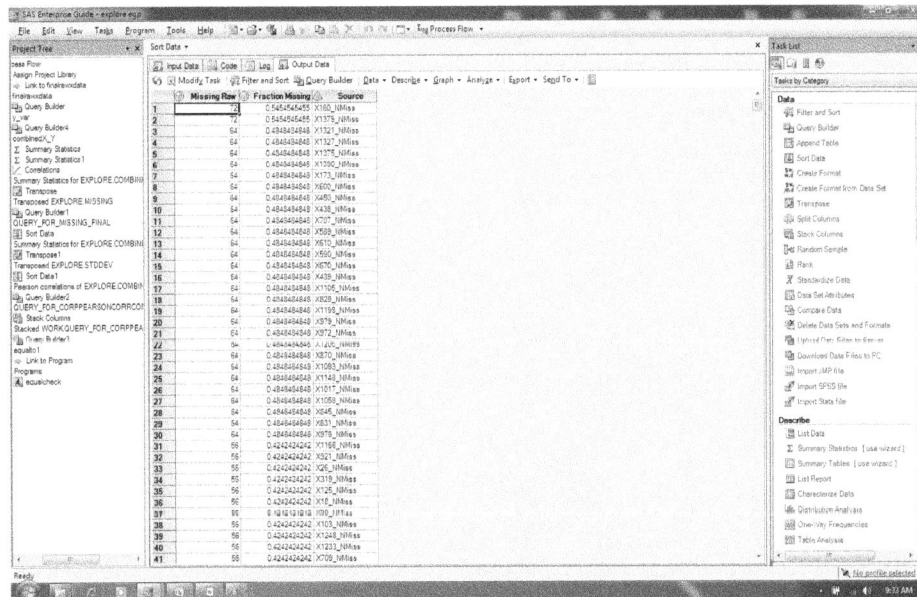

Figure 12.13 has a list of those X variables that have Standard Deviations of 0 indicating they do not change over the time period of interest. All Xs with Standard Deviations of 0 are dropped from the analysis as they offer no information for forecasting as they are constant.

Figure 12.13: Constant X Variables

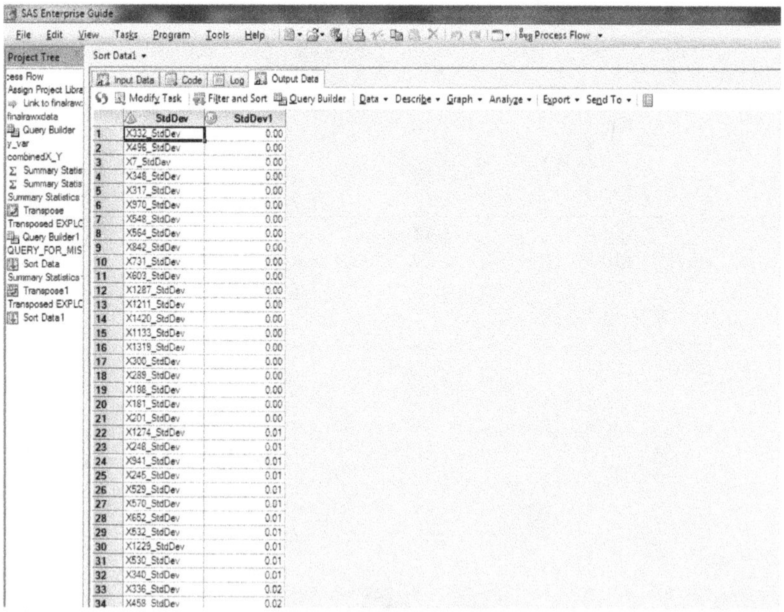

	StdDev	StdDev1
1	X332_StdDev	0.00
2	X496_StdDev	0.00
3	X7_StdDev	0.00
4	X348_StdDev	0.00
5	X317_StdDev	0.00
6	X970_StdDev	0.00
7	X548_StdDev	0.00
8	X564_StdDev	0.00
9	X842_StdDev	0.00
10	X731_StdDev	0.00
11	X603_StdDev	0.00
12	X1287_StdDev	0.00
13	X1211_StdDev	0.00
14	X1420_StdDev	0.00
15	X1133_StdDev	0.00
16	X1319_StdDev	0.00
17	X300_StdDev	0.00
18	X289_StdDev	0.00
19	X198_StdDev	0.00
20	X181_StdDev	0.00
21	X201_StdDev	0.00
22	X1274_StdDev	0.01
23	X248_StdDev	0.01
24	X941_StdDev	0.01
25	X245_StdDev	0.01
26	X529_StdDev	0.01
27	X570_StdDev	0.01
28	X652_StdDev	0.01
29	X532_StdDev	0.01
30	X1229_StdDev	0.01
31	X530_StdDev	0.01
32	X340_StdDev	0.01
33	X336_StdDev	0.02
34	X458_StdDev	0.02

Some Xs actually measure the same phenomena. Or, in the case of Global Insights, if you are not careful, data for Baseline, Pessimistic, and Optimistic might be pulled on the same variable. All three would be measuring the same variable but in different scenarios. Thus, the simple correlation analysis is useful to detect and eliminate these phenomena. Choosing which variable to drop when having correlations of 1 might need to be addressed with your client as they could have a preference.

Figure 12.14: Correlations of 1 or Very Near 1

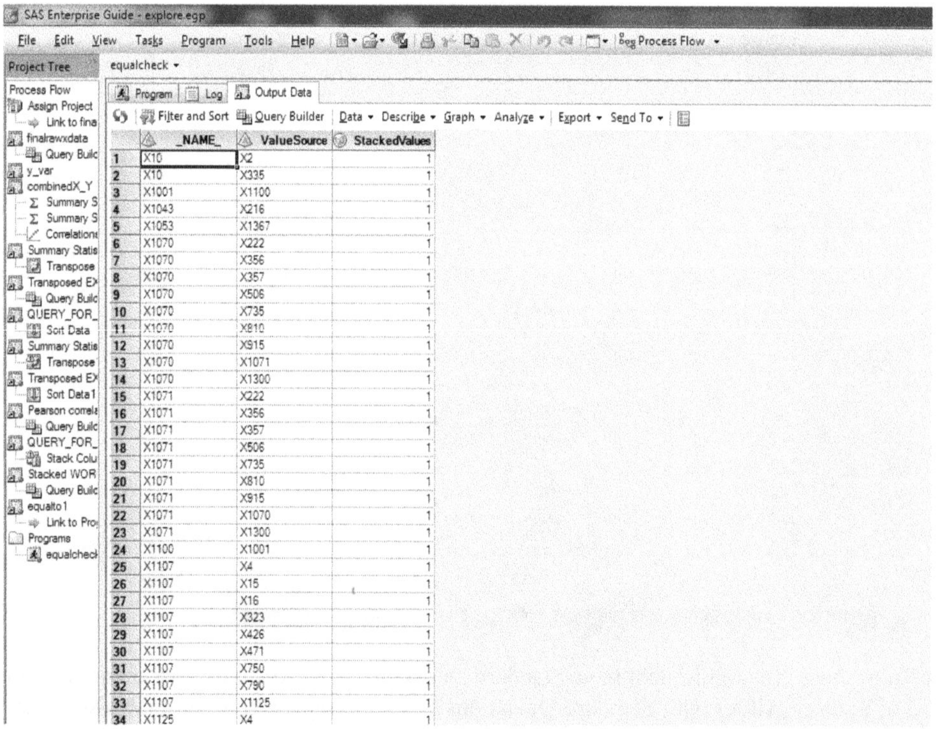

	NAME	ValueSource	StackedValues
1	X10	X2	1
2	X10	X335	1
3	X1001	X1100	1
4	X1043	X216	1
5	X1053	X1367	1
6	X1070	X222	1
7	X1070	X356	1
8	X1070	X357	1
9	X1070	X506	1
10	X1070	X735	1
11	X1070	X810	1
12	X1070	X915	1
13	X1070	X1071	1
14	X1070	X1300	1
15	X1071	X222	1
16	X1071	X356	1
17	X1071	X357	1
18	X1071	X506	1
19	X1071	X735	1
20	X1071	X810	1
21	X1071	X915	1
22	X1071	X1070	1
23	X1071	X1300	1
24	X1100	X1001	1
25	X1107	X4	1
26	X1107	X15	1
27	X1107	X16	1
28	X1107	X323	1
29	X1107	X426	1
30	X1107	X471	1
31	X1107	X750	1
32	X1107	X790	1
33	X1107	X1125	1
34	X1125	X4	1

Having done the exploratory analysis, the Xs have been reduced to 1121 from 1441 (1443 original variables less Date and the Y). Some of these variables still need imputing due to missing data.

12.7 X Variable Imputation

There were 13 X variables that needed backcasting. That is, their missing data was less than 25% and was in the initial time periods. We use SAS Forecast Server for this process. It is quite simple. Since 75% of the data is in fact available and is available for the most recent 75% of the time horizon, we feel very comfortable using univariate ARIMA for backcasting. We are simply trying to maintain the trend, seasonality, and any cycle in the early part of the data. The process goes like this. First, reduce the data set to just those Xs of interest, maintaining the original data variable as well. Set up a new date variable reflecting the most recent time having the missing data. Figure 12.15 is the data set before sorting to reverse time. Sort the data in reverse time. Save the data set in this new format into a proper SAS data set for SAS Forecast Server (Figure 12.16). Import the data into SAS Forecast Server and call all of the Xs the dependent variables. Use univariate ARIMA for modeling. In this case we generally do not bother with a holdout. Once satisfied with the SAS Forecast Server models, simple re-sort the data set with the forecasts into the original Date sequence and put the data back into the original data set.

Figure 12.15: Data for Backcasting Before Sorting

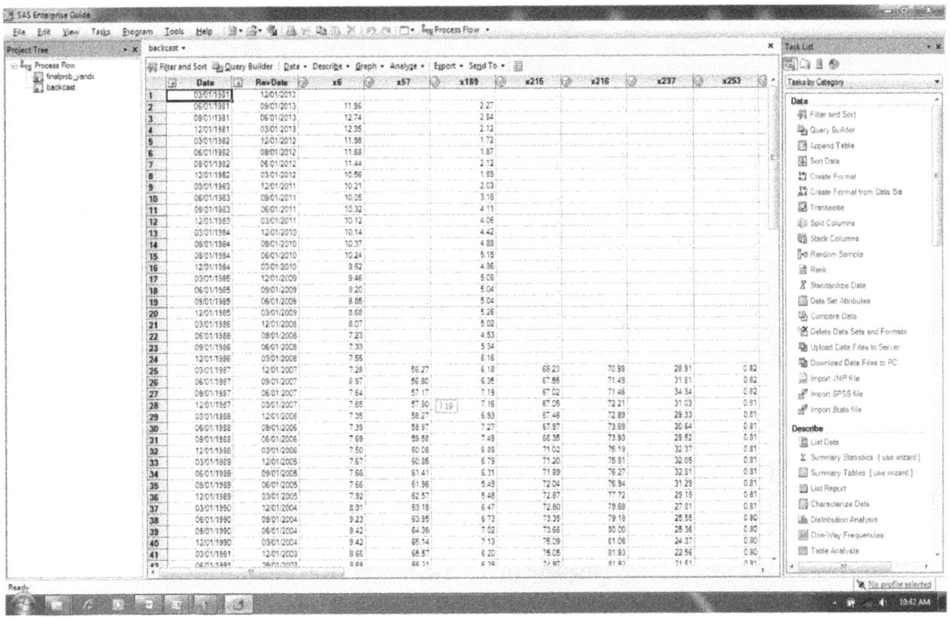

Figure 12.16: Data Ready for Backcasting

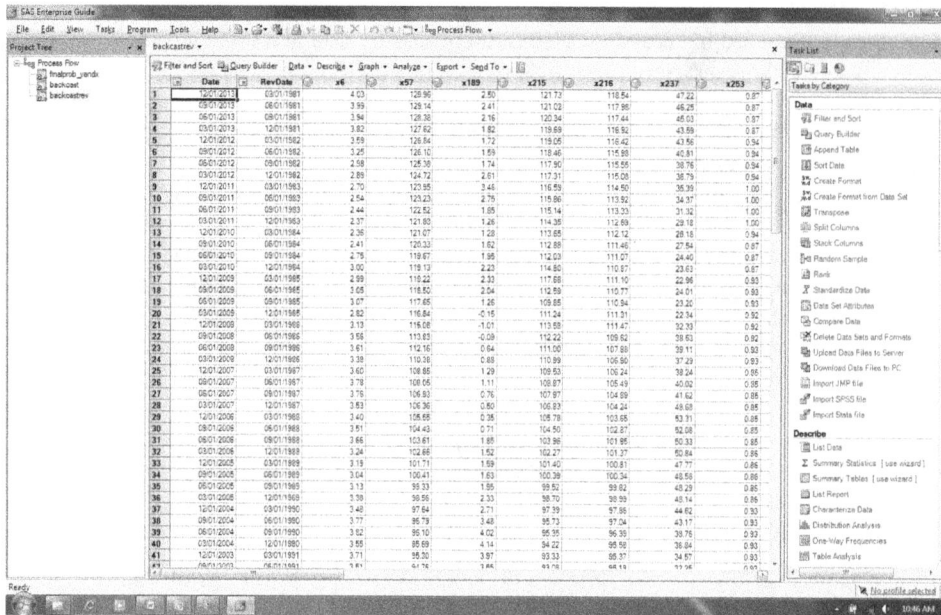

Having already chosen RevDate for the data variable for forecasting, we now declare all of the Xs as "Dependent" variables so that we can backcast them all at the same time, as shown in Figure 12.17. In Figure 12.18 we are setting up the preliminary options for diagnosing seasonality and outliers, using the median forecast and allowing SAS Forecast Server to choose the transformation on the dependent variable automatically. The intermittency tests and the minimum number of points for certain types of models are irrelevant in this case since we know we have data for all the time periods for each X.

Figure 12.17: Declaring all Xs as Dependent Variables

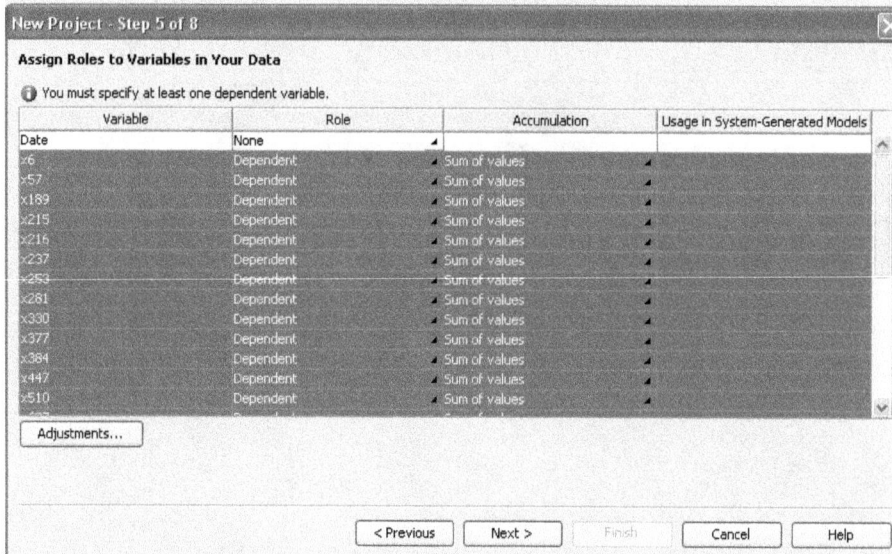

Figure 12.18: SAS forecast Server Preliminary Options

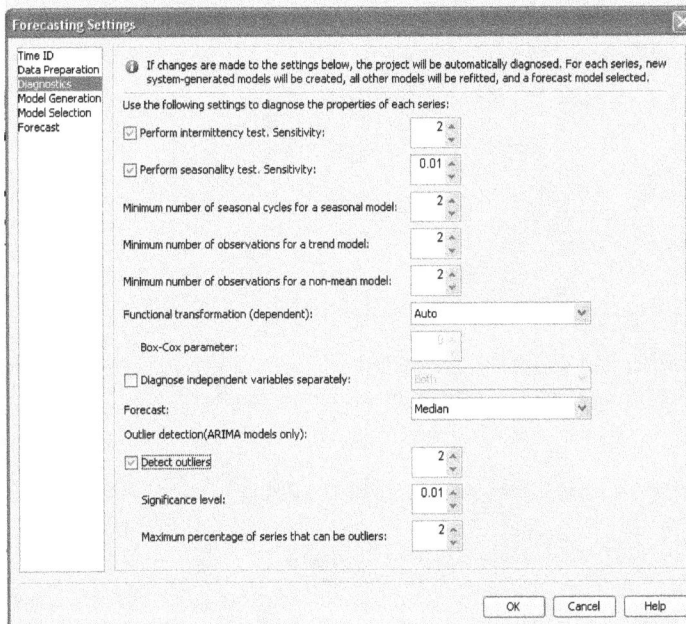

In Figure 12.19 we are setting the options for actually doing the backcasting. We prefer to force the models as ARIMA to alleviate the typical "shadow effect" that exponential smoothing models produce and to ensure the best chance of capturing trend, cycle, and seasonality. Figure 12.20 shows the summary for the 27 Xs to be backcast. We first see that all but a few have MAPEs less than 10, all are ARIMA (as we forced them to be), many had outliers detected and no models failed to be estimated.

Figure 12.19: SAS Forecast Server Model Options

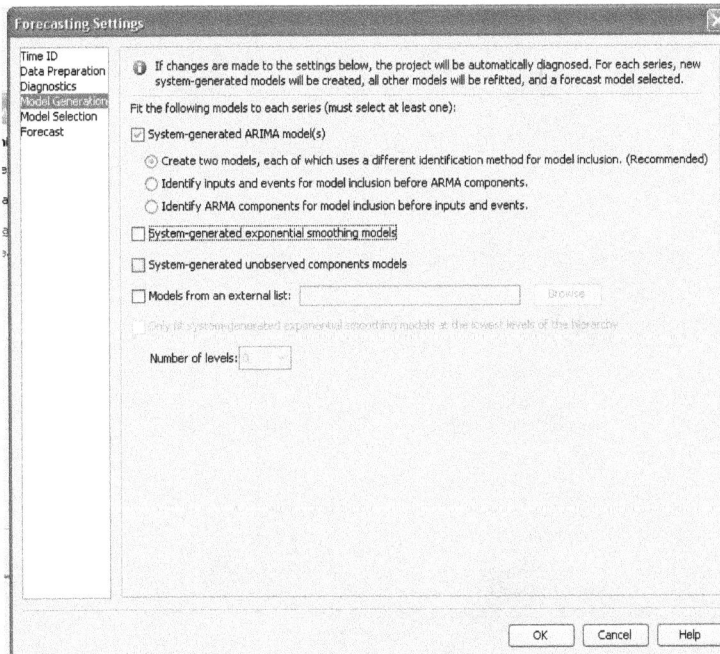

Figure 12.20: Summary

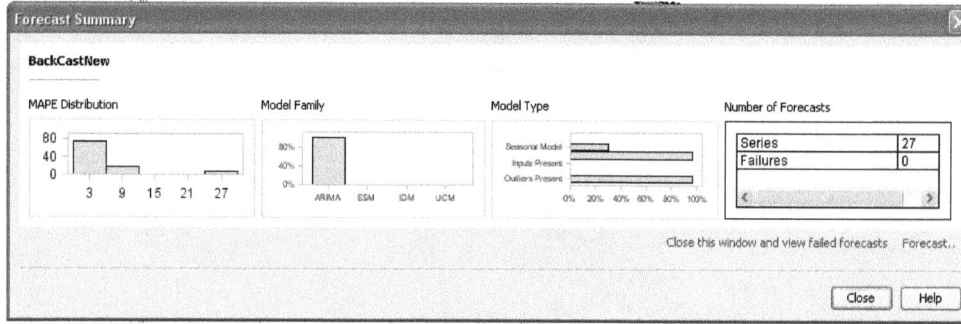

Figures 12.21 and 12.22 are examples of backcasted variables. To the left of the graph we can see each of the MAPEs for the 27 backcasted variables. The highlighted variable is shown in the graph.

Figure 12.21: Example of Backcasted X

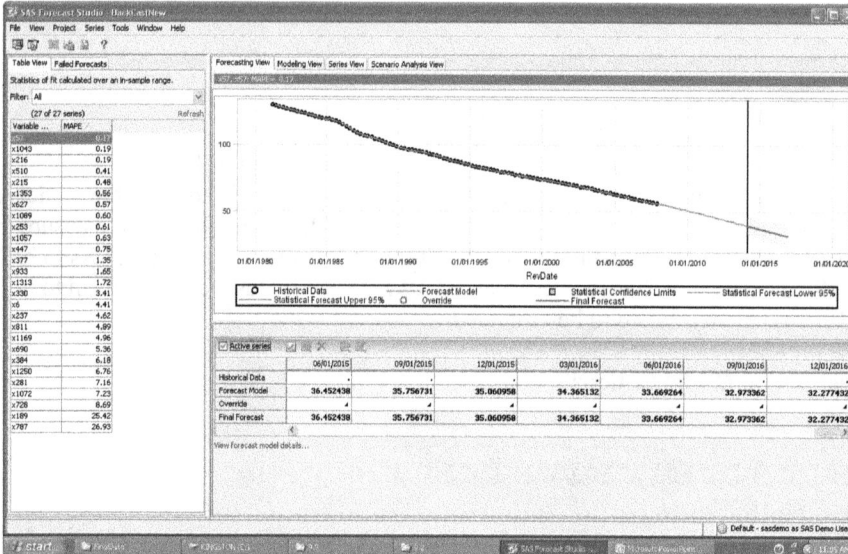

Figure 12.22: Example of Backcasted X

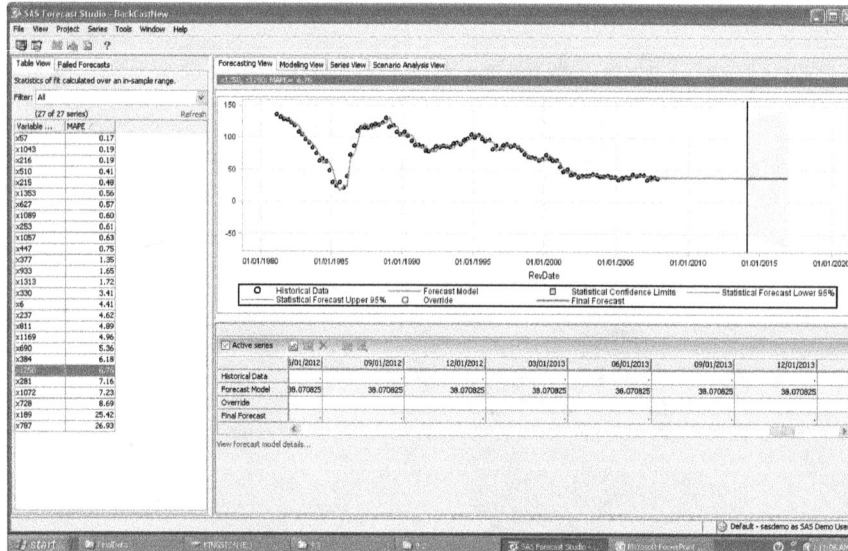

To conclude the imputation phase, the forecasted values are substituted for the missing values and the data set is re-sorted in the original date order. The variables with their backcasted values are then returned to the original data set. Concerning any individual data points missing within the series, SAS Forecast Server's options are used to replace them and are shown later.

12.8 Variable Reduction and Selection

In this phase of the analysis, we are starting with 1121 X variables. This number of Xs is far too many for SAS Forecast Server to operate on. Therefore, a two-step process is in order. First, the number of Xs has to be reduced and then, based on this reduction, a subset of these variables is selected for SAS Forecast Server to work with. This first reduction is done using a combination of similarity analysis along with variable clustering as described in Chapter 7. Figure 12.23 shows the SAS Enterprise Guide flow containing the similarity and variable clustering code. Figure 12.24 shows the actual SAS code for similarity and variable clustering.

Figure 12.23: SAS Enterprise Guide Flow for Variable Reduction

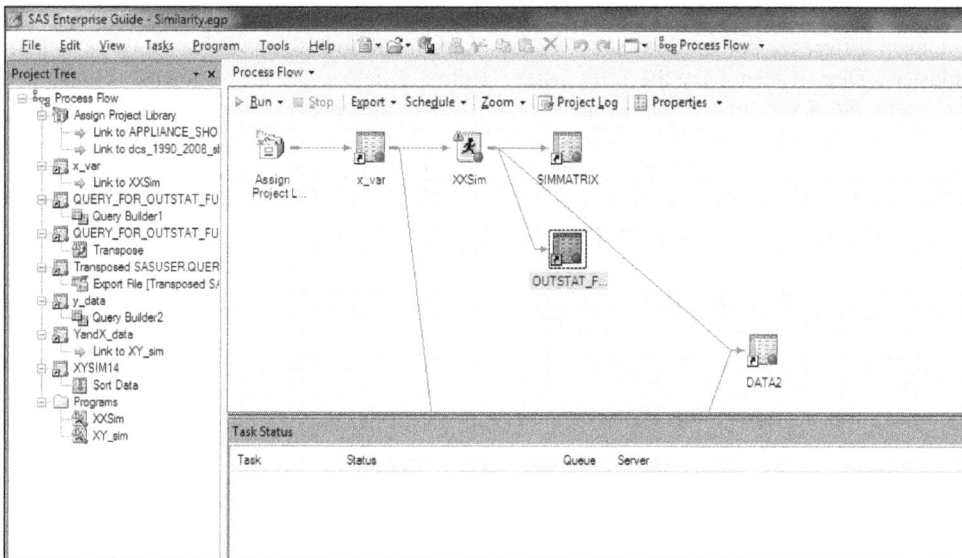

Figure 12.24: SAS Code for Similarity Analysis and Variable Clustering

Since the variable clustering algorithm does not solve directly for an optimal number of clusters, and since the number of variables will be further reduced, the number of clusters chosen is normally around 20% of the number of Xs (here potentially 200). But, it is important to make sure the clusters are homogeneous. That is, the correlation range is generally in the .9 to .8 and reduces somewhat continuously across the variables in the cluster. To demonstrate this phenomenon, a range of cluster solutions is given below (10, 40, 75, 150, 200).

Table 12.1 shows the minimum and maximum correlation for the X variable with the highest correlation to the centroid of the cluster along with the number of single variable clusters.

Table 12.1 Cluster Solution Comparison

Number of Clusters	Minimum	Maximum	Number of Single Variable Clusters
10	0.507	0.874	0
40	0.584	0.939	0
75	0.724	1	1
150	0.869	1	9
200	0.895	1	19

The 75-cluster solution still has lower minimum correlations than desired, so moving to 150 or 200 is probably best. There is not much difference in the minimum correlations from 150 to 200, and there are twice as many single variable clusters. Thus, it seems like the 150-cluster solution is the best for this problem. Therefore, for each of the 150 clusters, the X variable with the highest correlation to the cluster centroid will be the variable that is moved to the next stage of the analysis. This is purely a statistical decision. At this point it is often a good idea to engage the business subject matter expert because they might prefer a particular X in the list for a given cluster. If the correlation to the centroid of the cluster is not that different from the best X variable, and both have forecasts from a source like Global Insights, then their preference is best.

Using the top 5 variables in each cluster, it becomes obvious what the cluster intent or theme actually is. For X251 (Cluster 38) the theme is motor vehicle sales, and more specifically trucks. For X742 (Cluster 71), the variables are more related to saving, interest rates, and higher-level discretionary spending. X785 (Cluster 74) seems to reflect the State and Local governments' ability to spend based on their receipts. Lastly, X1279 (Cluster 14) reflects exports and non-residential construction, which both indirectly reflect industrial spending.

X251 (Cluster 38)

- Sales of all new trucks. Source: BEA. Units: millions annual rate
- Industrial production—motor vehicles and parts. Source: FRB. Units: 2007=100
- Unit sales of new light trucks. Source: BEA. Units: millions annual rate
- Unit sales of new light vehicles. Source: BEA. Units: millions annual rate
- Unit sales of new light domestic trucks. Source: BEA. Units: millions annual rate

X742 (Cluster 71)

- Gross saving. Source: BEA. Units: billions of dollars annual rate
- Sales of new heavy and medium trucks. Source: BEA. Units: millions annual rate
- Prime discount rate—Federal Reserve Bank of New York. Source: FRB. Units: percent per annum
- Real Consumer Spending -Recreational services. Source: BEA. Units: billion 2005 dollars annual rate
- Real Consumer Spending -Leasing. Source: BEA. Units: billion 2005 dollars annual rate

X785 (Cluster 74)

- State and local government social insurance tax receipts. Source: BEA. Units: billions of dollars annual rate

- State and local government employer-paid social insurance tax receipts. Source: BEA. Units: billions of dollars annual rate

- Other labor income (fringe benefits). Source: BEA. Units: billions of dollars annual rate

- State and local net interest payments. Source: BEA. Units: billions of dollars annual rate

- Effective state and local government social insurance tax rate on wages and salaries. Source: IHS Global Insight. Units: - decimal fraction

X1279 (Cluster 14)

- Exports of factor services. Source: BEA. Units: billions of dollars annual rate

- U.S. international transactions—income receipts from the rest of the world. Source: BEA. Units: billions of dollars annual rate

- Real exports of factor services. Source: BEA. Units: billions of chained 2005 dollars annual rate

- Gross private investment in nonresidential construction Building and Other. Source: IHS Global Insight (total-mining and utilities). Units: billions of dollars annual rate

- Federal budget surplus—NIPA basis. Source: BEA. Units: billions of dollars annual rate

- Nonresidential investment-structures—commercial and health care. Source: BEA. Units: billions of dollars annual rate

- Private investment-nonresidential, non-farm buildings. Source: IHS Global Insight. Units: billions of dollars annual rate

Having completed the variable reduction step, the next step of the process is variable selection. As described in Chapter 7, three different time series variable selection methods will now be used, similarity (X-Y), cointegration, and cross correlation function analysis. Table 12.2 contains a summary of the analysis derived using the SAS Enterprise Guide flow in Figure 12.25.

Table 12.2: Variable Selection Analysis

X_var	Coin_Y	XYSim_Y	Abs(LagCORR_Y)	Best	Pick
_x1023	0.1512487	0.3235767	0.6476772	1	1
_x1059	0.0501944	0.3574313	0.6884455	1	1
_x1311	0.9288972	0.3591984	0.8026373	2	1
_x1119	0.8377197	0.3608021	0.7809169	2	1
_x112	0.7114264	0.3616429	0.7638551	2	1
_x299	0.918025	0.3635842	0.8039425	2	1
_x1401	0.4871511	0.3675027	0.6423653		
_x1076	0.3595368	0.3684917	0.5308634		
_x37	0.4929427	0.3686142	0.7436871	2	1
_x236	0.274021	0.3735979	0.540229	1	1
_x405	0.6419131	0.3756427	0.7435856	2	1
_x1237	0.8082668	0.3767561	0.725936	2	1
_x251	0.6465527	0.3786432	0.7549783	2	1
_x1129	0.2835249	0.3801517	0.6554995	1	1
_x1387	0.4772896	0.3806968	0.617184		
_x796	0.8612461	0.3871107	0.7138978	2	1
_x244	0.4155125	0.3897994	0.625084		
_x39	0.9287392	0.3900057	0.7866294	2	1
_x547	0.2241301	0.3932483	0.5763677	1	1
_x267	0.6684609	0.4017867	0.6781281	2	1
_x98	0.9621871	0.4051987	0.7779253	2	1
_x916	0.024157	0.4052213	0.4169243	1	1
_x742	0.8521528	0.4097431	0.7522278	2	1
_x812	0.7537492	0.412634	0.6876199	2	1
_x560	0.1932381	0.4127743	0.490175	1	1
_x1396	0.987245	0.4156912	0.7111687	2	1
_x1235	0.8125586	0.4157532	0.780293	2	1
_x1406	0.8667983	0.4193117	0.7168955	2	1
_x1192	0.9879722	0.4199711	0.7359344	2	1
_x227	0.6674007	0.4204923	0.5447775		
_x307	0.9999579	0.4227097	0.7407838	2	1
_x785	0.9942017	0.427841	0.7247526	2	1
_x167	0.9976324	0.4298321	0.6846007	2	1
_x1167	0.9931081	0.4303521	0.7720413	2	1
_x1160	0.529658	0.4313959	0.5763836		
_x223	0.9882791	0.4320334	0.7770826	2	1
_x797	0.9203266	0.4326965	0.6510546	2	1
_x1250	0.2818367	0.4348595	0.6647641	1	1
_x948	0.9971148	0.4392408	0.7654264	2	1
_x1326	0.8747022	0.4398601	0.6919678	2	1
_x868	0.8221225	0.4443366	0.5742418		
_x1349	0.6901579	0.4446143	0.7190697	2	1
_x892	0.9986562	0.4453802	0.640237		
_x1279	0.9705853	0.4527356	0.6767041	2	1
_x105	0.9971841	0.4550817	0.6823598	2	1
_x517	0.9963528	0.4557623	0.7378036	2	1
_x87	0.992426	0.4567416	0.711599	2	1
_x451	0.9999442	0.4592807	0.7451752	2	1
_x613	0.7827147	0.4606164	0.4667177		
_x403	0.9998313	0.4609271	0.7497567	2	1
_x428	0.9814056	0.4619457	0.7761062	2	1
_x895	0.9985936	0.462069	0.6694238	2	1
_x710	0.9744768	0.4630021	0.7416002	2	1
_x772	0.9055489	0.4640899	0.6865074	2	1
_x1103	0.9999596	0.4662272	0.6885741	2	1
_x1320	0.9997196	0.4694437	0.7596356	2	1
_x197	0.6840477	0.470412	0.53182		
_x54	0.9994697	0.4705299	0.7491572	2	1
_x1271	0.9827311	0.4710417	0.6031985		
_x1400	0.9563082	0.4715611	0.652673	2	1

Figure 12.25: Special Similarity, Cointegration and Cross Correlation Function Analysis

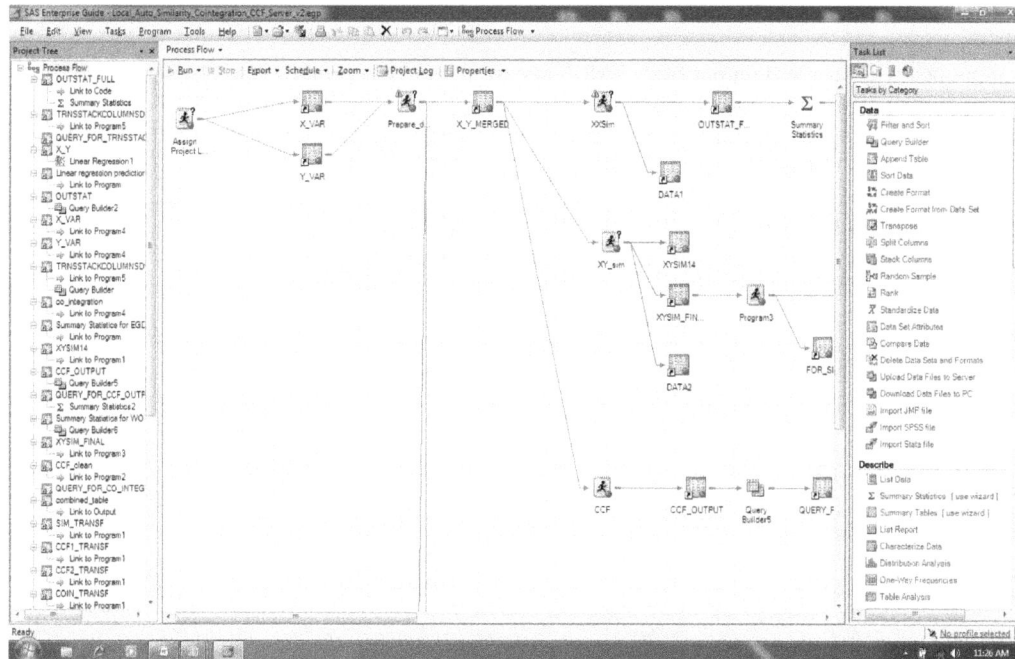

The cut-off values for highlighting in the table were as follows: less than .3 for cointegration (this is actually a P-value), greater than .65 for CCF (simple Pearson-Product-Moment correlation at different lags in the X) and then simply the rank order from lowest to highest for similarity. The column "Best" is based on the simple algorithm that if all three are highlighted or cointegration and similarity are highlighted with CCF being close to .6. Those that have a 2 in the "Best" column were instances where combinations of two of the three criteria were met. The "Pick" column is where Best is 1 or 2. This results in 50 Xs to use in the next step.

To review, this final set of 50 Xs were chosen based on the above algorithm applied to the three time series–based supervised variable selection methods described above. These three methods were applied to the 150 Xs that were selected out of the 1121 Xs using the unsupervised variable reduction method called variable clustering. The latter was based on the similarity matrix amongst the original 1141 Xs. These final 50 X variables will be presented to SAS Forecast Server for modeling.

12.9 Modeling

Now that the 1441 X variables, first reduced to 1121, have been reduced to 50 Xs for modeling, the following best practice modeling framework will be introduced. First, a general exploratory analysis is conducted, which includes analyzing for the time series components of the Y itself. Next a baseline model is set up. In this case there was no previously developed model, so an exponential smoothing model was used as the baseline. All models need a frame work for comparison (Gilliland 2008). This is where holdout and out-of-sample points come in. There are enough data points in this problem to warrant using both a holdout and an out of sample. One other best practice used to help ensure the "winning" model is stable is to actually change the number of points used in the holdout and out-of-sample sets. In this case, the base study was done using eight holdouts (two years) and four out of sample (1 year). Once the exploratory analysis is done, SAS Forecast Server is then allowed to automatically build models. That is, SAS Forecast Server will select the proper transformation on the Y, discern stationarity of the Y and Xs (choosing to apply differencing or not), detect seasonality, select the Xs inclusive of lags and functional form of the transfer function with the Y, look for outliers as well as level shifts, and lastly apply the proper adjustments to maintain normal i.i.d. errors. The modeler does have various choices to make for comparing models. As mentioned earlier, different holdout and out-of-sample scenarios can be set up. Plus, different groups of Xs can be presented to SAS Forecast Server. For example, the top 50 variables chosen in the variable selection process or just the top cointegration or similarity or cross correlation function variables. Lastly, SAS Forecast Server also allows a choice of Try to Use or Use if Significant when considering the Xs. For reference SMAPE (Symmetric Mean Absolute Percent Error) on the holdouts is being used for model selection.

All of the output below is standard output from SAS Forecast Server. The process starts by setting up the project in an eight-step process. The first step is simply naming the project, identifying which variable is the TimeID, and then specifying the hierarchical variables, if any. Step 1 is shown in Figure 12.26. Then you identify the data source, as shown in Figure 12.27.

Figure 12.26: Naming the Project

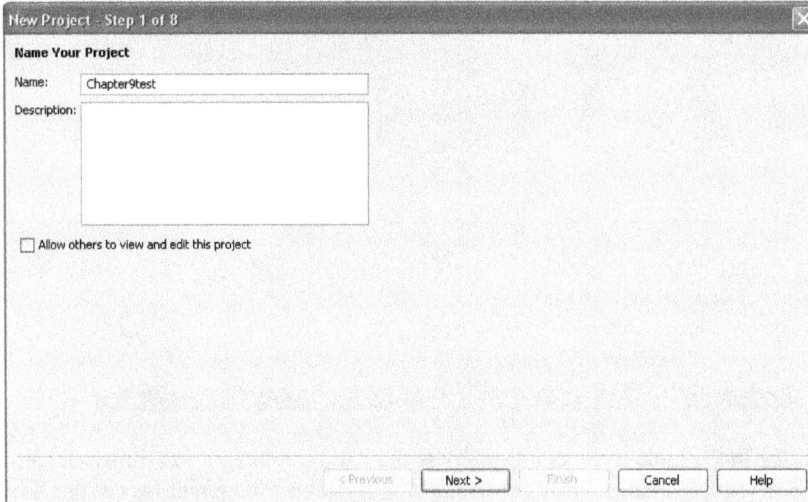

Figure 12.27: Identifying the Input Data

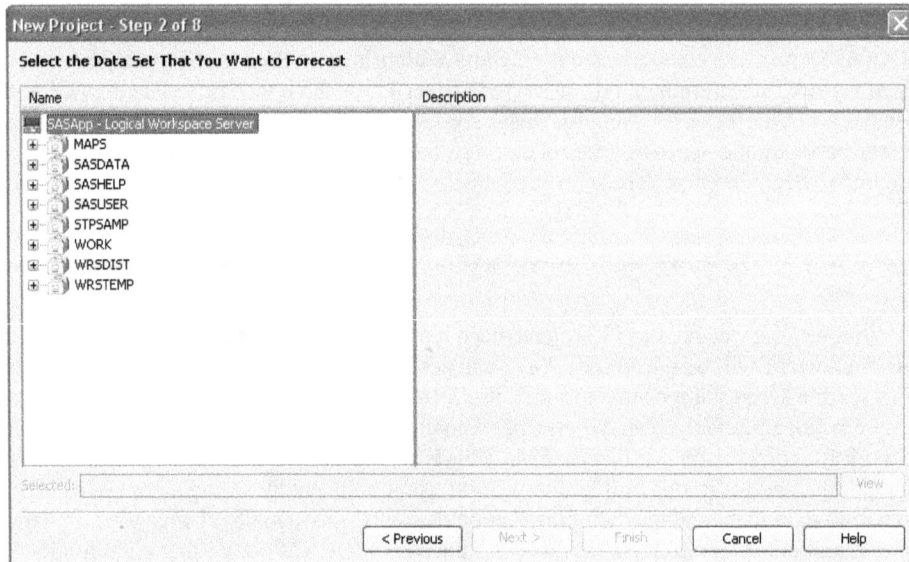

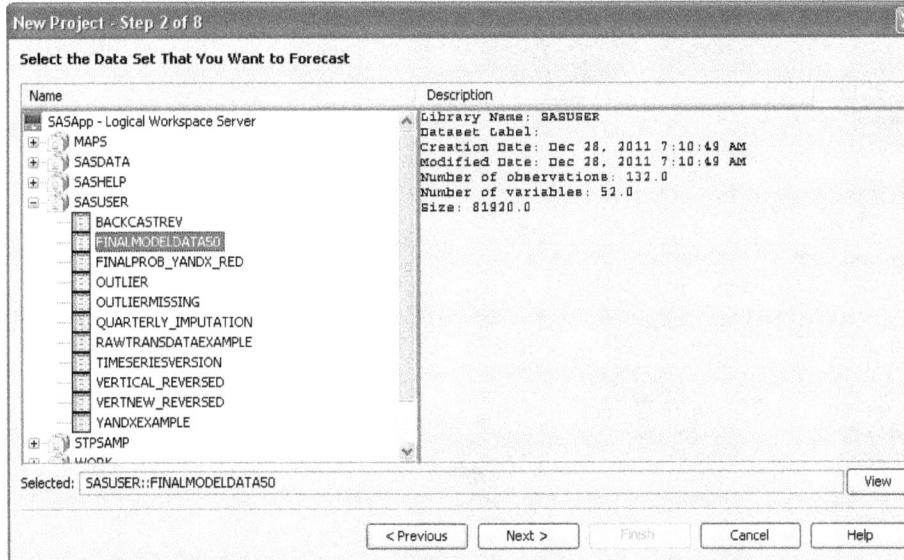

In the next series of steps the Date variable, Y variable and X variables are all identified. Figure 12.28 shows how the Date variable is identified. Options include setting the Interval of the data (Monthly, Quarterly, Annual, and so on), setting the Multiplier (the multiple of the interval), setting the Shift (the offset for the interval) and finally setting the Seasonal Cycle Length. Figure 12.29 shows how the Y and X variables are identified. The Role selection sets the type of variable (Dependent [Y] or Independent [X]). The Accumulation option enables you to specify numerous accumulation types such as averaging or summing along with many others. For the Usage in System-Generation option you can specify Try to Use or Use if Significant.

Figure 12.28: Identifying Date Variable

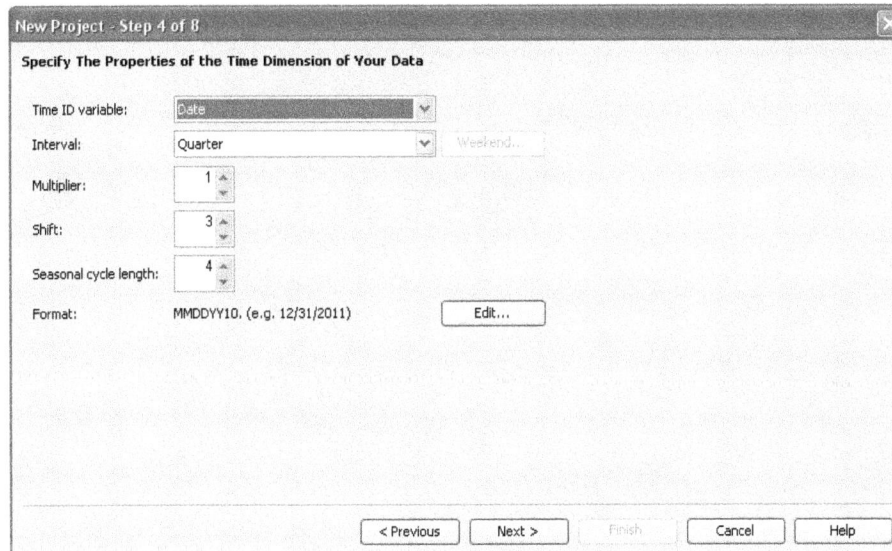

Figure 12.29: Identifying the Y and X Variables

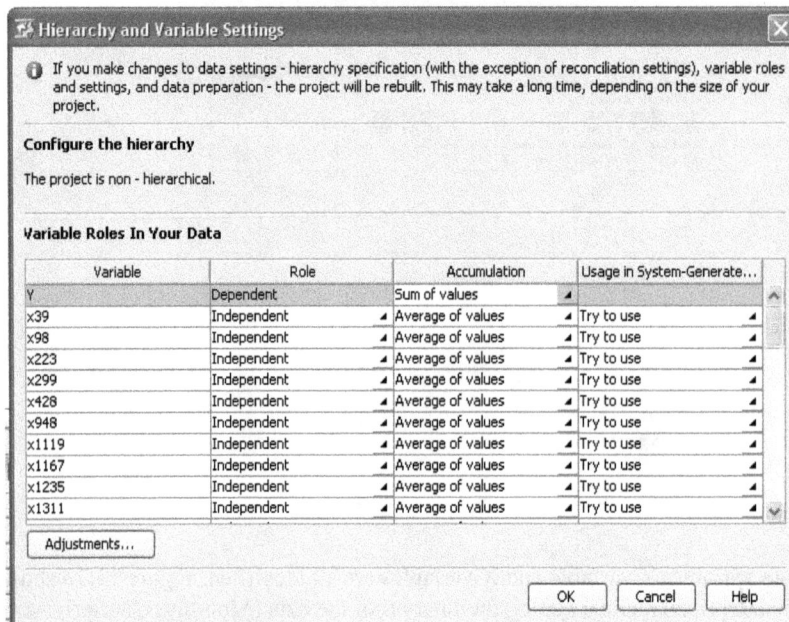

The last step in the process is to specify all of the Forecast Settings. The first of these steps is the Data Preparation step, shown in Figure 12.30. In this step, SAS Forecast Server enables you to impute missing values within the time series as well as leading and trailing missing values or zeros.

Figure 12.30: Missing Data Imputation

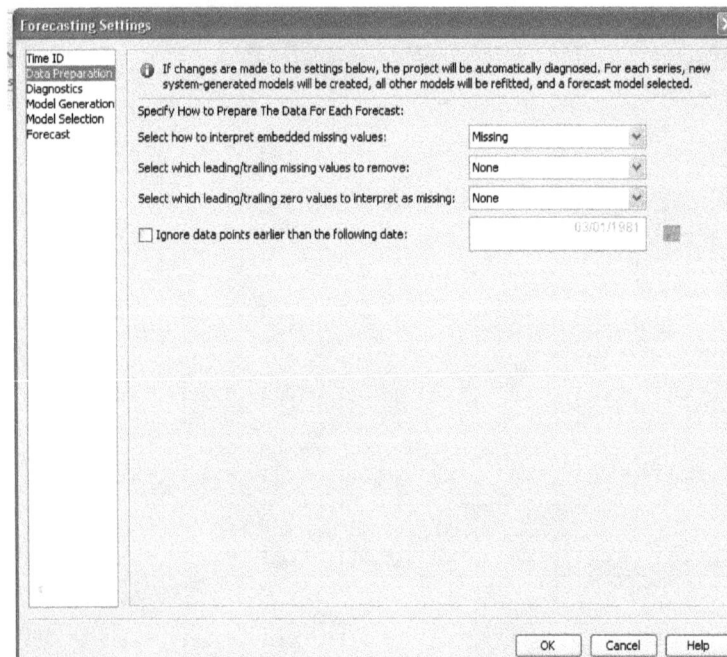

Various diagnostics are now set up as shown in Figure 12.31. First, in the case of hierarchical forecasting, the lowest level of the hierarchy might not have complete data sets and thus a special model called the Intermittent Demand Model must be used. The first option provides for the control of the intermittent data diagnostic. The second option provides for the control of the seasonality diagnostic. The next three options set the minimum number of data points for various model structures (again, this is more appropriate for hierarchical forecasting, and this problem is not hierarchical). Next, the transformation for the Y variable can be chosen. Normally, this is left to Automatic. Forecasting a Mean or Median can be selected next, and Median is the norm here. Lastly, sensitivity options can be set for detecting outliers, and the settings selected are the standard.

Figure 12.31 Diagnostic Settings

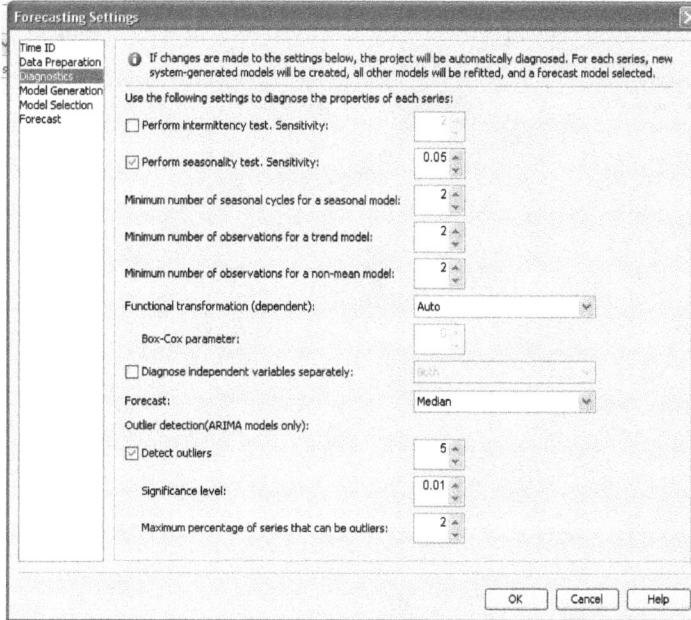

After setting Diagnostics, the Model Generation selections are made. There are three families of models available: ARIMA(X), Exponential Smoothing, and UCM (Unobserved Components Models). For the ARIMA(X) models, three different options are available as can be seen in Figure 12.32. The "recommended" option is normally used.

Figure 12.32: Model Generation

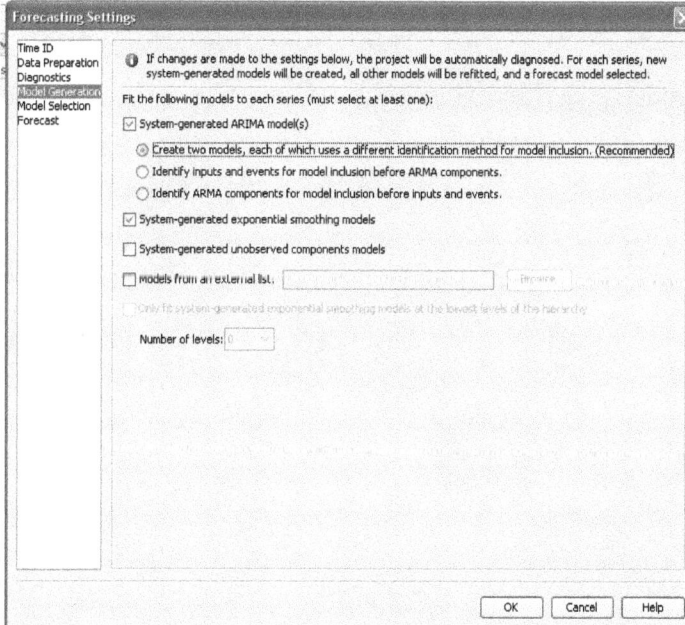

In order to provide for Model Selection, it is a best practice to use a holdout. SAS Forecast Server diagnoses the model form, inclusive of selecting the Xs to be used in the model, and then it estimates the parameters for the diagnosed form. Lastly, SAS Forecast Server forecasts the Y over the holdout space, calculates the fit statistics (in this case, SMAPE) and picks the model that performs the best over the holdout set. Figure 12.33 indicates the holdout set selection size (in data points or % of data or both)—here we have specified 8 (or a maximum of 25%) and selected the fit statistics for the SMAPE model.

Figure 12.33: Model Selection

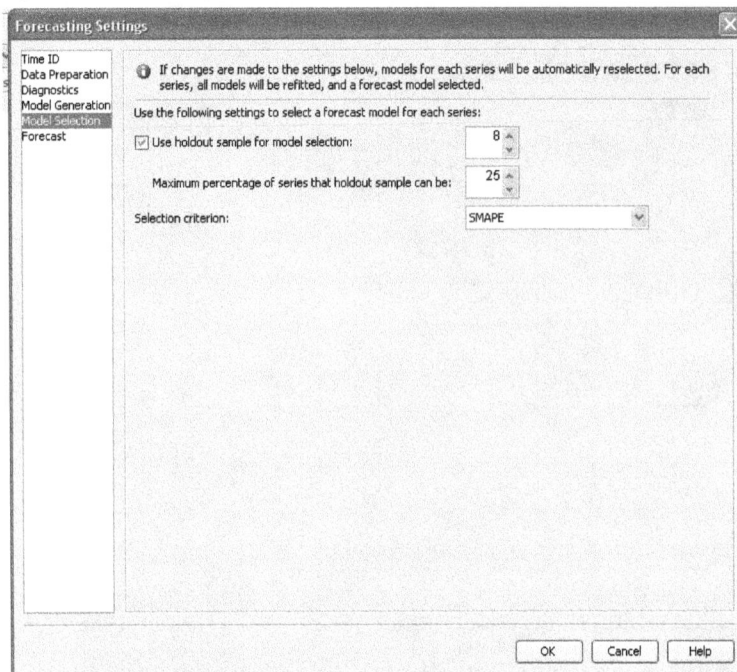

The last Forecast Settings options are for the forecast itself, as shown in Figure 12.34. Here the number of periods to forecast is set along with the number of data points for an out-of-sample fit statistic calculation, if that is desired. Lastly, SAS Forecast Server offers a significance level choice for the Confidence Bands. An out-of-sample fit statistic calculation is a true test of model adequacy since the holdout set is used to select the best model in the first place.

Figure 12.34 Forecast Options

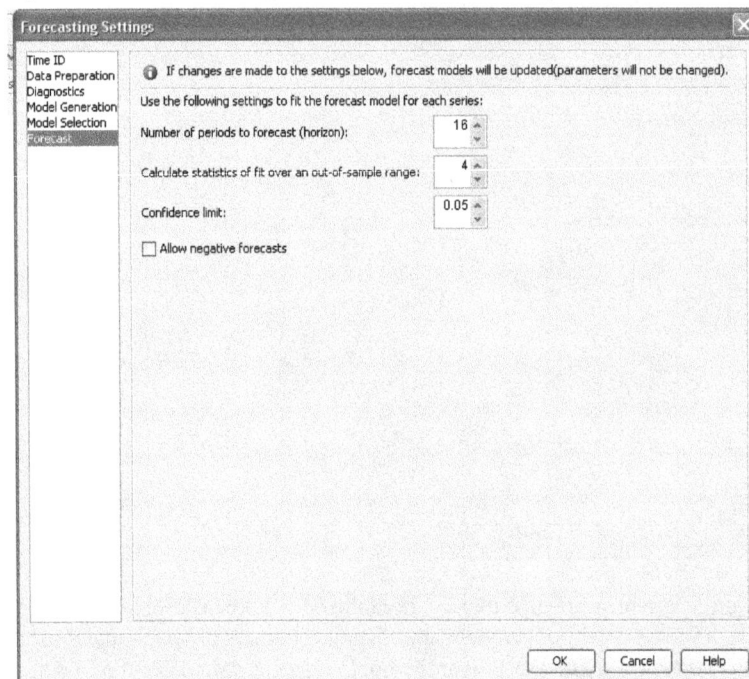

Developing models in SAS Forecast Server is done using the Modeling View shown in Figure 12.35. SAS Forecast Server offers automated model form diagnosing, estimating, and forecasting. The following equation is the general exogenous variable model that SAS Forecast Server diagnoses, estimates, and forecasts.

$$f(y_t) = \mu + \sum_{k=1}^{t-1} d_k f(y_k) + \sum_{i=1}^{N} \sum_{k=1}^{t} \beta_{i,k} f_i(x_{i,k}) + \sum_{k=1}^{t} w_k \varepsilon_k$$

This automatic model diagnosing algorithm includes:

- Automatic Functional Transform Detection, $f(y_t)$
- Automatic Difference Filter Order Detection d_k
- Automatic Constant Term Detection μ
- Automatic Disturbance Filter Order Detection w_k
- Automatic Transfer Function Filter Order Detection $f_i(x_{i,k})$
- Automatic Exogenous input and Calendar Event Selection $x_{i,k}$

See Leonard (2009) for details.

The following process is a best practice for model exploration and development. First, look at the simple exploratory plots to understand the general form of the Y. Next, use SAS Forecast Server to develop the base components of the Y (trend, cycle, seasonality). From here, select the holdout (HO) and out-of-sample (OOS) framework (8 and 4 respectively), and then develop a baseline model, such as a simple univariate Exponential Smoothing model. Next, save the various SAS Forecast Server automated models offered by SAS Forecast Server. Then, set up different sets of Xs for SAS Forecast Server to select from. Finally, to test the robustness of various models, change the HO and OOS data set sizes and run the same models. One option SAS Forecast Server offers is to save either an automated or custom-built model. In this way, under various variable groupings and HO and OOS selections, these custom and automated models are still available for testing.

Figure 12.35 shows various models that were developed using the process above. The names of the models reflect the type of model and their options. Thus, ARIMAX, UCM, and ESM are the types of models (the three main types available in SAS Forecast Server). Then, TTU and IS reflect the options for the Xs (Try To Use and If Significant). Lastly, COINT, SIM, and CCF reflect the type of X variable selection used. The Modeling View platform in Figure 12.35 shows the standard output for each model. There is a Plots pull-down menu (Figure 12.36) as well as a Tables pull-down menu for access to more details about the models. In the left-hand pane the SMAPE for the in-sample data is given. The holdout SMAPE is listed on the far right for each model listed. The out-of-sample SMAPE is given below the Modeling View tab.

Figure 12.35: Modeling View

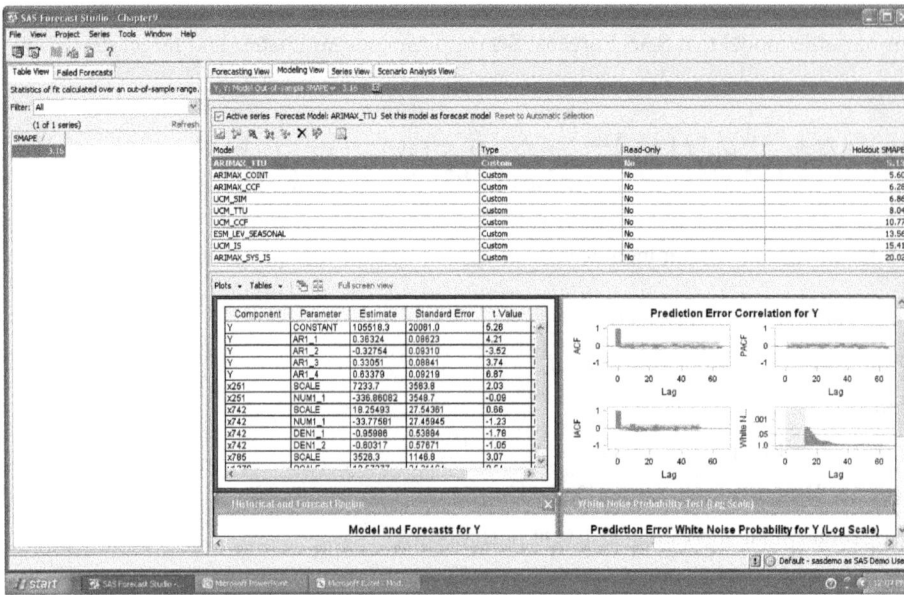

Figure 12.36: Plot Pull-Down Menu

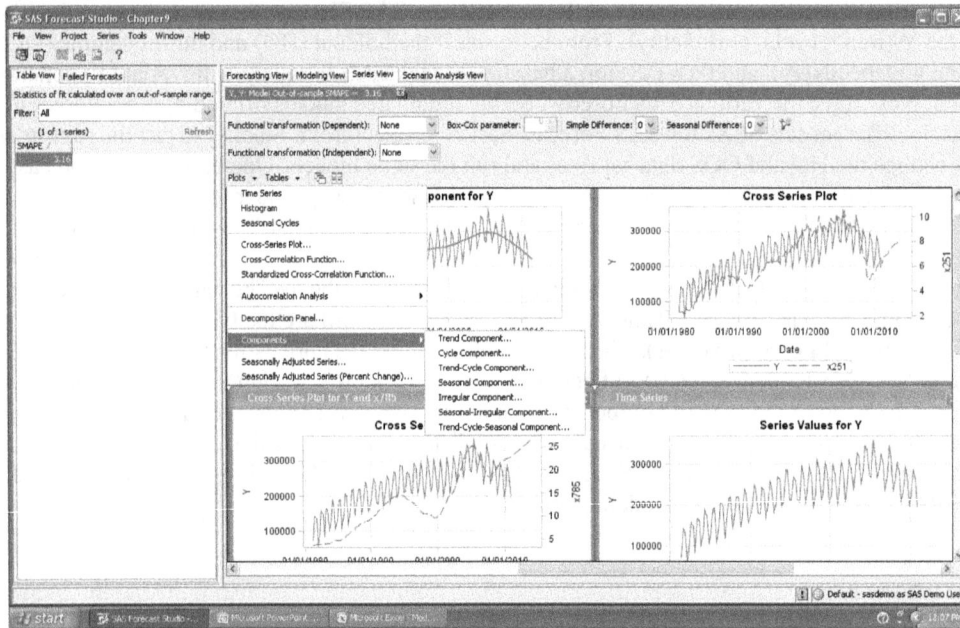

Figure 12.37 is simply the Y variable plotted over time while Figure 12.38 is the distribution of the Y variable. After a general increasing trend, there seems to be a couple of cycles along with a current downward trend. There is obvious seasonality. The Y seems to be normally distributed.

Figure 12.37: Exploratory Plots

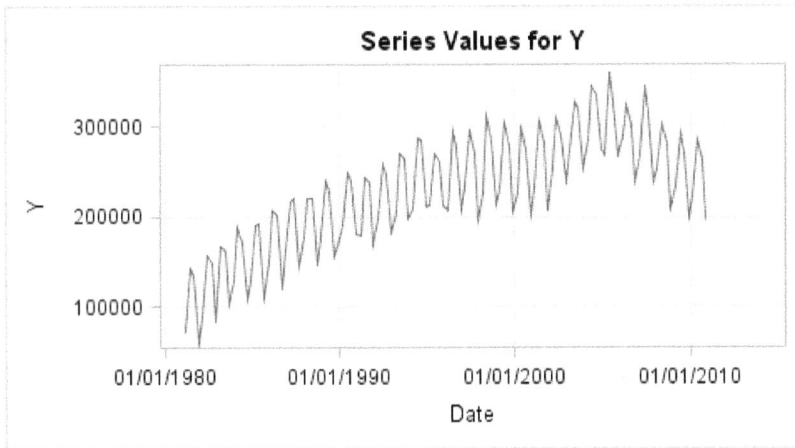

Series Values for Y

Figure 12.38: Distribution of Y

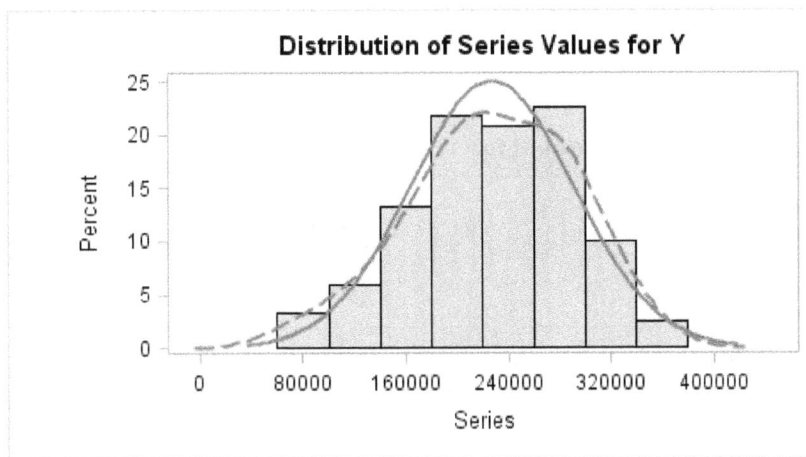

Distribution of Series Values for Y

Figure 12.39 represents the trend component, verifying the early upward trend and the current downward trend. Figure 12.40 is the cycle component confirming a large cycle in 2000 and some smaller ones earlier and more recently.

Figure 12.39: Trend Component

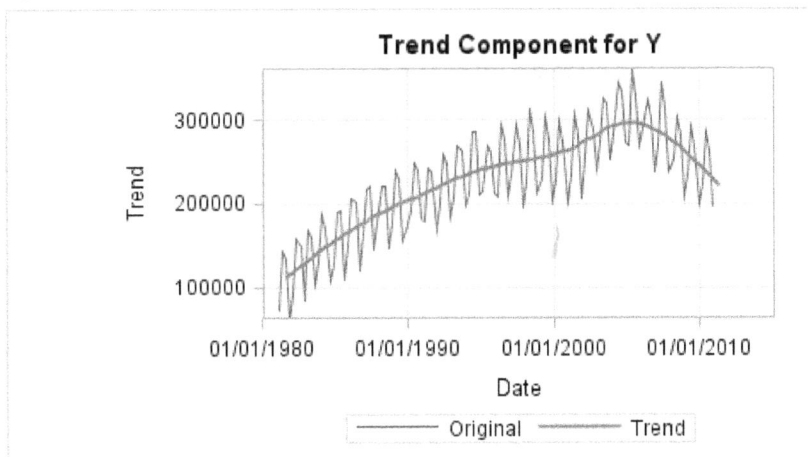

Trend Component for Y

Figure 12.40: Cycle Component

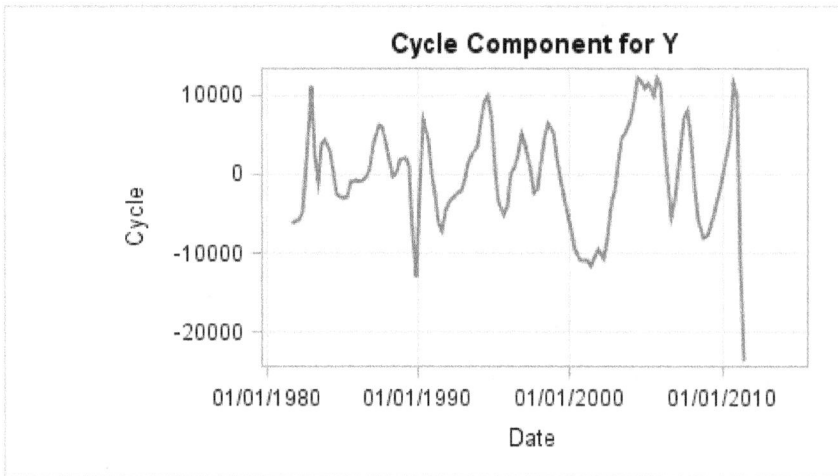

Figure 12.41 shows that the 2nd quarter is high with the 4th quarter is low at about a 15% seasonal affect. Figure 12.42 shows the seasonally adjusted series over the original series.

Figure 12.41: Seasonality

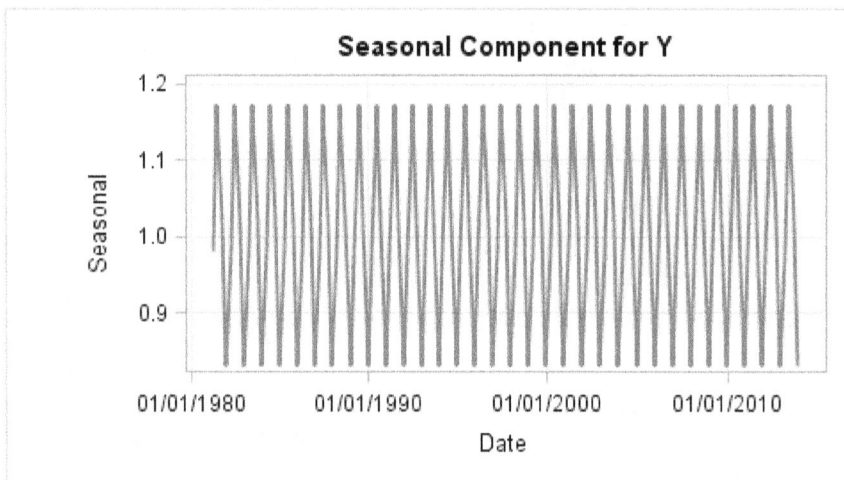

Figure 12.42: Seasonally Adjusted Series

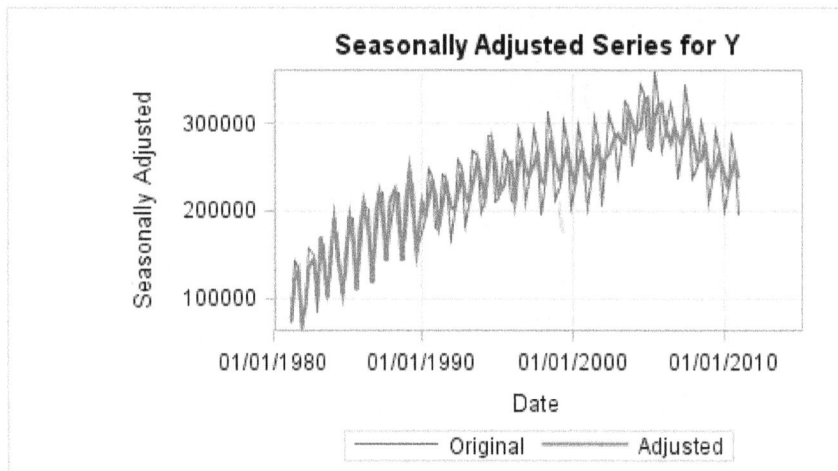

Lastly, Figure 12.43 shows the trend-cycle and seasonally adjusted series overlaid on the original series indicating the history might be a good indicator of the future.

Figure 12.43: Trend-Cycle-Seasonally adjust series

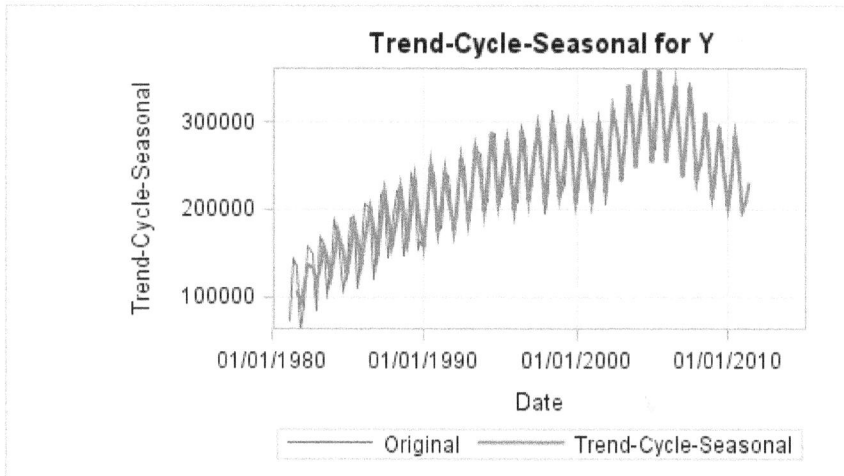

Figure 12.44 gives the time series correlation structure of the Y. Previous chapters gave details about how to interpret this structure to and how to propose the AR, I and MA structure for the Y portion of an ARIMAX model. SAS Forecast Server does this automatically. It is shown here for reference. The ACF, PACF, IACF, and white noise plots all show significant structure.

Figure 12.44: Exploratory Correlation Structure

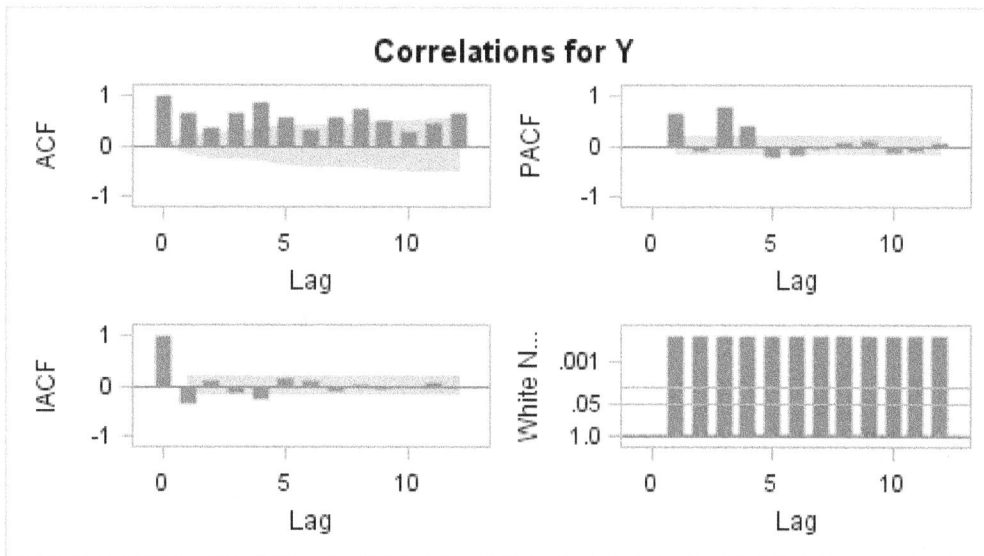

The following models were all set up in the Modeling View in SAS Forecast Server and are all automated models chosen by simply changing either the Try to Use or If Significant options on the Xs or subsetting the Xs that SAS Forecast Server tries to use.

- ARIMAX _TTU is an ARIMAX with the Try to Use option selected on all Xs.
- ARIMAX _SYS_IS is an ARMIAX with the use If Significant option selected on all Xs.
- ARIMAX_COINT is an ARIMAX with the Try to Use option selected on just the top cointegration Xs.
- ESM_Lev_Seasonal is an ESM model with level and seasonal components.

- UCM_CCF is a UCM model with just the top cross correlation variables with the Try to Use option selected,
- UCM_SIM is a UCM model with just the top similarity variables with the Try to Use option selected.
- UCM_IS is a UCM model with all Xs with the use If Significant option selected.
- UCM_TTU is a UCM model with all Xs with the Try to Use option selected.

Other models were attempted, specifically trying to reduce those listed above by removing in-significant Xs, but, all attempts lost ground in predictive capabilities. Figure 12.45 compares all of these models to one another. In general, the forecast trend is similar, slightly up.

Figure 12.45: Overlay Plot for All Proposed Models

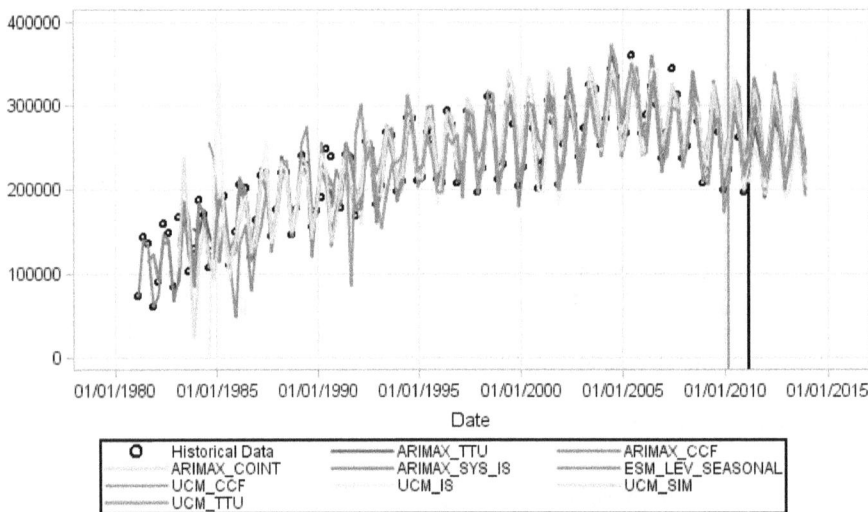

Figure 12.46 is the Overlay of the Top 5 models based on the holdout SMAPE. The forecast trend is still the same. Thus, until the in-sample, holdout, and out-of-sample statistics are reviewed, it is difficult to tell which model is best, although the UCM_SIM (darker gray line) does overshoot some of the earlier points.

Figure 12.46 Top 5 Overlay Plot

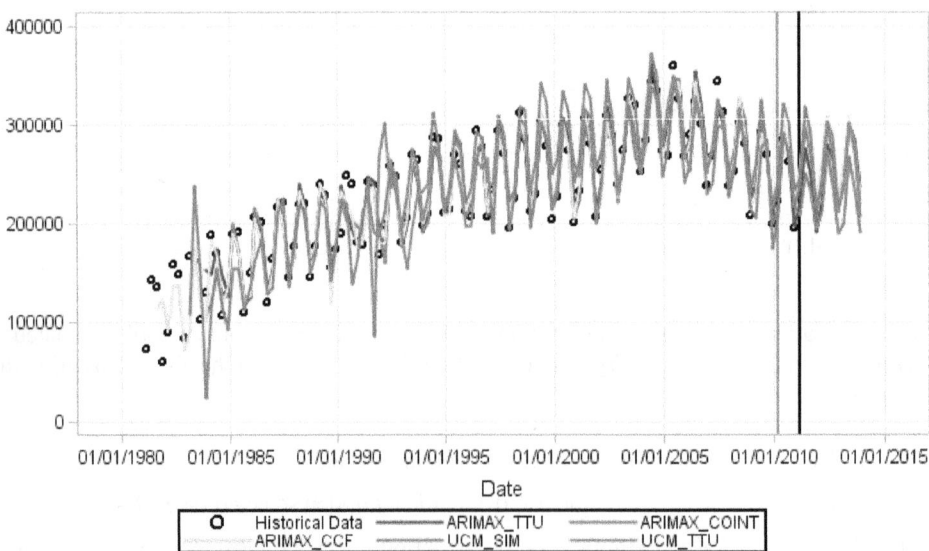

In order to make a more formal comparison, the in-sample statistics (along with the SMAPE for holdout and out of sample for various HO and OOS sizes) were collated into Table 12.3. The ARIMAX_TTU model (all 50 Xs made available using the Try to Use option) shows large decreases in RMSE (Root Mean Square Error) and large increases in R-square. The SBC (Schwartz Bayesian Criteria) and SMAPE are the lowest for the ARIMAX_TTU model. Thus, in all cases the ARIMAX_TTU model has the best in-sample fit statistics. So, the next question is, how does this model do in the holdout and out-of-sample scenarios? The initial model selection scenario was a holdout of 8 and an out of sample of 4, and in this scenario, the ARIMAX_TTU model is the best again. In the HO 6 and OOS 4 scenario, ARIMAX_TTU is the best HO and 3rd best OOS. In the HO 8 and OOS 6 scenario, ARIMAX_COINT is first in HO but ARIMAX_TTU is 3rd in HO and 1st in OOS. So, in the end it seems that the original best model held its own and remained the best.

Table 12.3: Top 5 Model Comparisons

Model	In-Sample				HO8 OOS 4		HO6 OOS4		HO8 OOS6	
	RMSE	Rsquare	SBC	SMAPE	SMAPE	SMAPE	SMAPE	SMAPE	SMAPE	
ARIMAX_TTU	13896	0.8434	152.629	5.13	5.17	3.16	3.16	7.24	8.34	4.75
ARIMAX_COINT	18122	0.7336	156.878	5.60	5.60	4.76	4.76	2.87	2.87	4.80
ARIMAXC_CCF	18786	0.7139	157.454	6.28	6.28	7.16	7.16	8.34	8.34	7.13
UCM_SIM	20878	0.6597	159.143	6.86	6.86	12.81	12.81	8.62	10.26	5.34
UCM_TTU	22317	0.6119	160.210	8.04	8.04	5.36	5.36	4.75	5.51	5.61

Figure 12.47 is the final ARIMAX_TTU model showing the holdout, out of sample, and forecasts. This model does very well on the 4 out-of-sample data points (3.16% error) as well as the 8 holdouts (5.17% error). Figure 12.48 shows the final ACF, PACF, IACF, and white noise plots for the ARIMAX_TTU model. All of the ACF type plots look reasonable. It would be preferable if all of the bars were below the .05 line on the white noise plot. Figure 12.49 contains the residual plot of which should look random. There are 4 points outside of the +/− 2 sigma bars and potentially some non-random patterns—some of which might have been indicated in the ACF plots for the model.

Figure 12.47 Final ARIMAX_TTU Model and Forecasts

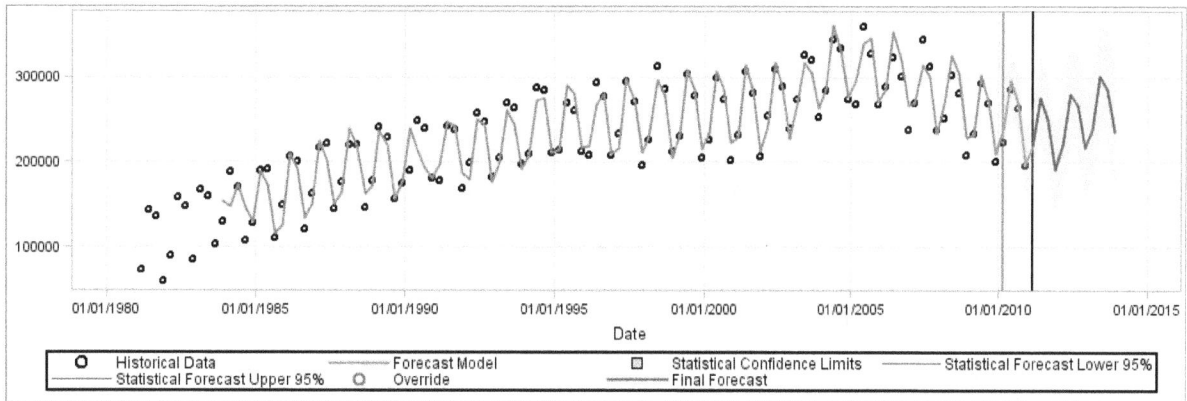

Figure 12.48: ARIMAX_TTU ACF Plots

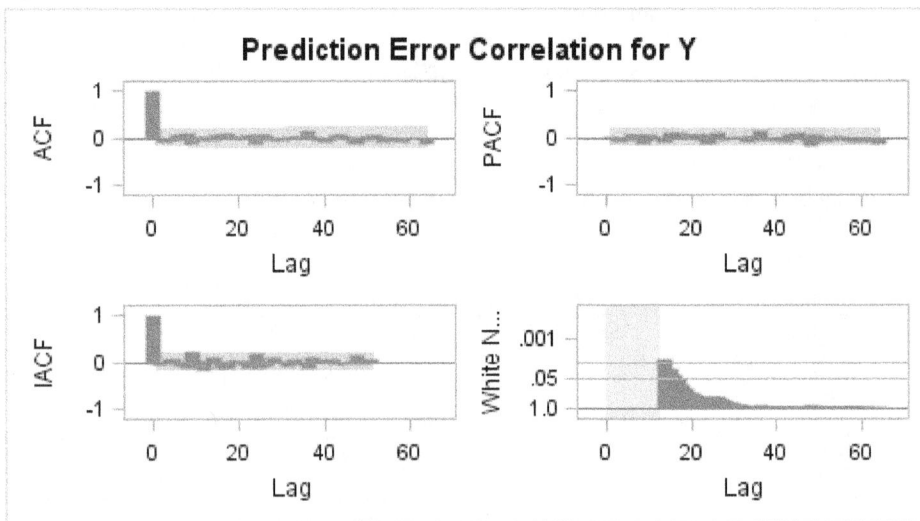

Figure 12.49: ARIMAX_TTU Model Residual Plot

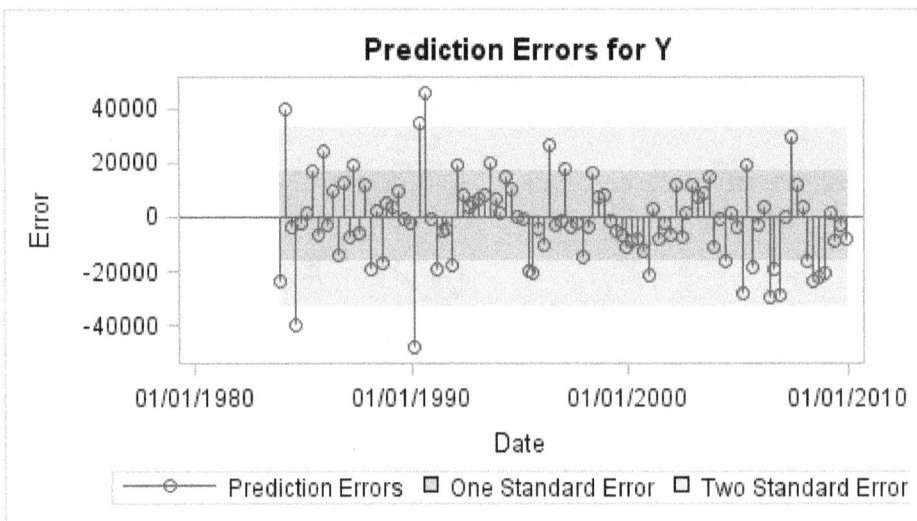

Table 12.4 gives the final parameter estimates, t values, and critical significance levels for each X term in the model. It is important to note that transfer functions have hierarchical effects. That is, a scale affect cannot be removed from the model if there is a significant numerator or denominator term. There is seemingly an insignificant numerator term for X251. Upon trying to remove this term, the fit statistics change quite dramatically. The same went for the scale term on X1279. It is critical note that, unlike transactional data (as explained in Chapter 1), there is never a situation where time series Xs are truly orthogonal to one another. Thus, the parameter estimates are related to one another, which is the cause of these phenomena. This is tough on business clients since the ideal situation is to have the best predictive model with the purest cause and effect model—a utopian situation rarely experienced in time series modeling.

Table 12.4: ARIMAX_TTU Parameter Estimates

Component	Parameter	Estimate	Standard Error	t Value	Approx Pr > |t|
Y	CONSTANT	105518.3	20061	5.26	<.0001
Y	AR1_1	0.36324	0.08623	4.21	<.0001
Y	AR1_2	-0.32754	0.0931	-3.52	0.0007
Y	AR1_3	0.33051	0.08841	3.74	0.0003
Y	AR1_4	0.63379	0.09219	6.87	<.0001
x251	SCALE	7233.7	3563.8	2.03	0.0453
x251	NUM1_1	-336.86082	3548.7	-0.09	0.9246
x742	SCALE	18.25493	27.54361	0.66	0.5091
x742	NUM1_1	-33.77581	27.45945	-1.23	0.2218
x742	DEN1_1	-0.95986	0.53884	-1.78	0.0782
x742	DEN1_2	-0.60317	0.57671	-1.05	0.2984
x785	SCALE	3526.3	1146.8	3.07	0.0028
x1279	SCALE	18.57277	34.31164	0.54	0.5896

Figure 12.50: Cross Series Plots

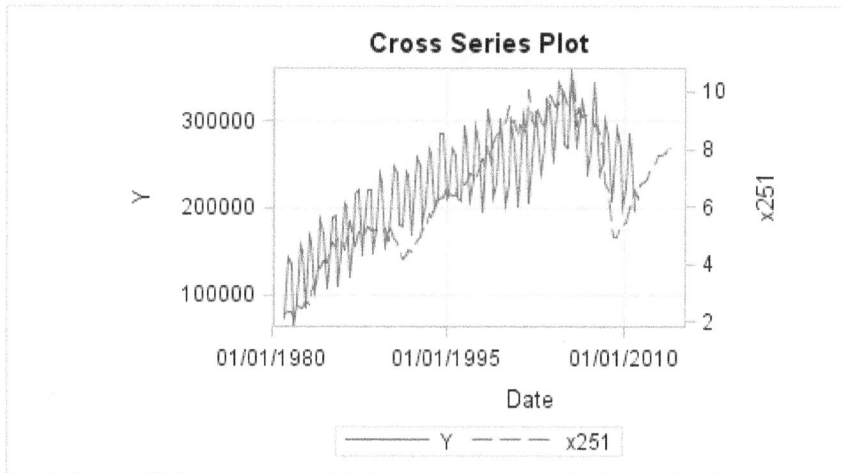

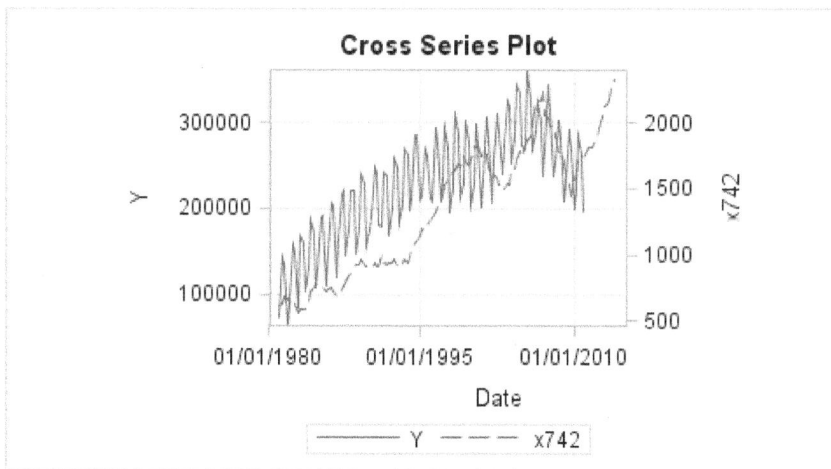

Cross Series Plot

Cross Series Plot

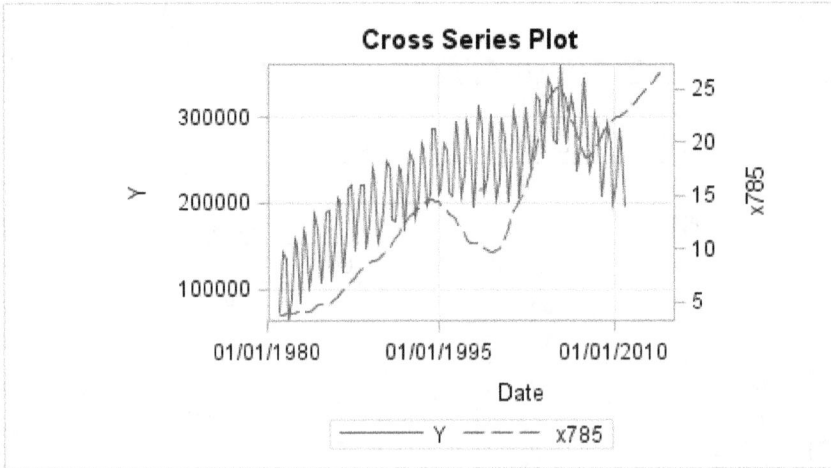

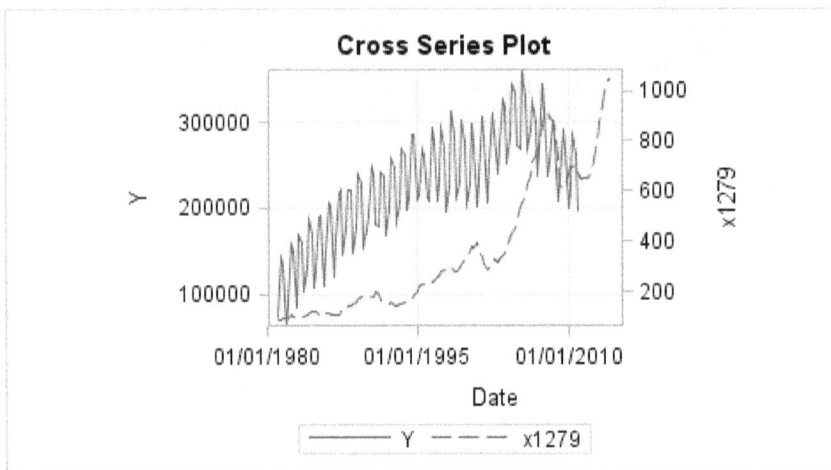

For completeness, the next best model should be investigated, ARIMAX_COINT. Table 12.5 contains the parameter estimates for this model. The critical significance levels look a little better. The ACF plots in Figure 12.51 show a little more pattern in the PACF, and this shows up a bit in the residual plot in Figure 12.52. Therefore, in the end, both models can be shared with the business client, but the ARIMAX_TTU model is still preferred.

Table 12.5: ARIMAX_COINT Model Parameter Estimates

Component	Parameter	Estimate	Standard Error	t Value	Approx Pr > |t|
Y	MA1_4	0.61168	0.24013	2.55	0.0125
Y	AR1_1	-0.31273	0.10283	-3.04	0.0031
Y	AR1_2	-0.7286	0.1063	-6.85	<.0001
Y	AR1_3	-0.29832	0.1047	-2.85	0.0054
Y	AR1_4	0.16725	0.12965	1.29	0.2002
Y	AR2_4	0.80287	0.19322	4.16	<.0001
x560	SCALE	-411.87917	216.55607	-1.9	0.0603
x560	NUM1_1	-254.19376	221.39217	-1.15	0.2538
x560	NUM1_2	-417.40529	215.10908	-1.94	0.0554
x1129	SCALE	83613.2	116067.6	0.72	0.4731
x1129	NUM1_1	332705	118506.8	2.81	0.0061
x1129	NUM1_2	-239202.7	114973.7	-2.08	0.0402

Figure 12.51: ACF Figures for ARIMAX_COINT Model

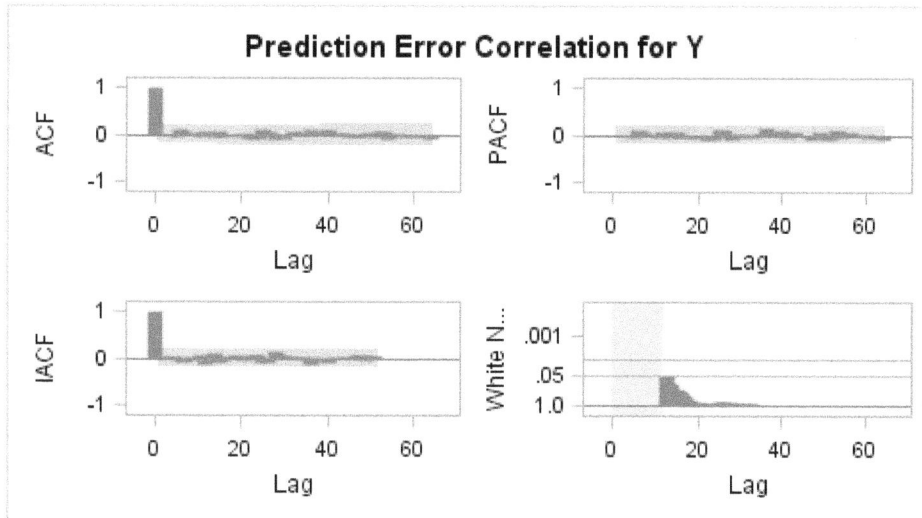

Figure 12.52: ARIMAX_COINT Model Residuals

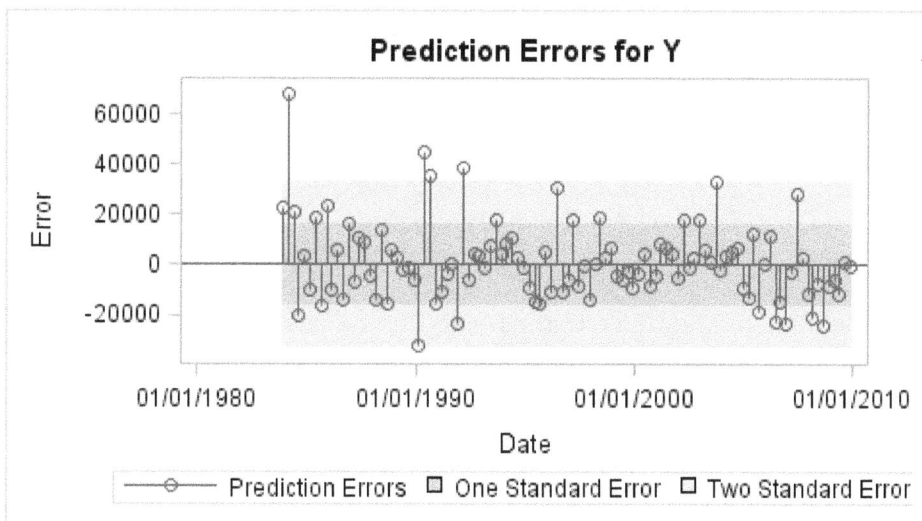

X560 was Consumer Spending -New autos. Source: BEA. Units: billions of dollars annual rate

X1129 was Marginal rate of investment tax credit on other information equipment. Source: IRS. Units: decimal fraction

12.10 Summary

Much work has been done to data mine the original 1441 Xs. First, we removed non-informative Xs (no data, no variance, and correlated perfectly with other Xs), and then we imputed missing data (either by backcasting or within the data). Next, we used rigorous time series–based approaches to reduce the number of Xs through similarity and variable clustering, and finally we selected the Xs for modeling through three different time series methods (similarity, cointegration and cross correlation). Given this reduced set of 50 Xs, SAS Forecast Server was used to explore and build the best ARIMAX model possible. The topics covered in Chapters 8-11 could be used to refine this model even further if that is needed.

Appendix A

1) Some background on denominator order polynomials. The simple relationship shown in equation 1 maps directly to the way a denominator polynomial order works. Consider the resemblance between equation 1 and the denominator order 1 polynomial shown in equation 2.

When $\quad |\omega| < 1$

$1 + \omega + \omega^2 + \omega^3 + \omega^4 ...$

$$\rightarrow$$

$$\sum_{i=0}^{\infty} \omega^i = 1 / (1 - \omega) \qquad\qquad 1$$

$$y_t = \gamma / (1 - \delta B) \cdot x_t \qquad\qquad 2$$

$$\rightarrow$$

$$= \gamma \cdot (1 + \delta B + \delta^2 B^2 + \delta^3 B^3 + ...)x_t$$
$$= \gamma \cdot x_t + \gamma\delta \cdot x_{t-1} + \gamma\delta^2 \cdot x_{t-2} + \gamma\delta^3 \cdot x_{t-3} ...$$

Where γ is the scale effect, and B is the backshift operator. Note, above $\Rightarrow |\delta| < 1$.

2) ARMA models as ratios of polynomials.

Consider the general ARMA (p, q) specification, below.

$$y_t = \varphi_1 y_{t-1} + \varphi_2 y_{t-2} + \varphi_3 y_{t-3} + ... \varphi_p y_{t-p}$$
$$+ \xi_t - \theta_1 \xi_{t-1} - \theta_2 \xi_{t-2} - ... - \theta_q \xi_{t-q}$$

Rearranging,

$$y_t - \varphi_1 y_{t-1} - \varphi_2 y_{t-2} - ... \varphi_p y_{t-p} = \xi_t - \theta_1 \xi_{t-1} - \theta_2 \xi_{t-2} - ... \theta_q \xi_{t-q}$$

Applying the Backshift operator, and rearranging some more;

$$y_t \cdot (1 - \varphi_1 B - \varphi_2 B^2 - ... \varphi_p B^p) = \xi_t \cdot (1 - \theta_1 B - \theta_2 B^2 - ... \theta_q B^q)$$

To save typing, use $(\varphi B), (\theta B)$ to denote general polynomials in the backshift operator.

$$\rightarrow y_t (\varphi B) = (\theta B)\xi_t \quad \rightarrow y_t = (\theta B) / (\varphi B)\xi_t$$

The ratio of polynomials creates a "filter" to translate variation in the white noise error term into variation in the dependent variable Y.

Appendix B

For and MA1:

$$y_t = \mu - \theta\varepsilon_{t-1} + \varepsilon_t$$

Covariance at lag 0 (variance of y):

$$\gamma_0 = (y_t - \mu)^2 = (\varepsilon_t - \theta\varepsilon_{t-1})^2$$
$$= \varepsilon_t^2 + \theta^2\varepsilon_{t-1}^2 - 2\theta\varepsilon_t\varepsilon_{t-1}$$

Cross terms drop because of the independence assumption: $E[\varepsilon_t\varepsilon_{t-k}] = 0 \qquad \forall k \; n.e. 0$

$$\gamma_0 = \sigma^2 + \theta^2\sigma^2$$
$$= \sigma^2(1 + \theta^2)$$

Covariance at lag 1:

$$\gamma_1 = (y_t - \mu)(y_{t-1} - \mu) = (\varepsilon_t - \theta\varepsilon_{t-1})(\varepsilon_{t-1} - \theta\varepsilon_{t-2})$$
$$= \varepsilon_t\varepsilon_{t-1} - \theta\varepsilon_{t-1}^2 - \varepsilon_t\theta\varepsilon_{t-2} + \theta^2\varepsilon_{t-1}\varepsilon_{t-2}$$

Cross terms drop.

$$\gamma_1 = -\theta\sigma^2$$

Covariance at lag 2 (and all higher lags):

$$\gamma_2 = (y_t - \mu)(y_{t-2} - \mu) = (\varepsilon_t - \theta\varepsilon_{t-1})(\varepsilon_{t-2} - \theta\varepsilon_{t-3})$$
$$= \varepsilon_t\varepsilon_{t-2} - \theta\varepsilon_{t-1}\varepsilon_{t-2} - \varepsilon_t\theta\varepsilon_{t-3} + \theta^2\varepsilon_{t-1}\varepsilon_{t-3}$$

Cross terms drop.

$$\gamma_2 = 0$$

We see that for MA1, innovations persist for one interval. This correspondence between order and persistence holds for all MA specifications; the length of persistence of innovations is given by the MA order.

Now, contrast this with an autoregressive model of order 1.

One characteristic of a stationary series is that its mean is independent of time: $E(y_t) = \mu$

$$y_t = \tau + \varphi y_{t-1} + \varepsilon_t$$

The mean of the process is seen by taking expectations of both sides of the equation:

$$E(y_t) = E(\tau + \varphi y_{t-1} + \varepsilon_t)$$
$$\mu \;\;= \tau + \varphi\mu \rightarrow \mu = \tau / (1 - \varphi)$$

The <u>covariance at lag 0</u> of the AR1 process is shown below. Note, in the following, it is assumed $\tau = 0$. This minimizes clutter, and, in practice, represents a re-scaling of the series to be centered around zero.

$$\gamma_0 = E(y_t)^2 = E(\varphi y_{t-1} + \varepsilon_t)^2$$
$$\rightarrow E(\varphi^2 y_{t-1}^2 + \varepsilon_t^2 + 2\varphi y_{t-1}\varepsilon_t) = \varphi^2 \gamma_0 + \sigma^2$$
$$\gamma_0 = \sigma^2 / (1 - \varphi^2)$$

The covariance at lags 1 and 2 of the AR1 process are derived next.

$$\gamma_1 = E(y_t y_{t-1}) = E(y_{t-1}(\varphi y_{t-1} + \varepsilon_t))$$
$$\rightarrow E(\varphi y_{t-1}^2 + y_{t-1}\varepsilon_t) = \varphi\gamma_0$$
$$\gamma_1 = \varphi\sigma^2 / (1 - \varphi^2)$$

$$\gamma_2 = E(y_t y_{t-2}) = E((\varphi y_{t-1} + \varepsilon_t)y_{t-2})$$
$$\rightarrow E(\varphi y_{t-1} y_{t-2} + \varepsilon_t y_{t-2}) = \varphi\gamma_1$$
$$\gamma_2 = \varphi^2 \sigma^2 / (1 - \varphi^2)$$

$$\Rightarrow \gamma_n = \varphi^n \sigma^2 / (1 - \varphi^2)$$

The series is stationary, so the AR1 parameter is less than one in absolute value. We could continue to derive the auto-covariance function for longer lags, but the above shows the general idea. Innovations in AR1 series persist, but diminish over time.

References

Achuthan, Lakshman, and Anirvan Banerji. 2004. *Beating the Business Cycle*. New York, NY: Doubleday.

Antunes, Cláudia, and Arlindo Oliveira. 2001. "Temporal Data Mining: an overview." KDD Workshop on Temporal Data Mining.

Arstrong, Scott J., ed. 2001. *Principles of Forecasting: a Handbook for Researchers and Practitioners.* Springer.

Azevedo, Ana, and Manuel Santos. 2008. "KDD, SEMMA and CRISP-DM: A Parallel Overview." *Proceedings of the IADIS.*

Banerji, Anirvan.1999. "The Lead Profile and Other Nonparametric to Evaluate Survey Series as Leading Indicators." *24th CIRET Conference.*

Barbazon, Anthony, Michael O'Neil, and Ian Dempsey. 2008. "An Introduction to Evolutionary Computation in Finance." *IEEE Computational Intelligence Magazine* 3:42–55.

Berry, Michael J. A., and Gordon S. Linoff. 2011. *Data Mining Techniques for Marketing, Sales, and Customer Relationship Management.* 3rd Edition. Hoboken, NJ: John Wiley & Sons.

Box, G. E. P., and G.M. Jenkins. 1972. *Time Series Analysis: Forecasting and Control.* rev. ed. Oakland, California: Holden-Day.

Breiman, Leo. 1996. "Heuristics of Instability and Stabilization in Model Selection." *The Annals of Statistics* 24(6):2350–2383

Breyfogle III, Forrest. 2003. *Implementing Six Sigma: Smarter Solutions Using Statistical Methods.* 2nd ed. Hoboken, NJ: John Wiley & Sons.

Brocklebank, John, and David Dickey. 2003. *SAS for Forecasting Time Series.* 2nd ed. Cary, NC: SAS Press.

Buzan, T. 2003. *The Mind Map Book.* 3rd ed. BBC Active.

Cabena, Peter et al. 1998. *Discovering Data Mining: From Concept to Implementation.* Prentice Hall.

Capps, Oral. 2008. *Advanced Topics in Applied Econometrics Course Notes.* Cary, NC: SAS Institute Inc.

Chase, Charles. 2009. *Demand-Driven Forecasting: A Structured Approach to Forecasting,* Hoboken, NJ: John Wiley & Sons.

Chattratichat, J. et al. 1999. "An Architecture for Distributed Enterprise Data Mining." *Proceedings of the 7th International Conference on High-Performance Computing and Networking.* London, UK: Springer-Verlag.

Cherkassky, Vladimir, and Filip Mulier. 2007. *Learning from Data: Concepts, Theory, and Methods.* 2nd ed. New York: John Wiley & Sons.

Clements, M. P., P. H. Franses, and N. R. Swanson. 2004. "Forecasting Economic and Financial Time Series with Non-linear Models. *International Journal of Forecasting* 20:169–183.

Cohen, M. and E. Nagel, 1934. *An Introduction to Logic and Scientific Method.* Oxford, England: Harcourt, Brace.

Croston, J.D. 1972. "Forecasting and Stock Control for Intermittent Demands." *Operations Research Quarterly* 23(3):289–303.

Dagum, Estea Bee, and Pierre A. Cholette. 2006. *Benchmarking, Temporal Distribution, and Reconciliation Methods for Time Series.* Lecture Notes in Statistics 186. Springer.

Dasu, Tamraparni, and Theodore Johnson. 2003. *Exploratory Data Mining and Data Cleaning.* Hoboken, NJ: John Wiley & Sons.

Davenport, Thomas, Jeanne Harris, and Robert Morison. 2010. *Analytics at Work: Smarter Decisions, Better Results.* Boston, MA: Harvard Business School Press.

Dickey, David, and W. Fuller. 1979. "Distribution of the Estimators for Autoregressive Time Series with a Unit Root." *Journal of the American Statistical Association* 74:427–31.

Dickey, David. 2008. Personal Communication.

Duling, D., H. Plemmons, and N. Rausch. "From Soup to Nuts: Practices in Data Management for Analytical Performance." *Proceedings of the SAS Global Forum 2008 Conference.* Cary, NC: SAS Institute Inc. Available at http://www2.sas.com/proceedings/forum2008/129-2008.pdf.

Duling, D. and W. Thompson. 2005. "What's New in SAS® Enterprise Miner™ 5.2." *SUGI 31 Proceedings.* Cary, NC: SAS Institute Inc. Available at http://www2.sas.com/proceedings/sugi31/082-31.pdf

Ellis, J. 2005. *Ahead of the Curve: A Commonsense Guide to Forecasting Business and Market Cycles.* Boston, MA: Harvard Business School Press.

Engle, R. F., and W. J. Granger. 2001. *Long-run Economic Relationship, Readings in Cointegration.* New York, NY: Oxford University Press.

Evans, C., C. Liu, and G. Pham-Kanter. 2002. "The 2001 Recession and the Chicago Fed National Activity Index: Identifying Business Cycle Turning Points." *Economic Perspectives* 26(3):26–43.

Fayyad, U et al., eds. 1996. *Advances in Knowledge Discovery and Data Mining.* AAAI Press.

Fildes, Robert, and Paul Goodwin. 2007. "Good and Bad Judgment in Forecasting: Lessons from Four Companies," *Foresight: The International Journal of Applied Forecasting* 8:5–10

Ghysels, E. 2010. *Mixed Data Sampling. Encyclopedia of Quantitative Finance.* DOI: 10.1002/9780470061602.eqf19008. John Wiley & Sons.

Gilliland, Michael. 2010. *The Business Forecasting Deal.* Hoboken, NJ: John Wiley & Sons.

Gilliland, Michael. 2003. "Fundamental Issues in Business Forecasting." *Journal of Business Forecasting* 22(2):1–13.

Glasserman, Paul. 2004. *Monte Carlo Methods in Financial Engineering.* New York: Springer-Verlag.

Glymour, Clark et al. 1997. "Statistical Themes and Lessons for Data Mining." *Data Mining and Knowledge Discovery* 1:11–28. Netherlands: Kluwer Academic Publishers.

Guyon, Isabelle and André Elisseeff. 2003. "An Introduction to Variable and Feature Selection." *Journal of Machine Learning Research* 3(3):1157–1182.

Hamilton, James D. 1994. *Time Series Analysis.* Princeton, New Jersey: Princeton University Press.

Han, J., M. Kamber, and J. Pei. 2006. *Data Mining: Concepts and Techniques.* San Francisco, CA: Elsevier, Inc.

Hand, David. 1998. "Data Mining: Statistics and More?" *The American Statistician* 52(2):112–18.

Harvey, A. C. 1989. *Forecasting, Structural Time Series Models and the Kalman Filter.* Cambridge, UK: Cambridge University Press.

Haykin, Simon. 1998. *Neural Networks: A Comprehensive Foundation.* 2nd ed. Upper Saddle River, NJ: Prentice-Hall.

Kalos, Alex, and Tim Rey. 2005. "Data Mining in the Chemical Industry." *Proceedings of the Eleventh ACM SIGKDD International Conference on Knowledge Discovery and Data Mining,* 763–769. Chicago, IL.

Kantardzic, Mehmed. 2011. *Data Mining: Concepts, Models, Methods, and Algorithms.* Piscataway, NJ: IEEE Press.

Koller, Daphne, and Mehran Sahami. 1996. "Towards Optimal Feature Selection." *International Conference on Machine Learning,* 284–292.

Kordon, Arthur. 2009. *Applying Computational Intelligence: How to Create Value.* Heidelberg: Springer.

Koza, John. 1992. *Genetic Programming: On the Programming of Computers by Means of Natural Selection.* Cambridge, MA: MIT Press.

Kurgan, Lukasz, and Petr Musilek. 2006. "A Survey of Knowledge Discovery and Data Mining Process Models." *The Knowledge Engineering Review* 21:1, 1–24.

Laxman, Srivatsan and P.S. Sastry. 2006. "A Survey of Temporal Data Mining." *Sadhana* 31, Part 2, 173–98.

Lee, Taiyeong, et al. "Two-Stage Variable Clustering for Large Data Sets." *Proceedings of the SAS Global Forum 2008 Conference.* Cary, NC: SAS Institute Inc. Available at http://www2.sas.com/proceedings/forum2008/320-2008.pdf.

Lehman, Thomas. 2010. Personal Communication.

Leonard, Michael, and Brenda Wolfe. 2002. "Mining Transactional and Time Series Data." *SUGI 30 Proceedings.* Cary, NC: SAS Institute Inc. Available at http://www2.sas.com/proceedings/sugi30/080-30.pdf.

Leonard, Michael, et al. "An Introduction to Similarity Analysis Using SAS." *Proceedings of the SAS Global Forum 2008 Conference.* Cary, NC: SAS Institute Inc. Available at http://www2.sas.com/proceedings/sugi30/080-30.pdf.

Leonard, Michael. 2009. "ARIMA training." SAS internal training white paper.

Lucas, Bob. 2008. Personal Communication.

Makridacis, Spyros, Wheelwright, Steven C., and Rob J. Hyndman. 1998. *Forecasting: Methods and Applications.* New York: John Wiley & Sons.

Matignon, Randall. 2007. Data Mining Using SAS Enterprise Miner. Hoboken, NJ: John Wiley & Sons.

Miller, Rupert G. 1981. *Simultaneous Statistical Inference.* 2nd ed. New York: Springer-Verlag.

Mitsa, Theophano. 2010. *Temporal Data Mining.* Taylor and Francis Group, LLC.

Montgomery, Douglas, Elizabeth Peck, and Geoffrey Vining. 2006. *Introduction to Linear Regression Analysis.* 4th ed. Hoboken, NJ: John Wiley & Sons.

Mundy, Joy, Warren Thornthwaite, and Ralph Kimball. 2011. *The Microsoft Data Warehouse Toolkit.* Indianapolis, IN:Wiley Publishing.

Nast, Jamie. 2006. *Idea Mapping.* Hoboken, NJ: John Wiley & Sons.

Nisbet, Robert, John Elder, and Gary Miner. 2009. *Handbook of Statistical Analysis and Data Mining Applications*, Burlington, MA: Elsevier.

Pankratz, Alan. 1991. *Forecasting with Dynamic Regression Models.* John Wiley & Sons.

Pearson, Ronald. 2006. *Mining Imperfect Data: Dealing with Contamination and Incomplete Records.* Philadelphia, PA: SIAM.

Pindyck, Robert S., and Daniel L. Rubinfeld. 1991. *Econometric Models and Economic Forecasts.* 3rd ed. New York: McGraw-Hill.

Pletcher, Tim, et al. 2005 "Business Intelligence Architecture." *CMURC Steering Committee Communications.*

Poli, Riccardo, William Langdon, and Nicholas McPhee. 2008. *A Field Guide to Genetic Programming.* Available at www.lulu.com.

Pyle, Dorian. 2003. *Business Modeling and Data Mining.* San Francisco, CA: Morgan Kaufmann.

Pyle, Dorian. 1999. *Data Preparation for Data Mining.* San Francisco, CA: Morgan Kaufmann.

Rey, Tim, and Alex Kalos. 2005. "Data Mining in the Chemical Industry." *Proceedings of the eleventh ACM SIGKDD*, 763–69.

Rey, Tim. 2009. "Large Scale Forecasting in Manufacturing Using Data Mining Techniques for Pre-Processing." Presentation, *SAS A2009 Analytics Conference.*

Sall, John, Lee Creighton, and Ann Lehman. 2007. *JMP Start Statistics: A Guide to Statistics and Data Analysis Using JMP.* 4th ed. Cary, NC: SAS Press.

Sankoff, David, and Joseph Kruskal. 1999. *Time Warps, String Edits and Macro Molecules: The Theory and Practice of Sequence Comparison.* CSLI Publications.

Sapankevych, Nicholas, and Ravi Sankar. 2009. "Time Series Prediction Using Support Vector Machines: a Survey." *IEEE Computational Intelligence Magazine* 4:24–38.

SAS Institute Inc. 2010. *SAS/ETS® 9.22 User's Guide.* Cary, NC: SAS Institute Inc.

SAS Institute Inc. 2011. *SAS® Forecast Studio 4.1: User's Guide.* Cary, NC: SAS Institute Inc.

Schubert, Sascha, and Taiyeong Lee., "Time Series Data Mining with SAS® Enterprise Miner™." *Proceedings of the SAS Global Forum 2011 Conference.* Cary, NC: SAS Institute Inc. Available at http://support.sas.com/resources/papers/proceedings11/160-2011.pdf.

Smits, Guido, et al. 2006. "Variable Selection in Industrial Data Sets Using Pareto Genetic Programming." In *Genetic Programming Theory and Practice III,* 79–92. New York: Springer.

Smits, Guido, and Mark Kotanchek. 2004. "Pareto-Front Exploitation in Symbolic Regression." In *Genetic Programming Theory and Practice II,* 283–300. Boston, MA: Springer.

Svolba, Gerhard. 2006. *Data Preparation for Analytics Using SAS®.* Cary, NC: SAS Press.

Tan, Pang-Ning, Michael Steinbach, and Vipin Kumar. 2005. *Introduction to Data Mining.* Addison Wesley.

Tsay, R. S. 2002. "Nonlinear Models and Forecasting." In *A Companion to Economic Forecasting.* Malden, MA: Blackwell Publishers.

Vapnik, Vladimir. 1998. *Statistical Learning Theory.* New York: John Wiley & Sons.

Wagner N., Z. Michalewicz, M. Khouja, and R. McGregor. 2007. "Time Series Forecasting for Dynamic Environments: The DyFor Genetic Program Model." *IEEE Transactions on Evolutionary Computation* 11(4):433–452.

Wallace, Thomas. 2004. *Sales and Operations Planning: The How To Book.* 2nd ed. T. F. Wallace Co.

Willemain, T. R., C. N. Smart, and J. H. Shocker. 1994. "Forecasting Intermittent Demand in Manufacturing: Comparative Evaluation of Croston's Method." *International Journal of Forecasting* 10:529–538.

Woodfield, Terry. 2008. *Forecasting Using SAS ETS Procedures; a Programing Approach Course Notes* (FETSP). SAS Institute Inc.

Zhang, Guoqiang, B. Eddy Patuwo, and Michael Y. Hu. 1997. "Forecasting with Artificial Neural Networks: The State of the Art." *International Journal of Forecasting* 14:35–62.

Index

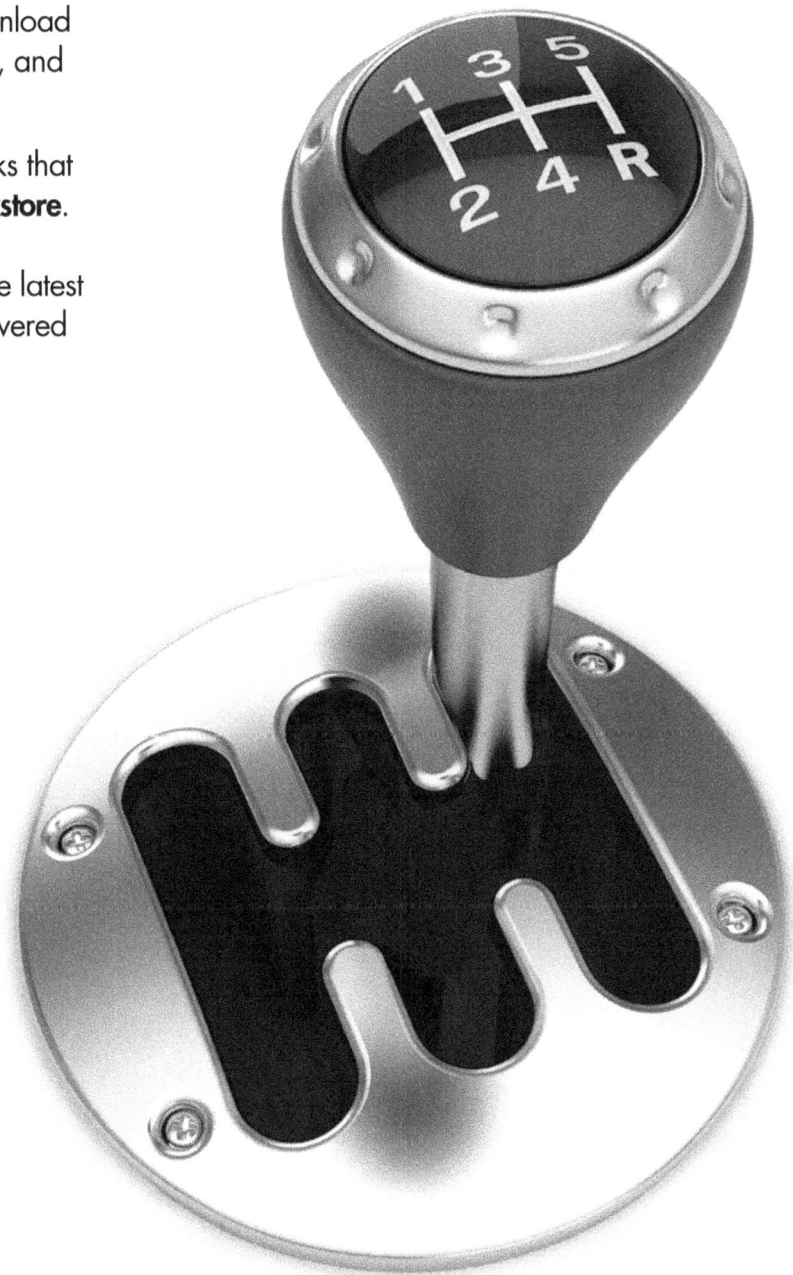